COMPANY-G
UNIVERSIT
NOTRE
COMPANY COMM
Ju

Company-G, Navy V-12 Unit, University of Notre Dame, Notre Dame, IN, July 31, 1944.
Company Commander: G.A. Bender. (Courtesy of Bernard J. Freed)

NAVY V-12 UNIT
Y of NOTRE DAME
DAME, IND.
ANDER' G. A. BENDER
ly 31, 1944

Bagby Photo Co.
South Bend

NAVY
V-12

TURNER PUBLISHING COMPANY

TURNER PUBLISHING COMPANY

Copyright © 1996
Turner Publishing Company.
All rights reserved.
Publishing Rights: Turner Publishing Company

Turner Publishing Company Staff:
Chief Editor: Robert J. Martin
Designer: Lora Ann Lauder

Library of Congress
Catalog Card Number: 95-60331
ISBN: 978-1-56311-189-1

Additional copies may be purchased directly from Turner Publishing Company.

This publication was compiled using available information. The publisher regrets it cannot assume liability for errors or omissions.

Photo: EX Fraternity Wabash College 1944-45. (Courtesy of William Thompson)

Table of Contents

Frontispiece

The person to whom I am most grateful for maintaining my interest in collecting data and information on the World War II U.S. Navy V-12 Program is my wife, Alice. Since my retirement in 1975, she knew of my desire to write a chronicle with its primary focus on Wesleyan University, Middletown, Connecticut, where I served as the Commanding Officer of its V-12 Unit.

Accordingly, on three separate occasions, we scoured the archives in search of information at Berea College, Berea, Kentucky; Dartmouth College, Hanover, New Hampshire; and Wesleyan University. Pertinent information to my own file of data motivated my writing, with Alice offering constructive criticism as she typed the manuscript.

Upon completion, my manuscript was submitted to the editor of *The Press* at Wesleyan University, entitled, *Wesleyan University in World War II-A Cooperative Venture in Education*. After much waiting, I was informed in a terse note: *"The University Press is closed."*

So, much of Part I in my Wesleyan manuscript has been adapted for inclusion in this publication. *Henry C. Herge, Sr.*

Preface

In late June 1945, it was a distinct pleasure for me to inform Dr. Alonzo G. Grace, Connecticut Commissioner of Education, of my affirmative response to his offer of the associate directorship of a study, "Implications of the Armed Services Educational Programs," to be sponsored by the American Council on Education (ACE), Washington, DC, and funded jointly by the Carnegie and Rockefeller Foundations.*

Highly honored by Dr. Grace's offer, I doubted seriously whether I might be released from active duty as the Commanding Officer of Wesleyan University's Navy V-12 Unit, which was still in full operation. Nevertheless, I initiated the request for release from active duty in order to join the ACE staff by September 1, 1945.

I was aware of the very important role played by the ACE in the early policy and decision-making process pertaining to the utilizations of colleges as contract institutions for training officer personnel. So, I remained hopeful and waited.

My release from active duty was approved by Admiral Randall Jacobs, Chief of the Bureau of Naval Personnel. Accordingly, I was granted terminal leave authorizing my replacement to report to Wesleyan University for the change of command.

In the period September 1, 1943 to December 24, 1945, it was my good fortune to visit 14 institutions in which Armed Services Training Units were operative. Thus, I was able to glean valuable data, information and literature for subsequent use in the ACE Commission Staff studies.

Also, while on active duty, but prior to the ACE Commission appointment, I was a recipient of temporary duty orders from the Director of Training, Third Naval District, to attend "inspections" of Navy V-12 Units. In all, there were four such "inspections" during my tenure as a Commanding Officer that were fruitful also for acquiring valuable information of Navy programs-in-action.

The combination of these experiences enabled me to prepare an ACE staff document, No. 10, entitled, *The War and Higher Education*, for the Commission members.

Now, more than five decades later, this nation stands much indebted to the Armed Services and the 131 contract colleges that participated in the preparation of highly qualified civilian youths, together with a large contingent of carefully selected enlistees from the Fleet and Marine Corps, that produced college-educated officer candidates for the Navy and Marine Corps.

The V-12 Program provided young men who grew up in the nation's severest depression with a new vision, especially those high school graduates who, at age 18, saw little hope of ever going to college.

When the V-12 Program terminated oin June 30, 1946 (page 363, *Navy V-12 Leadership for a Lifetime* by James Schneider), it had produced more than 60,000 Navy and Marine Corps officers. Today, records show that many trainees remained on active duty and became the highest ranking officers while others in civilian life, during the post-war period, achieved national prominence in business, industry and government.

In boastful pride, trainee alumni can point today to their success in being selected for admission to the Navy V-12 Program in April 1943. The nationwide qualifying examination was taken by 60 percent of all young men in the draft age category, which had been lowered to age 18. While 125,000 were in the V-12 program, only 117,295 passed the first national test and selected the Navy. An additional 10,000 came from the Fleet and active Marine corps.

In tribute to those trainees who became commissioned officers and saw action in the Fleet from Normandy to Saipan, Iwo Jima or Okinawa, or the Pentagon, and to those who made a career in the Navy and Marine Corps - High Praise! It is likewise appropriate that the V-12ers who stayed in the U.S. Naval Reserve in the post-war period deserve acclaim, inasmuch as many were "called up" for fighting in Vietnam while still others fought in Korea if they had become U.S. Navy Reserves. *Henry C. Herge, Sr.*

Endnote:
The Commission on Implications of Armed Services Educational Programs began its work in July 1945. It undertook to identify features of the wartime training and educational programs worthy of adaptation and experimentation in peacetime civilian education of any and all types and levels. It also undertook to make available to the public well-considered answers to the questions: What should education in America gain from the experience of the vast wartime training efforts? What are the implications for education and the national culture and strength, now and in the future?

Acknowledgments

The Commission on Implications of Armed Services Educational Programs began its comprehensive studies in July 1945. Subsequently, the Commission published 10 separate studies before it terminated in 1949.

One of these studies was focused primarily on the U.S. Navy V-12 Program:*

Part I of this study was strictly my authorship. It focused primarily on Navy contract training programs in colleges and universities during World War II.

The American Council on Education (ACE) was both the sponsor and publisher of the 10-volume series of the Commission. It is fitting, therefore, to note that portions of the report cited above were adapted for use in this manuscript and to thank the ACE Manager of Publications, James Murray, for permission to do so with an "appropriate credit line."

Acknowledgment is due Raymond J. Connally, who served as Chief, Contracts Unit, Purchase Control Section, Procurement Branch, War Department, for his comprehensive in-depth report of all the contract relationships between the United States Government and civilian colleges and universities during World War II. Connally's focus on the financial relationships between the military and civilian colleges remains pertinent as an important reference source in the event of another conflict with hostile nations.

The source of much useful information was the Wesleyan University, Middletown, Connecticut, Olin Research Library, where Elizabeth Swaim, archivist, was most cooperative. Finally, the author is grateful that materials dealing with the Navy College Training Programs at Dartmouth College, Hanover, New Hampshire, were made available by Kenneth C. Cramer, archivist in Dartmouth's Baker Library. *Henry C. Herge, Sr.*

Endnote:
*HERGE, HENRY C., et al. Wartime College Training Programs of the Armed Services. Washington, DC. The American Council on Education, 1947.

Dedication And In Respectful Tribute To Arthur Stanton Adams

To accomplish its purpose as an instrument of force against our enemies, the Navy has built a vast and formidable fleet. To meet the personnel needs of that new fleet, the Bureau of Naval Personnel (BuPers) undertook the unprecedented programs of procurement, training and distribution of personnel. All three programs have kept pace with the rapid growth of the Navy's ships and facilities to ensure a balanced development at maximum efficiency in all stages. *Rear Admiral Randall Jacobs, USN, Chief of Naval Personnel.*

Admiral Jacobs' statement depicted well the task faced by the Navy at the time this nation entered World War II. Certain individuals warrant attention for their leadership in fulfilling the Navy's mission. It is appropriate to cite one officer in particular, Captain Arthur Stanton Adams, who earned the appellation, "Father of the Navy's V-12 Program," a title bestowed upon him by his Navy peers as well as by University administrators and college faculty members across the nation.

In the fall of 1942, as Admiral Jacobs indicated, the Armed Services were expanding rapidly. Although the Army, Army Air Corps, Navy and Marine Corps were training and assigning inductees for a vast array of specific duties, the manpower projections indicated the urgent need for a more systematic way to assure a flow of qualified commissioned officers.

This history of the Navy V-12 College Training Program of World War II focuses on its evolution, operation and contribution in the war effort. It narrates the Navy's early experiences in sponsoring several programs for college students who were granted inactive duty status while they remained under instruction. These were the Navy V-1, V-5, and V-7. College students were not applying for these programs in keeping with the Navy's expectations. Something needed to be done to bring order to what many civilian critics called "a confused, unorganized state."

The Secretary of War, on September 8, 1942, took action that hastened decisions. He ruled that all students who were of Selective Service age would be summoned to active duty at the end of the semester.

It was at this juncture that the Navy searched for a person of high competency and proven ability who could plan the ways and means whereby the reservoir of trainees already enrolled in the several "V-Classification" programs would be salvaged and incorporated into a single, cohesive program. In the search, the Navy recalled to active duty Dr. Arthur S. Adams to be the officer-in-charge of the Administration Division of BuPers. His credentials clearly indicated his eligibility to cope with this complex situation.

Arthur Adams' boyhood home was Winchester, Massachusetts. His Navy career began as a Midshipman in the United States Naval Academy where he acquired the nickname, "Beany," possibly because he was intelligent and articulate as well as balding.

Shortly after his commissioning as an Ensign, he chose the submarine service and attended the Submarine School. In 1918, he was ordered to California, where he attached to the Mobile Submarine Division. He had intended to make the Navy his life career but he contracted tuberculosis and was retired for physical disability as a Lieutenant (j.g.) on November 25, 1921.

As a convalescent, Arthur Adams enrolled for his Master's Degree in Physics at the University of California. He enrolled next at the Colorado School of Mines, where he earned his Ph.D. Degree in Metallurgical Engineering. Following graduation, he remained at Mines as a junior faculty member, later achieving full professor rank during his 19-year tenure. During this period, he also served as assistant to the president.

Just as the threat of war was nearing in 1940, Dr. Adams was invited by Cornell University to become Assistant Dean of Engineering. His tenure there was interrupted by the U.S. Navy.

The search committee for chief planner-administrator for the Navy's College Training Programs spotlighted Dr. Adams. He was recalled from retirement status on November 12, 1941, and promoted from a Lieutenant (j.g.) rank to Lieutenant Commander and subsequently, in BuPers, "spot promoted" to Captain, USNR.

During Captain Adams' administration, first in the Training Section and then in the Administration Division at BuPers, he exhibited outstanding leadership and wisdom in skillfully redesigning the muddled college training program.

Captain Adams' first major test came on May 14-15 at the national conference at Columbia University, convened by the Navy, consisting of college and university representatives of 131 contract institutions and prospective commanding officers of Navy V-12 Units. In the two-day session, Captain Adams held center stage as ranking Naval officers interpreted all aspects of the new program. The most fitting of all was a statement by Captain Adams setting forth the basic principles upon which the Navy V-12 was designed.

"This is in no matter of commandeering the colleges. It is in no sense a matter of dictating to the colleges, of remodeling education or, indeed, is it a matter of preserving the American tradition of education and, particularly, it is not a way of providing a grand opportunity for worthy young people to get a good education at government expense. It is none of these things. It is for the benefit of the Navy to win the war."

"The program is for the needs of the Navy. It is my guess, if I am not wholly out of touch with the youth of America, that we are going to have real difficulty with some of these lads in getting them to go to college. They want to go out and shoot Japs tomorrow. So do I. However, once they are in the service, their choice in the matter will not operate to determine just what they will do, and if they have the capabilities for developing themselves into capable leaders, it is to the best interest of the Navy that they do so, and they will do so. It will be your college instructors' high responsibility, and perhaps your grave concern, to see that their morale is held at the highest peak and that they are constantly realizing that they are not in college to benefit themselves. They are there to make themselves more useful to our very urgent needs."

Captain Adams and several officers from BuPers carefully explained the comprehensive plan, which he stated was designed cooperatively by many civilian educators and Navy training personnel. He explained in detail the method for the procurement of male candidates through a screening process approved by the ACE/Navy Advisory Council on Education and BuPers. President Franklin D. Roosevelt made public announcement of the program on December 7, 1942.

In concluding the conference, Captain Adams was brilliant in summing up the salient points underscored by the speakers and in replying to a multitude of questions posed by the university administrators. His command of the situation led to spontaneous applause. The enthusiasm he engendered was expressed in terms of praise for the program's soundness and the possibility it held for the accomplishment of the Navy's goal for training officer candidates.

The writer attended this conference, elated to be part of the Navy's plan, still unaware of his next assignment but ready for the challenge and grateful to Captain Adams for his vision, his tact and genteel manner in restraining a few cantankerous college administrators.

Through his calm efforts during the next three years, Captain Adams, with the cooperative assistance of numerous civilian educators and Navy personnel, trained 150,000 junior officers for the expanding Navy.

When the program terminated on June 30, 1946, Captain Adams retired again. This time, he departed BuPers with the Legion of Merit. He returned to Cornell, a layman again, this time as the University Provost. It was while in this position he accepted, in 1948, the call of the trustees of the University of New Hampshire to be its 11th president. His short-term presidency ended in 1951 when he accepted appointment to the presidency of the American Council on Education (ACE) in Washington. The Council, throughout the war, was considered to be "the capstone of higher education in America." Dr. Adams continued in leadership position and served with great distinction as the ACE spokesman for 11 years.

Not fully ready for retirement, Dr. Adams accepted an appointment as president of the Salzburg Seminar in American Studies. He retired in 1965 and returned to Durham, New Hampshire and to the house he built on Cedar Point, Great Bay, to sail his boat, a love he acquired in his midshipman days.

Dr. Adams continued to be prominent in retirement. He became a popular commencement speaker, particularly at institutions which, in World War II, held Navy V-12 contracts. In keeping with custom, he became the recipient of honorary degrees, totaling 28 in his lifetime.

Dr. Adams died November 18, 1980, at 84 years of age. He was accorded full military honors at the National Cemetery in Arlington, Virginia.

To Arthur Stanton Adams, more than any other individual in the Navy during World War II, was due the honor and credit for blueprinting and administering the V-12 College Training Program, the most comprehensive educational undertaking in American higher education. Indeed, his skillful coordination and clear-sighted ability to anticipate potential problems endeared him to college presidents and Navy officers alike, for he was, indeed, "a man for all seasons."

In tribute to him, President Leonard Carmichael of Tufts University said:

"Captain Adams was of the opinion that American colleges had heretofore overlooked the wealth of student material who, because of lack of certain 15 units, or failure to take certain prescribed courses, or because of economic and/or guidance errors were in commercial or vocational curricula, were college calibre and possessed innate ability to do superior academic work.

Experience of some men from the Fleet and Marine Corps, also, from civilian status procured for V-12 was that they possessed unusual leadership qualities and a burning desire to make good academically in spite of their shortcomings. Incentive was undoubtedly a factor but capacity, plus the willingness of faculty in tutorial help, were reasons for success.

Through the college (V-12) program, many deserving American boys were able to take advantage of training at the college level who would not have been able to do so otherwise because of financial limitations of their families."

Foreword

What A Naval Officer Should Be

It is by no means enough that an officer of the Navy should be a capable mariner. He must be that, of course, but also a great deal more. He should be, as well, a gentleman of liberal education, refined manner, punctilious courtesy, and the nicest sense of personal honor.

He should be the soul of tact, patience, justice, firmness and charity. No meritorious act of a subordinate should escape his attention or be left to pass without its reward, if even the reward be only one word of approval. Conversely, he should not be blind to a single fault in any subordinate, though, at the same time, he should be quick and unfailing to distinguish error from malice, thoughtlessness from incompetency, and well-meant shortcoming from heedless or stupid blunder. As he should be universal and impartial in his rewards and approval of merit, so should he be judicial and unbending in his punishment or reproof of misconduct. *John Paul Jones. From a statement to the Maritime Commission, 1775.*

Prologue

The Swift Momentum of Events Leading to World War II

The typical V-12 trainee alumnus, if asked today concerning a particular on-campus event that occurred in the early 1940s while in his Apprentice Seaman's uniform at one of the 131 contract colleges or universities, can recall quickly the time and place.

The commentary that follows will provide a record of events, both national and international, that a typical V-12 trainee alumnus may recall but vaguely, if at all.

This Navy V-12 Program history focuses on the significant events that led to the utilization of American institutions of higher learning for the specific purpose of supplying officer candidates in the several fields of specialization. It also records the rigid admission requirements, curricula leading to those specializations and the assignment of graduates to Naval Reserve Officer Candidate Schools prior to commissioning in the United States Navy and Marine Corps.

This history is intended as a tribute to the Navy V-12 Program of World War II and every trainee who became an officer. The Epilogue ends with a poetic toast to all who participated!

Franklin Delano Roosevelt, a physically handicapped ex-Governor of New York, was elected President, the first of his four terms in 1932. Beginning with his inaugural ceremony, he exhibited outstanding ability in contending with difficult crises that molded U.S. history.

At the inception of his administration until his sudden death on April 12, 1945, Mr. Roosevelt was a vigorous and commanding Chief-of-State. He possessed high quality leadership ability in influencing people and a keen ability in political debates.

The President came to Washington in the wake of a national crisis - the worst in U.S. history. On his arrival, he had to cope with the devastating effects of the banking system's collapse of October-November 1929 that was the inception of the Great Depression. A total of 5,504 banks closed, creating a shock wave felt in every town and city across the country.

The newly installed President knew that he must do something constructive and do it quickly to stem the high tide of depression, mounting unemployment and dwindling industrial production.

With his charm that brought solace to many citizens, President Roosevelt delivered a memorable "New Deal" inaugural address containing phrases of encouragement to dispell despair and to stimulate hope among his listeners and the unemployed. He

exhorted his vast audience, with typical composure, to be patient: "....the only thing we have to fear is fear itself."

Thus, with a new administration in power, along with the thrust he gave the banking system, private business and re-employment, Mr. Roosevelt inspired optimism.

The President's first "hundred days" in office (March 6-June 16, 1933) were a period during which the Congress remained in session to work cooperatively on his carefully drafted program of relief legislation. His array of proposals gave rise to the repeated creation of new federal agencies, staffed by selected businessmen whom he trusted.

The President's critics became vocal but his reaction, always objective, signified that businessmen possessed the skills to deal directly with industrial production. In addition, he pointed out that these persons were temporary and without federal civil service status.

Until Mr. Roosevelt's arrival, the Congress had appeared to lack the leadership to provide measures to correct the ever-mounting problems inherent in the serious depression. Forthwith, the President addressed the sky-rocketing prices, the rationing of food supplies and scarce commodities. Of special concern to him were the poorly equipped Army and Navy and the absence of any "blueprint" to guide the nation if it be drawn into war.

On April 14, 1938, Mr. Roosevelt delivered to Congress more legislative proposals, which he explained in detail that same evening during his radio address to the nation. It is obvious that his public messages were productive in passage of legislation.

On June 21, 1938, the Emergency Relief Appropriations Act was enacted into law. This Act was a reversal of the administration's deflationary policy. Surprisingly, it authorized the sharp expansion of the WPA rolls from one and a half million to three million; a three million dollar recovery and relief program; the "de-sterilizing" of more than three billion dollars in vaulted U.S. Treasury gold; easy reconstruction finance corporation loans; and a new policy for the Federal Reserve System authorizing a "loose money policy."

Initially, the President's foreign affairs position was strict isolationism, reflecting the wishes of a majority of voters but, later, as the war clouds abroad blackened the skies in Europe, it changed to neutrality. Soon after, however, the nation's unpreparedness moved Mr. Roosevelt, on May 17, 1938, to

sign the Naval Expansion Act that authorized a gradual enlargement of the U.S. Navy, with a marked increae in the number of capital ships and creation of a two-ocean Navy "in 10 years."

Now, the frequency of the President's "fireside chats" to the nation increased as he sought to calm the fear of the populace who opposed any involvement in another European war. Isolationists expressed their major concern: "Will our neutrality be respected if the U.S. begins shipping needed supplies to Britain and France, and possibly to other nations?"

The President, himself, soon recognized that America was vulnerable to attack inasmuch as the Atlantic Ocean could not serve as a shield to Axis' attacks, so he sought legal measures to defend the nation. Whatever he announced, he provoked more criticism from frenzied isolationists across the country. Then, he declared publicly in another "fireside chat" the need to provide full aid to Britain.

Before long, the Nazi war machine began to roll as an invincible force. It began to occupy nation after nation. On March 31, 1939, the British Prime Minister, Winston Churchill, announced in Parliament that if the German armed forces invaded Poland, then Great Britain would declare war on Germany.

With a startling swiftness, however, the Nazi air and ground forces swept across half of Poland and in rapid succession conquered the Netherland, Belgium and Luxemburg, and entered Norway on a pretense of an acute shortage of iron ore.

As the European crisis mounted, President Roosevelt declared the United States was in a state of limited neutrality. His fireside chats continued to be his most efficient means of communicating with the general public.

The tide of concern in America continued high and the courage of Mr. Roosevelt's dedication and talent of leadership lay in his skill in stating the issues in clear phrases understood by the average man:

"That was how he told the American people of his decision that aid to Britain should no longer be cash-and-carry, but should be lent or leased, that the U.S. should take over the role of arms supplier, inject itself into the war against Hitler, drop all pretense of neutrality, take the risk that Hitler would see this as a declaration of war. It was his way of selling a program that, however necessary and desirable, was totally fraudulent, as he described it. But it worked, even though Senator Taft, his voice sounding like chalk on a dry blackboard, replied immediately: 'Lending arms is like lending chewing gum. You don't want it back.'"**

As the European crisis worsened, Mr. Roosevelt declared a state of limited neutrality for the United States. His radio speech delivery, always calm and informal, was effective in allaying fears of his countrymen. Yet, he knew only too well, as did the top officials in the War Department, that America was indeed ill-prepared to fight a war, even if attacked.

It was during this time period the Japanese military forces were waging an aggressive war campaign in Asia. They invaded and occupied vast areas of land in China and South Pacific, Thailand, Singapore, and small island nations fell before the vastly superior air and land forces of the Japanese.

In Europe, the Axis military forces swept across France to the English Channel, their primary objective, in order to split apart the military forces of England and France. Having achieved that objective, the Nazi high command then directed attention to their conquering Paris and the remainder of France.

Mr. Roosevelt, having already proclaimed the limited national emergency, then asked the Congress to repeal the arms embargo. Congressional compliance thus gave the President an incentive for creating a long list of new government agencies to stimulate private industries, banks and agriculture in mass production of much needed supplies for Allies abroad as well as for consumption here at home.

In short order, the doubtful "new dealers" in Congress started to voice their skepticism about the sudden mushrooming of those new agencies which Mr. Roosevelt kept establishing. In Congress, the Republican opposition also voiced concern. Why was Mr. Roosevelt recruiting personal friends, top-level executives from private corporations, to run these agencies?[1]

The Washington Press Corps' media game questioned the need for these agencies if they were being established to perform the functions that were the jurisdiction of established departments in federal government. Metropolitan newspapers also questioned his abuse of power, whether the President was engaged in becoming a dictator.[2]

On November 4, 1939, President Roosevelt urged Congress to repeal the Neutrality Act, and to authorize U.S. merchant ships carrying munitions into European war zones "to be armed!" The Congress complied with his request but, in doing so, the U.S. was, in effect, "at war without a formal declaration of war."

Promptly, the Atlantic and North Sea lanes became danger zones for all U.S. shipping. The first casualties were the U.S. destroyer, *Kenny*, followed by the U.S. destroyer, *Reuben Jones,* while on convoy duty, the victims of U-boats.

By 1941, the Battle of the Atlantic intensified, with numerous sinkings. Using "wolf pack" tactics, Nazi Ü-boats sank a total of Allied country and U.S. ships totaling 668,000 tons.

Thus, U.S. involvement in war appeared imminent, so much so that Mr. Roosevelt began frequent conversations with Prime Minister Winston Churchill of Great Britain. Their secret letters and highly classified messages were numerous. President Roosevelt had already been informed by Albert Einstein, in October 1939, of the scientific possibility of developing an atomic bomb.

The historic records now show that Dr. Albert Einstein informed Mr. Roosevelt of the A-bomb feasibility in a letter dated August 2, 1939, and that the U.S. research endeavors to do so were initiated on June 15, 1940, with the Presidential appointment of Dr. Vannevar Bush as Chair of the newly established National Defense Research

Committee. Great Britain was not invited to participate in the research.

As the decade of the 1940s began abroad, the powerful Nazi Luftwaffe began its intensive air embarkment of Great Britain. Italy, now an Axis Ally of Germany, completed its invasion of Egypt and Albania, while the Nazi battle of Britain reached a climax during August, September and October 1940, as the Luftwaffe waged its all-out offensive against British land installations and its shipping, planning for an early land invasion.

On September 15, 1940, the British defenses destroyed "56 Nazi war planes." Winston Churchill, the Prime Minister, glowingly reported differently by his maintaining the Nazi air losses "totaled 185 planes." Mr. Hitler's plans for the land invasion of the British Isles were promptly abandoned - a major defeat for Nazi Germany.

These major events occurred prior to the re-election of President Roosevelt to his Third Term. He immediately took administrative steps after his Inaugural Ceremony to establish another agency, the Office of Production Management, to coordinate all defense efforts as well as to accelerate essential war supplies for Great Britain. With William S. Knudsen as the director of O.P.M., the United States became widely known as the "Arsenal of Democracy" worldwide.

The U.S. defense measures were further enhanced markedly in September, 1940 with the approval of the nation's Selective Training and Service Act. This peacetime registration was compulsory for all males, able-bodied U.S. citizens, between the ages 21 and 25, and for their induction for military training to last one year. This registration to draft males produced an eligible roster of more than 16 million men.

United States Isolationism - Neutrality - Involvement

Perhaps the most exacting draft (Winston Churchill) ever dictated...was not a speech but a letter. The July 31, 1940, cablegram of 4,000 words addressed by "Former Naval Person" to President Roosevelt pleaded for destroyers. These ships, "so frightfully vulnerable to air bombing," were crucially needed "to prevent seaborne invasion." To Churchill, the fate of Western civilization hung on the acquisition of these destroyers. "Mr. President, with great respect, I must tell you that in the long history of the world this is a thing to do now.'"*

Soon the resourceful Roosevelt found a way, in isolationist America, to release 50

destroyers in exchange for the United States' use of British naval bases in the West Indies. To Churchill, it was more than destroyers that were promised - it was deliverance. The first step had been taken in the road toward U.S. involvement. Only with the total commitment of the vast American resources would the balance be tilted against the Axis powers.

Endnotes:
*James C. Humes. Churchill, Speaker of the Century. New York: Stein and Day, 1981, p. 198.
**David Brinkley, Washington Goes to War. (New York: Alfred A Knopf, 1988). p. 51.
[1] Among the succession of new government agencies

established were the following: Office of Production Management, Agricultural Adjustment Administration, National Recovery Administration, War Resources Board, War Manpower Administration, War Shipping Administration, Works Progress Administration, Export-Import Bank, Civil Works Administration, War Industrial Board, Farm Credit Administration, Reconstruction Finance Corporation, Public Works Administration, Federal Bank Deposit Insurance Corporation, Tennessee Valley Authority, Homeowner's Loan Refinancing Corporation and others.
[2] The President's rebuttal occurred only within a closed cabinet meeting at which he stated that Federal Department personnel were not the caliber - just not qualified to meet the complex demands a war would require.

Campus of Duke University, Marine V-12, 1943. Front L to R: Major Walter G. Cooper, USMCR and Warrant Officer Blanchard. (Courtesy of Walter G. Cooper)

Navy V-12

A Tribute To Navy V-12 Trainees of World War II

By: Henry C. Herge, Sr.

I
All Aboard!

Yes, the Program took teenagers - poor depression-ridden kids,
Who had no college tuition, but Uncle Sam did!
Just to dream of college was much beyond belief.
Where could one get the money? One had to be a thief!
Social prestige, not a factor, for admission to good ole Yale;
One had to have the stamina and the will to prevail!
Some came from the fleet - or Marine Corps, too,
While "Old Salts" washed aboard straight from the briny blue!
You had to pass that very hard test,
To give you entree to the Navy's best.
Which made your parents stand and cheer,
While you trained to become an Officer!

II
On Retention

Oh, trainee, were you enrolled fifty years ago, plus some,
Did sheer fatigue and comfort cause you to succumb,
To slumber as "Reveille" announced the "break of day!"
And you lay dreaming of————, what can I say?
Oh, demerits, demerits seemed to just come your way,
And give you restrictions, but what could you say?
Why the skipper's reprimand made you weep
'Cause you couldn't help it - you did have to sleep!
Weekend "shore leave" for you, sheer delight,
And you hoped it would be with a Susie Bright!
As "leave" for you usually meant a date!
That was, of course, if you were not too late.
Math, the sciences, physics courses, all touch, you must agree;
While "Warplane Recognition" slides were just too quick to see
And "Courts and Boards" were harder than you thought,
Especially when a 4.0 was the grade you really sought!
Semester by semester and you found you still were there,
Not "bilged" like some lads for whom you really cared;
The question rose where you would go next,
It kept you and dear ones generally perplexed!
Then, your time to be graduated finally rolled around;
Your mom and dad came and you they found;
Your status exemplary, for they really knew,
Your plans for the future would in time come true!

III
Refrain

On leaving V-12, you went in a group
To Princeton and Colgate and Kansas and Duke;
You went a "trainee" to Navy's ROTC School,
Four months more - the general rule!
When you were commissioned, you became a man.
With that stripe of an Ensign, who rightfully can
Lead men into battle, be it thick or thin,
You had strength of character and manly discipline.
Well, now war is over, but you shouldn't forget
That your V-12 preparation was your strongest asset;
So, isn't it fitting to at least thank your host
Who in your time of need gave you the most?
    The United States Navy V-12
    College Training Program!

Chronology
Pathway to Global War
1932-1945

I. Focus on America
1932-1936

June 17, 1932 - Governor F.D. Roosevelt (New York) is nominated for President of U.S. by the Democratic Party at the National Convention, Chicago.

November 8, 1932 - Democrats gain majority vote and control of Congress as nation hails F.D. Roosevelt, "New Deal" President.

December 1932-March 1933 - U.S. in state of serious economic depression as banking system reveals huge deficits. On President's Inaugural Day some banks close - others were tottering.

March 6, 1933 - President declares a four-day bank holiday.

March 12, 1933 - President delivers first "fireside chat" to nation via radio and explains legal measures being taken to stem the nation's financial crisis and unemployment.

March 9-June 6, 1933 - President's "first one hundred days" - a critical period with deep depression continuing. Congress remains in session to enact social and economic legislation proposed by the "New Deal" President, F.D.R. Widespread isolation sentiment among voters keeps U.S. aloof to any European involvements.

January-December 1934 - Mood of cooperation remains in Congress with passage of relief measures to combat unemployment and to stimulate nation's economy.

April 8, 1935 - Enactment of a National Works Program that enrolls 3,400,000 persons at its peak in 1936, and 8,000,000 persons prior to its June 30, 1943 termination. "WPA" stood for Works Projects Administration, and "NYA," National Youth Administration, which was a part of WPA, concentrated on youths (16-25) not enrolled in colleges. NYA reaches its peak in 1940 when 750,000 youths enrolled.

August 14, 1935 - The President signs the Social Security Act to provide unemployment compensation, old age security and a variety of social service benefits for the destitute, homeless, the crippled and delinquent children.

June 9-11,1936 - The Republican Party, in its national convention, is critical of the "New Deal" Administration and of the President for his usurping Congressional powers, citing FDR's social legislation and creation of new agencies that are new government establishments. Democratic Party Convention, June 27, 1936, supports the President by heralding his courageous leadership for social reform and economic stimulation.

II. Isolation vs. Preparation
1936-1940

November 3, 1936 - In the national election, Mr. Roosevelt carries every state with an overwhelming majority (except in Maine and Vermont), giving the Democratic Party a large majority of 77-19 in the Senate and 328-107 in the House.

January 20, 1937 - President Roosevelt is applauded enthusiastically during his Inaugural Address for defending his efforts to stem the depression and increase employment.

January 3, 1938 - Mr. Roosevelt addresses Congress and reveals his deep concern about world conditions. In accenting the need for a strong defense, he also accents hemisphere solidarity.

May 17, 1938 - Naval Expansion Act passes.

June 28, 1938 - F.D.R. pressures Congress to enact armament appropriations, including $8,800,000 for anti-aircraft supplies; $6,080,000 for defense (industry) materials; and a huge appropriation for an extended program of Naval ship construction.

September 5, 1939 - President Roosevelt announces U.S. neutrality in the European War under the Neutrality Act of 1937.

September 8, 1939 - President Roosevelt proclaims a limited national emergency and, on September 21st, he asks Congress to repeal the arms embargo.

November 4, 1939 - Arms embargo is repealed and Congress authorizes the export of arms munitions to belligerent nations (allies).

June 28, 1940 - The Alien Registration Act passes Congress, giving additional strength to previous legislation governing deportation of aliens.

July 20, 1940 - Congress enacts legislation authorizing a two-ocean Navy with an appropriation totaling $4 billion for construction of capital ships and aircraft carriers.

September 17, 1940 - First peacetime Compulsory Military Service Act requires registration of all male U.S. citizens between 21-35 years to register.

November 5, 1940 - President Franklin D. Roosevelt is re-elected for his third term.

III. Neutrality vs. Involvement
July 1941-August 1941

March 11, 1941 - Lend-Lease Act passes with a special provision for $7 billion in military supply *credits* for Great Britain. This provision is extended to Russia in November.

April 9, 1941 - Denmark and U.S. agree to U.S. occupation of Greenland "for the duration," and on July 7, 1941, U.S. armed forces occupy Iceland to prevent German takeover.

May 27, 1941 - F.D.R. declares an unlimited emergency and orders the German and Italian Consulates in Washington to close. Axis powers reciprocate promptly by closing the United States Embassies.

July 24, 1941 - F.D.R. freezes all Japanese credits in the U.S., and Japan retaliates by freezing all U.S. and British funds.

July 26, 1941 - F.D.R. "nationalizes" the Philippines' armed forces and places General Douglas MacArthur in command.

August 18, 1941 - U.S. Selective Service Act is amended by Congress to require all draftees to serve an additional 18 months.

IV. Hitler's Rise To Power
1932-1945

1906-1913 -Vienna boyhood is aimless and lost as he dreams of his leadership. Deep-seated hatred of church and Jews takes form:

The ideology that gripped Hitler in his early Vienna years was Pan-Germanism, the belief that all the Germans of Europe should form one nation. Hitler detested the Hapsburg Empire, of which he was a subject, because the former rules, German though they were, had accorded political equality to the Empire's non-Germans - Poles, Serbs, Hungarians, Italians, Slavs and Czechs - whom the German-Austrians had once dominated. Hitler nurtured an abiding hatred for the Czechs who, commercially and intellectually, were the most successful of the Empire's minorities. When he came to inquire why the non-Germans had displaced the Germans from dominance, however, he identified none of the minorities as "the villains of the piece," yet another people altogether - the JEWS!*

1914-1941 - Adolph Hitler (1889-1945). His youth and early manhood are turbulent. As a soldier in World War I, prisoner and "would-be political leader," his anti-semitism is strong and vocal. He very slowly begins to gain as the obstacles to public recognition fade away and he gains stature among Storm Troopers and the Nazi Party members.

In Munich, Hitler suffers public rejection, November 9, 1933, with the "Beer Hall Putsch," and his arrest and imprisonment. While in prison, Hitler pens *Mein Kampf*, which defines his bigoted political philosophy: the advocacy of Germany's right to govern "inferior peoples," and his hatred of the Jewish people.

In December 1924, Hitler is firmly resolved to regain political status as the recognized leader of the Nazi Party. He employs every opportunity to spread his political philosophy while he is intent on gaining partisan support in Parliament.

The severe economic depression in the United States and in Europe is the barrier to early attainment of his ambition. Unemployment and financial turmoil in Central Europe spread widely. As a result, Hitler, 1925-1930, is frustrated and spends a major portion of his time toward improving his strident oratory and skill in arousing his audience emotionally.

He begins to gain the attention of the elderly military of World War I. Field Marshall Paul von Hindenburg, the idol of German voters and popular President of the new Republic in the elections of 1925 and 1932, is aware of Hitler's anti-communism stance.

January 1933 - Adolph Hitler, having won the Chancellorship, succeeds in gaining absolute dictatorial powers and the "Nazification" of the government. His announced goal is now "world domination" and the extermination of the Jews!

June 1933 - Hitler becomes "der Fuhrer" and with adequate votes in Parliament, after the "blood purge," he also succeeds with passage of the "Enabling Act," whereby he, alone, makes the law. So, with the stroke of his pen, World War II is a reality! Der Fuhrer, having created strong animosities, is able now to tear down, in quick time, all respect for human life and social institutions. Worldwide aggression becomes Hitler's philosophic objective as he teams up with Italy and Japan in the Tripartite Pact, on December 11, 1941, and declares war on the United States.

V. The World At War
World War II Involves Every Major Power
1931-1945

September 18, 1931 - Japanese troops, in violation of the Kellogg-Briand Pact and League of Nations, invades Manchuria and all of Southern Manchuria by June 14, 1932, and creates "Manchuko," a puppet state. Japan retains control despite League of Nations' diplomatic intervention efforts.

December 12, 1937 - United States-Japanese relations deteriorate with sinking of *USS Panay*, a river boat.

October 21-25, 1938 - Japan declares "New Order" in Asia with occupation of Nanking, Canton, Hankow by November 5, 1938.

September 1, 1939 - Poland is invaded by Nazis.

September 3, 1939 - Britain and France declare war on Germany.

September 27, 1940 - Germany, Italy and Japan sign three-power pact for military and economic alliance.

September 1940 - Japan launches campaign to dominate and control all of Southeast Asia and the South Pacific Island nations, including Indo-China, Thailand, the Philippines, Hong Kong, Malaysia.

January 22, 1941 - Nazi German forces start invasion of East Russia.

September 4-October 30, 1941 - Terror on U.S. Atlantic shores: German U-boats sink Allied ships at an unprecedented rate. Submarines are sighted from Maine to Florida, Office of Civilian Defense orders "black-outs" of shoreline towns and cities while civilian air raid wardens patrol the streets from dark to dawn. President Roosevelt issues order to "shoot on sight" while U.S. Naval vessels are on convoy duty. He also requests Congress to amend the 1939 Neutrality Act to allow United States merchant ships to have armed guards while transporting war supplies to Allies.

October 11, 1941 - *USS Kearney*, a destroyer, is torpedoed near Iceland by Nazi U-boat.

October 30, 1941 - *USS Reuben James* is sunk while on convoy duty in the North Atlantic.

November 17, 1941 - The 1931 Neutrality Act is repealed and revised by deleting all restrictive phraseology.

December 7, 11, 25, 1941 - Japanese air and naval forces attack Guam, Philippines, Wake, Hong Kong, Singapore and Midway Islands.

December 7, 1941 - Japanese, in a sneak attack, bomb U.S. Fleet at Pearl Harbor Naval Base.

December 8, 1941 - "Yesterday, December 7, 1941 - a date which will live in infamy - the United States of America was suddenly and deliberately attacked by naval and air forces of the Empire of Japan." *Franklin D. Roosevelt.*

December 8, 1941 - The U.S. Congress declares war on Japan; Great Britain declares war on Germany and Italy.

December 11, 1941 - Germany and Italy declare war on the United States, and Congress declares war on both nations.

December 19, 1941 - Congress extends military conscription to all male citizens between ages 20 and 44.

January 17-27, 1943 - Churchill and Roosevelt plan Allied invasion of Europe.

November 28-December 1, 1943 - Churchill, Stalin and Roosevelt convene and plan final war strategy vs. the Axis powers.

June 6, 1944 - Allies invade Normandy.

December 16, 1944 - Battle of the Bulge.

April 12, 1945 - President Roosevelt dies.

May 1, 1945 - Axis forces in Italy surrender.

May 8, 1945 - President Truman announces surrender of Germany via radio as U.S. celebrates "V-E Day."

August 17, 1945 - "Unconditional Surrender Ultimatum" terms are drafted at Potsdam Conference by Truman, Stalin and Churchill.

August 6 and 9, 1945 - U.S. drops two atomic bombs on Japan: (1) Hiroshima, and (2) Japanese Naval Base, Nagasaki.

August 14, 1945 - Japanese accept peace terms of the Allies. U.S. celebrates "V-J Day" as Japan surrenders.

Endnote:
* Keegan, John. "The World at War," U.S. News and World Report. August 28, 1989. P. 38.

1942-1945

Africa and Europe are the centers of hot war as U.S. Forces in Asia-Pacific Theater gradually turn back the Japanese air and naval forces.

Navy V-12 students at Milligan College. Both served aboard USS YP-619, Guam 1946. L to R: Al Gillespie and Carl Chance. (Courtesy of Carl Chance)

LCT Skippers relaxing in the Philippines - 1945. (Courtesy of J.H. Curtin)

Invasion of Southern France, August 1944. (Courtesy of P.B. Dalton)

Chapter I

Designing The Role For American Colleges

Education and the National Defense

Early in 1939, the ominous clouds of war cast dark shadows across the Atlantic upon the peace-loving people in America.

Addressing the need to create, quickly, a more adequate national defense, President Franklin D. Roosevelt declared, on September 8, 1939, that "....a limited national emergency exists."

As the "Arsenal of Democracy," the United States pledged war shortage materials for those European nations which so valiantly attempted to withstand the merciless onslaught of the Fascist powers.

Quickly, blueprints for mobilization were drafted and deficiencies in the national defense were brought into focus. Shortcomings were alarming! In rapid succession, the nation's munitions industries were pouring forth instruments of warfare intended for the Allies in Western Europe. The United States was not at war!

In the summer of 1940, the American Council on Education (ACE)[1] presented a document containing cautions and democratic process recommendations to the President of the United States.

The following are excerpts taken from that mandate:

All the agencies of education must be utilized for the most effective meeting of any national emergency.

Emergency programs should not interfere unduly with the regular work of the schools and higher institutions.

An undue insistence upon regimentations of thought and action, including distortion of textbooks and other materials of instruction, and the uncritical use of materials of propaganda, should be assiduously avoided.

The responsibility for administrative control of the agencies of education (should) continue in the hands of the educational officers of the schools and institutions of higher learning...curricula of schools and colleges...(should) be worked out cooperatively and be prepared under the direction of educators assisted by special committees charged with the formulation of general principles and the determination of basic areas for the preparation of such supplementary curricular materials as an emergency may make imperative...

Every effort should be expended for the preservation of the democratic process. The present international conflict has developed a clear unanimity of belief that upon (us) has now been laid, as perhaps never before,

the responsibility of maintaining and refining the essential process of democracy through their effective operation in the United States.[2]

To the members of Congress and the President, it was increasingly apparent the United States was ill-prepared to fight a technological war. It soon became a matter of mounting concern for members of the fourth estate as well.

Personnel Needs for Winning the War

On January 6, 1941, President Roosevelt alerted the entire nation when he addressed a Joint Session of Congress. He said, in part:

"We must all prepare to make the sacrifices that the emergency - as serious as war itself - demands. Whatever stands in the way of speed and efficiency in defense preparations must give way to the national need."

In short order, the press, radio and syndicated writers were addressing defense issues and commenting on the military master plan for complete mobilization of the nation's manpower for the armed forces and essential industries.

Selective Service Act of 1940

During the summer of 1940, the Congress passed the first peacetime Selective Service Act. On September 16th, the President affixed his signature and it became the law of the land. Immediately, every physically and mentally qualified male citizen between the ages of 21 and 36 became subject to classification for induction into the armed services - or for a deferment by virtue of his job classification.

Immediately, educators nationwide became concerned because a large proportion of college students were within the draft age population. The law, however, contained a clause for the deferment of any *bona fide* student until the end of the academic year 1940-1941, if he should request it of his local draft board. The Congress had inserted this clause in the Act to permit a flow of high caliber youths into the armed forces, yet conserve the institutions which supplied the training. University administrators began to visualize empty rooms in halls of ivy in the foreseeable future.

A second provision in the Selective Service Act allowed additional deferments for those men whose "employment in industry, agriculture or other occupations or employment, or whose activities in other endeavors were found to be necessary to the maintenance of the national health, safety or interest."

It was this latter provision in the Act that draft regulations were modified to allow for the deferment of students who were "in training and preparation for an essential

occupation." Twenty occupations and professions, wherein personnel shortages had developed, were listed in the official directive that served to guide Selective Service Boards. For example, instructors in certain areas as well as those preparing in engineering, medicine, dentistry, chemistry and physics were among those who might qualify for deferment.

But there were challenging opportunities opening for officer commissions. Sheer patriotic fervor moved many young students, as well as instructors, who might otherwise have remained deferred, to enlist.

Period of Vacillation

The delay of higher education and the armed services to define jointly the role of the colleges in the period of Limited National Emergency (beginning September 1939, until the spring of 1943), when the military announced the Army Specialized Training Program (ASTP) had been termed "the period of vacillation" for American higher education.

Some constructive action was beginning to take place within certain college and university laboratories in that selected personnel and facilities were being contracted for technological war research. Army and Navy ROTC units were competing on certain campuses for student enlistments. The National Roster of Scientific and Specialized Personnel (established in July 1940) was listing all highly qualified professionals in an array of specialized employment fields; and the Selective Service System was busily engaged in imposing numerical quotas of manpower for both civilian and military phases of national defense. ACE, however, was maintaining a constructive liaison between higher education and the armed services through its several standing committees.

Then, suddenly, on December 7, 1941, the Japanese bombed Pearl Harbor. Overnight, the U.S. citizenry were united nationwide. Patriotic fervor ran at high pitch within industry, commerce, business and higher education - all clamoring to assist in the war effort. The voice of higher education finally began to be heard as it pressed for specific policies and programs and for training men for military duty and professional service.

Power Struggle Within

In retrospect, one may question seriously why there was no announcement for setting in motion a comprehensive plan for an effective utilization of colleges and universities within the President's national address on the national emergency.

Investigation has now revealed that serious internal power struggles were operative. Through its several standing committees, the ACE had been quick to register its desires to be of service, motivated perhaps by a fear that as men were syphoned off by draft quotas, many institutions would be closed for the duration, and other institutions might become defunct, never to open again!

The War Manpower Commission (WMC) had made it very clear that its *legal role* included "responsibility for the appropriate utilization of manpower to meet civilian as well as military needs." The War and Navy Departments, on the other hand, felt it was *their* preprogative to determine *their* needs, including college-trained officers. Thus, the inertia and confusion within the government caused needless delays and frustrations among those who represented educational interests, particularly college administrators. At this juncture, the WMC re-emphasized *its* legal prerogatives for deciding the *total manpower needs* of the nation, and expressed its concern for educating scientists, doctors, dentists, technicians and an array of specialists required in civilian life. The WMC again stated it would fulfill its *role* for training specialized personnel for the military-a view that was obviously in contradiction with that held by the Army and Navy.

With the failure of the Congress to appropriate funds to aid college students to continue their studies during deferment, the chairman of WMC called upon the U.S. Office of Education "to draft a program for the feasible use of degree-granting institutions during the war period." This prompted the Army and Navy to join together!

Higher Education Takes the Initiative

Less than a month after the entry of the United States into the war, the most representative gathering of administrators in higher education was convened in Baltimore, Maryland, under the joint auspices of the U. S. Office of Education and Defense. While in session, January 3-4, 1942, the delegates pledged the total resources of American higher education to the war effort. The gathering recommended specifically the urgent need:[3]

(a) to determine the immediate needs of manpower and womanpower for the essential branches of national service - military, industrial and civilian; (b) to determine the available facilities of colleges and universities to prepare students to meet the needs; and (c) to appraise the ultimate needs in professional personnel for long-time conflict and for the postwar period in order that a continuous and adequate supply of men and women trained in technical and profes-

sional skills and in leadership to meet both immediate and long-range needs shall be maintained.

The Second Baltimore Conference on Higher Education and the War was convened on July 15, 1942. The Conference went on record reaffirming the earlier stand taken in support of the resolutions adopted in its January session and deplored the failure of the armed forces for their....lack of any adequate, coordinated plan for the most effective utilization of higher education toward winning the war...[4]

The Office of Education had formulated its recommendations, which were submitted to the chairman, War Manpower Commission. Made public at the same time were a series of resolutions adopted by the Second Baltimore Conference. The chairman, in turn, created a committee to review the U.S. Office of Education recommendations and the resolutions of the Baltimore Conferences. The WMC continued to be obdurate in its stand! It would not relinquish its control to any other agency of government. It and it alone would be responsible for developing a unified plan for college programs!

Finally, the internal rancor must have evaporated; for the question of control came to an abrupt halt in August 1942, when the WMC, the Army and Navy Personnel Board invited the ACE to appoint a committee on the Relationships of Higher Education to the Federal Government. This committee's creation proved to be the catalyst that quelled the forces that tended far too long to disturb the administrators of higher education.

The committee, at its creation, consisted of 12 presidents of the nation's most prestigious institutions, chaired by President Edmund E. Day of Cornell University.[5]

At this stage, both the Army and Navy procured, either as officers or as civilian consultants, many men and women directly from higher educational institutions to assist in laying the ground work for college training programs. With them they brought creative ideas, long dreamed about but never tried, together with educational experience and professional wisdom. By working together, they pooled their knowledge and suggestions. Theory and practice paid dividends; for never before in American higher education had there been such a golden opportunity nor so much funding available to put into operation such comprehensive and effective programs! Unlike the typical college or university system, the Army and Navy spared no expense in their training programs.

Representatives of the ACE were in agreement with the WMC in developing programs that would meet, speedily, civilian as well as military needs. However, it is appropriate to note that civilian educators and university administrators maintained a

deep-seated distrust of any governmental agency's curriculum efforts or utilization of facilities. They recalled their World War I unhappy experiences. (Note concluding paragraphs of this chapter.)

The WMC's position was that all students must prepare themselves for active, competent participation in the war! The ACE Committee subscribed to this declaration to a degree; and its rejoinder was as follows: (1) the WMC's statement was of little merit until it is translated into a program based upon factual data for types of service; (2) the number of persons required in various fields of specialization; and (3) the extent and nature of the training required.

In addition, the ACE Committee concurred in principle that all able-bodied men should be draft registered, but it qualified that stand by stating:[6]

"We assume the military authorities will (1) take into consideration over-all needs; (2) recognize that the needs of war, industry, and civilian life in certain specialized fields may be as important as those in the armed forces; (3) set up necessary programs; and (4) allocate trained manpower to war production and other essential civilian activities as well as the armed forces.

Thus, the period of vacillation came abruptly to a close! Already, the Bureau of Naval Personnel (BuPers) was actively engaged in long-range planning for a large scale use of America's colleges and universities since it became increasingly clear that the Navy V-1 College Program could not meet the need for young officer candidates demanded by the growing number of amphibious craft and other types of marine construction.

Differences and factions disappeared. The conciliation gave to the armed services the *sole* responsibility for the establishment, training and administration of training programs under military auspices. It was deemed unwise to create new training facilities so long as it was possible to use civilian institutions already available.[7]

The Joint Army-Navy-Manpower Committee moved rapidly in its certification of approximately 475 institutions as *eligible* to receive contracts.

It is of interest to note that the ACE continued to play a vital role in decision-making. Minutes of the Navy Advisory Educational Committee show that the ACE was strongly opposed to recruit training as a prerequisite to college unit assignment. This stand was accepted by the Navy and endured throughout World War II.

Emergence of Plans

The ACE Committee met with appointed representatives of both the Army and Navy.

Its first pronouncement indicated that a plan for the training of officers and specialists should provide for the selection of candidates on "...a broad democratic basis...irrespective of their economic status."

The committee report called also for a selection process by state and regional boards consisting of representatives of the Army and Navy and civilian educators. In addition, it called for the creation of a joint Army-Navy-Civilian Board to set quotas for contract colleges and universities selected for training.

Fortunately, it asked also for the utilization of as many institutions as seemed fitting under the conditions outlined. One feature of the report would allow each enrollee in a college program to register his own preference for an institution. His acceptance would be granted if the admissions process concurred.

In the preliminary phase for projecting a sound program of officer training at civilian colleges and universities, the Navy adopted a policy in order to avoid any semblance of being dictatorial or unreasonable in the exercise of its authority, but later on there were a few instances during the operation of the program that required tact or a compromise when a college tradition conflicted with Navy requirements.

In time of war, any consideration given to an enrollee for his right to select a college or the right of an institution to accept or reject him, reflected the rare wisdom of civilian educators on the committee. The recommendation, though modified later, remained in effect for pre-medical and pre-dental candidates once the college programs became operative.

Two significant aspects of the committee report called for a 12-month academic year, and for curricula that would be written jointly by Army and Navy personnel *and* civilian faculty members of institutions selected for specialization programs. In colleges where no ROTC (Army or Navy) was operative, it was recommended the officers-in-charge should (if possible) be selected from the faculties therein.

Oddly enough, the committee made no effort to keep policy matters within its control, inasmuch as the programs recommended were only generally defined; but the Navy V-1 Program had already demonstrated by that time that the military and the colleges could work together harmoniously. Later, the Navy V-12 Program confirmed the assumption.

The media, nationwide, headlined the announcement. It heralded the V-12 Program as a "cooperative venture in education" and praised its purposes enthusiastically, with the belief that its goal to produce college-educated officer candidates for

the Navy and Marine Corps could be accomplished.

The need for careful communication between the civilian educators in higher education and Navy officials became increasingly apparent. There were many unanswered questions that college administrators wanted answered, and professors wanted clarifications pertaining to prescribed curricula.

So, it was fortunate that before the program was made operative a national conference was convened at Columbia University in New York City on Friday and Saturday, May 14 and 15, 1943. Invitations were sent to the administrators of the 131 colleges and universities under contract from Vice Admiral Randall Jacobs, USN, the Chief of Naval Personnel. The invitations specified his own attendance, along with several Naval officers from within the Bureau, who were to be the administrators of the V-12 Program. In addition, the Admiral announced that the "Commanding Officers-to-be" of the 131 Officer Training Units would be in attendance. The Admiral's invitation also contained the suggestion that pertinent questions be submitted well in advance of the Conference so that they could be included in the Conference Agenda and made the focal point of open discussion periods during the two-day meeting.

Vice Admiral Randall Jacobs was the conference key-noter. He apologized for the delay in launching the V-12 and explained why an entire year had elapsed in setting up the program:

"Gentlemen, we are about to embark on an education program that will have important effects on American colleges, on the Navy, and most important of all, on the lives of thousands of this nation's finest young men. We must educate and train these men well so that they may serve their country with distinction, both in war and in peace. We must increase the temper of our education and our training. Everyone in this room, and the hundreds of others whom we represent, share the responsibility for the success or failure of this venture. I am confident that by pooling our resources of experience, judgment and energy, we shall start the program in the right way and carry it forward to final achievement of which we all may be proud....we decided that a plan of such magnitude, affecting the academic work of some 150 colleges and universities (including medical, dental supply, and theological schools) and the education of many thousands of students, should start only after the most thoughtful consideration, both by the Navy and by leading educators of the country."[8]

Admiral Jacobs then underscored three comments which would govern the Navy

in its dealings with the 131 contract colleges and manifold trainees:

1. The delays were warranted because every step in the V-12 Program had to be weighed carefully.

2. The Navy was making no promises it could not fulfill.

3. The Navy had to adapt its curriculum requirements in such manner that the least disruption would be created.

From the time public announcement was first made pertaining to the use of American colleges and universities until the selected start-up date, namely, July 1, 1943, for training Navy officer candidates at 131 colleges, was a bare six months.

In additional comments, Admiral Jacobs cautioned the educators assembled as follows:

"In working out the relations between the academic and the military portions of this program, specific questions will arise on every campus that will tax the patience and ingenuity of the college administrators, and the Naval officers in charge. I ask now that every effort be made by all parties concerned to solve such problems locally in a spirit of mutual confidence and full cooperation. The Bureau of Naval Personnel will be glad to offer suggestions, but in each college the final responsibility must rest on those immediately in charge.

This is a *college* program! Its primary purpose is to give prospective Naval officers the benefits of college education in those areas most needed by the Navy. We desire, insofar as possible, to preserve the normal pattern of college life. We hope that the college will give regular academic credit for all or most of the Navy courses, and we desire that college faculties enforce all necessary regulations to keep academic standards high.

We are contracting not merely for classroom, dormitory and mess hall space and for a stipulated amount of instruction, but for the highest teaching skill, the best judgment, and the soundest administration of which they are capable. We desire our students to have the benefits of faculty counseling, of extracurricular activities - in short, the best undergraduate education the colleges can offer."

The conference moderator was Captain Arthur S. Adams who, in the Training Division of the Bureau of Naval Personnel, was to be the officer-in-charge of the Navy College Training Program, V-12. Among his preliminary remarks as moderator, Captain Adams stated that the Navy had no intent of "taking over the colleges." Instead, he pointed out that the Navy sought initially to take full advantage of each institution's academic resources; second, to make maximum use of each institution's experience and knowledge of procedure by having each decide the length of the college day, the time for classes, scheduling of exercises, meals, recreation, division of time between classroom and laboratory as well as the textbooks for use in prescribed courses. Captain Adams also emphasized the role of the contract colleges was paramount in counseling V-12 trainees and in maintaining academic standards.

It was fitting that long before the termination of the V-12 Program, Captain Adams' colleagues in uniform had named him "*Father of the V-12 Program.*" For attendees at the two-day conference, it is stated that all questions pertaining to the program were answered fully and understandably and thereafter did not need to be countermanded at a later date.

The wording of the Navy's policy statement of World War I was the antithesis to the language in World War II when the government informed American higher education with Student Army Training Corps units as follows:

"During the war you are no longer degree-giving institutions but rather short-course training schools for the specific purpose of preparing officers for the Army and Navy."

The Columbia Conference was but one of many that was convened nationally in order to keep contract colleges fully informed of the Navy's mission and of problems that came into being. The V-12 Bulletins, as they appeared, also were intended to give advance notice of any changes in quotas, curtailments or procedures. These combined efforts resulted not only in good public relations, but also in maintaining communication.

The Effects of War on Higher Education

"All This in Time of War is Necessary and Desirable...."

"We are now going through a period when the whole higher education program is being shaken down and when, therefore, we might expect a desideratum of the really important considerations. One thing certainly stands out clearly. In a period of war, the federal government very largely determines what happens to colleges and universities. It tells them what students it may have. It chooses which institutions it will use for specialized training and fixes the curriculums of the institutions. Its actions deeply affect all sources of income, including income from endowment. It takes away such members of the faculty as it wishes. For those who remain, it determines hours of teaching and rates of compensation. Certainly, if this war continues for any considerable length of time, the colleges and universities are going to have extensive experience with federal control of higher education."[9]

At the close of World War II, college administrators of contract institutions had much reason to be thankful. The dour prophecy[10] of Dr. George F. Zook did *not* occur. In fact, their reactions to the prediction were directly different, as any university administrator will attest.

Endnotes:
[1] *The American Council on Education is a* council *of national educational associations; organizations having related interests; approved universities, colleges, and technological schools; state departments of education; city school systems; selected private secondary schools; and selected educational departments of business and industrial companies. It is a center of cooperation and coordination whose influence has been apparent in the shaping of American educational policies as well as in the formulation of American educational practices. Many leaders in American education and public life serve on the commissions and committes through which the Council operates.*
[2] *Zook, George F. "How the Colleges Went to War,"* The Annals of the American Society of Political and Social Sciences, *Vol. 231 (January 1944), p.1; and* American Council on Education, Education and National Defense, *The Council, (June 1940), pp. 11, 12 and 14.*
[3] *For a complete transcript of the conference, see the report published by the American Council on Education, entitled* Higher Education and the War, *The Council, 1942.*
[4] *Second Baltimore Conference,* Higher Education and the War Bulletin, *Number 31 (July 24, 1942), The Council: Washington, D.C.*
[5] *Other members included President James P. Baxter, Williams College; Dean Paul H. Buck, Harvard University; President W. H. Cawley, Hamilton College; President Carter Davidson, Knox College; President Rufus C. Harris, Tulane University; Dr. Guy E. Snavely, Association of American Colleges; Chancellor William P. Talley, Syracuse University; Professor Ralph W. Tyler, University of Chicago; Dr. George F. Zook, ACE ex officio; and Dean T. R. McConnell, University of Minnesota.*
[6] *War Manpower Commission* Report of the Special Committee on the Utilization of Colleges and Universities for the Purpose of the War, *WMC, August 1942.*
[7] *Ibid.*
[8] *Navy Department, Bureau of Naval Personnel, Conference on the Navy V-12 Program at Columbia University, May 14-16, 1943, NavPers 15012 (1943), p.3.*
[9] *George F. Zook, "Summary of the Effects of War on Institutions of Higher Education,"* Higher Education Under War Conditions, *Vol. XV, 1943, P. 158, University of Chicago.*
[10] *Editor's note: "Aye, aye, sir! How wrong you were!"*

Chapter II

Inception of Navy College Training Programs

Decentralization of Control

The major portion of this Chapter is an adaptation of Chapter V in Wartime College Training Programs of the Armed Services. *By permission; see Acknowledgements.*

Although over-all plans were publicly released in December 1942, it was not until May 19, 1943, that the Navy Department announced that nearly 80,000 young men would be called to active duty and assigned to study under the new college program on July 1, 1943. This release stated that of the 80,000 total, about 15 percent were to be officer candidates for the Marine Corps, a few hundred were to be assigned to service with the Coast Guard, and the remainder were to be future Naval officers.

Literally hundreds of thousands of officers and men were being indoctrinated or were receiving basic training at Naval training schools and contract units throughout the country. The need for decentralization was apparent, as the following indicates:

Accordingly, a director of training was appointed in each of the Naval Districts as a member of the Commandant's staff. These directors of training carried no authority of their own, but served a very useful purpose as service agents of the Bureau, interpreting Bureau directives and regulations to the various schools in the district, giving advice concerning total problems, making periodic inspections, serving as a clearing house for all district training activities, and thereby saving the Bureau much time.

When the V-12 Program was established, the district directors were given the general supervision of the units within their respective districts, and were ordered to make an inspection of every contract school at least once each term (four months). The visits were not solely for the purpose of inspection, but also to discover local procedures that were singularly effective in the administration of the program. These were to be reported to the Bureau and passed along to other commanding officers for their adoption.[1]

Swarthmore College, Dec. 1943. L to R: Chu Sha-Ping. Ch'ang Yii Kuei, Lu Chin-Ming. Kneeling: V-12 Francis C. Tatem. (Courtesy of F. C. Tatem)

Drew University, V-12, Dec. 1943. "The Third Floor Gang." Rear Row L to R: Dean Staats, Dave Shimmel, Bernard Kott, Krona Krause, Jess Byers. Front Row L to R: Stanley Williamson, Frank Miller. (Courtesy of Dean Roy Staats)

U.S. Navy V-12. University Professor Arnold demonstrating applicatios of the slide rule in Mathematics Class at Wesleyan University, Middletown, Connecticut, 1943. (Courtesy of H.C. Herge)

On July 1, 1943, the V-12 Program was launched at 131 colleges and universities, exclusive of medical, dental, and theological schools.

The program and its flexibility: Great difficulties had to be overcome in the early stages of quota planning but the comprehensiveness and flexibility of the V-12 Program made possible numerous adjustments necessary where projected plans were not in line with current needs. So comprehensive was the program that men formerly classified V-1 and V-7, no matter what their majors - pre-medical and pre-dental students, students preparing for commissions as supply, deck, or engineering officers, Marine Corps Officer candidates, and a few from the Coast Guard - all were included in the same program....The fact that there was a common core of training, which even the pre-medical and pre-dental students received in slightly different sequence, made possible the transfer of men from one curriculum to another whenever it was necessary.[2]

Transitional Programs

When the official announcement of the Secretaries of War and Navy on the armed services' plans for the use of college facilities was released on December 12, 1942, several Navy training programs were already in operation at colleges and universities. The story of their mission, scope, successes and shortcomings, and final absorption by the V-12 Program is best described in the Navy training history, from which the following sections are adapted, by permission.[3]

In the 1920s and 1930s the Naval Academy and Naval Reserve Officer Training Corps furnished the Navy with the young officers needed to fill its ranks. NROTC units had been established in 27 colleges or universities by 1940, and in 1940-1941 had a total enrollment of 3,096 students who took courses in Naval Science and Tactics taught by Naval officers, along with their other college work. As the numbers involved were small compared to the demands of war and only a few colleges had units, comparatively little administrative experience was gained that was applicable to the establishment of a college program vastly expanded under the urgency of war.

At the outbreak of the Second World War, the Bureau of Naval Personnel continued to approach the twin problems of procurement and training in terms of World War I. No plans were drawn for training personnel in anything like the numbers that were soon to be needed...Yet, by the end of that year, the number of officer candidates alone equalled the total officer and enlisted strength of the Navy in the summer of 1939...

The need for midshipmen's schools to train young Reserve officers for the fleet was recognized in the spring of 1940 and during that year Reserve Midshipmen's Schools were commissioned on the *Prairie State*, the old *USS Illinois*, which had been tied up in the Hudson River and converted into a training ship; at the Naval Academy; and on the Chicago campus of Northwestern University. A Supply Corps School was established the next year at the Harvard Business School, and several other universities, schools for training diesel officers had

been placed under contract. Thus, prior to Pearl Harbor, the Navy had made some use of the physical equipment and plants available on the campuses of the country, and in a few instances the skills of the faculty as instructors had also been employed.

The great development in the use of colleges and universities for training programs, however, came after December 7, 1941. Numerous training activities, for both officer and enlisted personnel, were established on campuses in 1942 to provide a wide variety of specialized training. Enlisted men were trained as radiomen, electrician's mates, signalmen, storekeepers, and operators of diesel engines. Officers were trained in meteorology, radar, military government, foreign languages, and bomb disposal, and attended "indoctrination" schools to the chagrin of those faculty members who had long looked upon "indoctrination" as the very antithesis of "education..." None of this training involved general education, and in a large proportion of the classes the instruction was by Navy personnel. College facilities were being used, but not, except to a very small extent, college faculties.

The Navy V-12 Program

Meanwhile, in June 1940, three days after the Franco-German Armistice had been signed, the V-7 Program was announced to further the flow of college men into the Reserve Midshipmen's Schools that were being planned. The original announcement called for the enlistment of 5,000 men as apprentice seamen, V-7, for one month's training afloat. At the completion of the cruise, the successful candidates were to be appointed Reserve Midshipmen and given a 90-day course of instruction leading to a commission as ensign, USNR.

Candidates were to have completed a minimum of two years of college, and were limited to the ages of 19 to 26. During the following year, 7,200 enlisted but only 4,600 succeeded in earning commissions.

Because of the large number of failures among the candidates of the first year, the educational requirements were raised to four years of college, including two semesters of mathematics. These higher requirements appear to have been justified as there was a 50 percent reduction in the rate of failure, but the reduction may also have reflected the greater incentive to win a commission that arose from the increasing probability that the United States would soon be at war. By November 1941, all quota restrictions on the enlistment of engineering students in the V-7 classification had been lifted as the rapidly expanding fleet called for an ever greater number of officers.

The Navy V-1 Program

Within a month of the announcement of the revised V-7 Program, the Navy created still another class of enlisted men who were to remain in an inactive status while undergoing college training. This new V-1 Program was open to men 17 and 19 years of age willing to attend college at their own expense until the completion of the equivalent of two academic years. Men in V-1 were to study a curriculum prepared by the faculty of whatever accredited college or university the individual chose to attend. The only limitation on the curriculum was that it must be acceptable to the Navy and must stress physical training, mathematics, and the physical sciences.

The Navy had hoped to procure "not more than" 80,000 men in V-1. Near the end of the third semester an examination was to be given on the basis of which approximately 20,000 were to be selected annually for transfer at the end of their fourth semester to V-5 and flight training. Another 15,000 were to be transferred to V-7 and continued in college at their own expense until a degree was received. Five thousand of these were to be engineering students. Those not selected for either V-5 or V-7 were to be called to active duty as apprentice seamen at the end of their fourth semester.

The only curricular requirement of the original V-1 plan was that the student follow a course of study drawn up by the faculty and approved by the Navy, which would emphasize physical science, mathematics, and physical training.

The department, in view of the fact that enlistments in V-1 were not what had been

hoped, became alarmed that such curricular would defeat the purpose of the entire program by discouraging all but a small number of the potential enrollees.

The V-1 Program was not a procurement success from the numbers point of view, but those college men who did enlist in it got more than an even break. Instead of getting something less than a 50-50 chance of entering an officer candidate class, all V-1 students who maintained themselves in good standing in a college or university through the academic year 1942-1943, were transferred to the V-12 Program which offered them more opportunities to gain a commission than had ever been promised. Throughout the entire transition to the V-12 Program, the Bureau was careful to see that every promise made to students in either of the two earlier programs was scrupulously met.[4]

The Navy V-5 Program

The Naval Aviation Cadets (V-5)[5] were men who had been procured by the Naval Aviation Cadet Selection Boards (NACSB) by a different procedure from that used subsequently in the selection of other V-12 students, and who were later transferred from V-5 to V-12 with the special designator "(a)." In their selection, different standards were applied. Motivation (a strong and persistent desire to fly, the construction of model planes, and other evidences of an interest in aviation) was given much emphasis. Physical standards were also high, but if a boy was strongly motivated and passed his physical examination, the intellectual requirements were less rigid than those required of V-12 apprentices. After the inception of the V-12 Program, instances were not uncommon where boys who were turned down for V-12 went across the street to the NACSB office and were enlisted in V-5, to be transferred into the V-12 Program at a later date as a V-12 (a).

When the V-5s were transferred to the classification V-12 (a), it was agreed by both DCNO (Air) and the Bureau of Naval Personnel that they should receive two terms of college training before assignment as aviation cadets. Apparently there was concern from the first that some of the V-5s lacked a suitable mathematical background to do college work, because an understanding was reached in a meeting of the Joint Procurement-Training Committee that the V-12 (a)s were to be subject to the same academic standards as the V-12s and also subject to the same separation procedures. Several meetings of the committee discussed the academic difficulties of the V-12 (a)s and there was good reason to do so, as rumblings of discontent were being re-

ported from several schools where they had been assigned. Some V-12 (a)s, whether interested or not, were not academically or intelligently prepared to do the work in the V-12 curriculum. In such cases, the maintenance of morale became a real problem. At the end of the first term, it was found that the academic attrition rate for those in V-12 was 5.86% and for those in V-12 (a), 9.62%.

In time, the V-12 (a)s adjusted themselves to the idea of eight months of college. Meanwhile, DCNO (Air) was beginning to discover that it had set its sights too high and that the aviation program would have to be trimmed down. One way to accomplish that, and at the same time raise the educational qualifications of the men, was to keep the V-12 (a)s in college for a third term.

The question of the third-term basic curriculum was finally settled in a meeting of the Joint Procurement-Training Committee. One of the representatives of training outlined objectives that were apparently acceptable to all: (a) give the aviation candidates such academic work as would help them become better Naval officers and not just better pilots, and (b) furnish training that would permit them, if it were ever desirable, to transfer to other curricula. In line with these objectives, the committee decided that all the V-12 (a)s should take the third-term basic curriculum and that no distinction should be drawn between V-12 and V-12 (a) Programs.

Naval Aviation Training: A Typical Pattern

Before the absorption of V-5 by the V-12 Program, a Naval Aviation Cadet (V-5), in order to achieve his "wings," was ordered to a chain of training activities. Originally, the pattern was as follows. However, the V-12 Program eliminated the first two schools:

1. Naval Flight Preparatory School (NFPS): Three months' formal ground training at any one of 20 colleges under strict Naval discipline, where the courses, except for Naval subjects, were civilian-taught and consisted of communications, navigation, recognition, physics, mathematics, theory of flight, power plants, and physical education.

2. CAA War Training Services (CAAWTS), or Preliminary Flight School: Three months' elementary and secondary flight training under the CAA in light aircraft, totaling 50 hours, and 240 hours of ground school instruction.

3. Pre-Flight School: The emphasis at each of five colleges was to develop qualities associated with first-class physical fit-

ness, quick thinking, iron discipline and teamwork. Cadets also took continuing courses in seamanship and gunnery.

4. Primary Flight School: Here the cadet was given 85 hours of flying instruction in Naval N2S and Stearman primary trainer land planes, and 168 periods of ground school work.

5. Intermediate Flight School: At this school (Corpus Christi or Pensacola) the cadet received 14 weeks' instruction in SNJ and SNV basic training planes, including use of equipment, take-offs and landings, formation, night and cross-country flying, and finally, instrument flying. Prior to being commissioned ensign or first lieutenant in the Marine Corps, the cadet was given advanced squadron training in the fleet.

6. Operational Training: When commissioned, the new officer received his official orders. On active duty, he received eight weeks' additional training in the type plane he would be flying. Thus, after 70 weeks, totaling 1,400 periods of ground training, 300 hours of flight training, the young Naval officer went into active combat.

Naval Reserve Officers' Training Corps (NROTC)[6]

The establishment of an NROTC in American colleges and universities was authorized in 1925. The original six institutions, where the course was put into effect in 1926, had increased to 27 prior to the outbreak of World War II. Until July 1, 1943, a basic course covering the first two years, and an advanced course extending through the junior and senior years, were offered. The curriculum, so far as the choice of subjects went, was based on the curriculum at the Naval Academy; but the actual scheduling of the courses was the individual concern of each professor of Naval Science and Tactics.

With little or no centralized control, the curricula at the various institutions authorized to conduct NROTC training grew up in virtual independence of each other. In a letter of September 4, 1941, the Bureau of Naval Personnel expressed its desire to standardize training to the point where all units would teach the same subjects in each of the four years allowed for the course, thereby bringing candidates to the same relative stage of training prior to graduation. Alternate type schedules were offered for comment; but no effort was made to prescribe how much time should be allocated to each subject beyond a recommendation that the general outline of instruction at the Naval Academy should be taken as a model.

On July 1, 1943, the Navy V-12 Program went into operation under wartime emergency legislation. The NROTC Program thereupon became amalgamated with it but was allowed to retain its own identity since it was a continuing program established by Congressional action and since it was always intended that, with the return of peace, the training of Naval Reserve officers should be put back on the pre-war basis. Since the first two terms of V-12 included no Naval indoctrination, other than a basic course in Naval Administration, NROTC students were required to complete, in an accelerated course of five semesters, the work which had hitherto been scheduled for eight semesters. The previous division of the course between basic and advanced levels was discontinued, and candidates, instead, indicated their preference for majors in General Line, Business Administration, or Engineering, within the Naval curriculum.

The V-12 Program[7]

The V-12 Program was distinctly a college program designed to give officer candidates the requisite and minimum education necessary to more specialized training as officers, and to give it on an accelerated schedule. Aside from the pre-medical and pre-dental students, there were three groups that had to be kept in mind in the preparation of curricula. The *first* included students who had been transferred from V-1 and V-7 with some college work already completed. Because of the promises that had been made at the time of their enlistment, they were permitted to continue the curricula they had been pursuing before their transfer to V-12 so long as they met certain minimal requirements and care was taken to send them to schools offering competent instruction in the field of their respective majors. *Bona fide* engineering students were permitted a total of eight terms of college work from their matriculation. Pre-medical, pre-dental, and pre-chaplain students were given the minimum time necessary to complete the requirements for admission to their professional training.

The *second* group of students included those who had never been enlisted in any of the earlier programs, but who also had advanced standing. They were allowed much freedom but were urged to include as much of the fully prescribed curricula as the number of terms allowed, and had to satisfy at least the minimum requirements established for the V-1s and V-7s. The trainees, in neither of these two groups, were classified according to the type of commission they would receive or the duty they would perform after commissioning until they were ready for Midshipmen's Schools.

Both of these first two groups of trainees, all of whom were advanced students, were called "irregular" and during the early part of the program furnished the unit administrative officers and college deans much work in adapting their former programs to the requirements of the Navy and in determining the number of terms of college work that had been completed. This use of time-spent-in-college rather than credits earned to fix the amount of time allowed was almost the only workable basis that could be used, however, where so very many institutions were involved. At all times, the Navy took the position that it could not be concerned with the interminable question of college credits, and in view of the extreme skepticism with which instutions of higher learning look upon the credits of sister institutions, the decision was sound.

The *third* group of students were those who entered the program as first term freshmen and who were required to follow a fully prescribed curriculum.[8]

All these "regular" V-12 trainees, with the exception of the pre-medical and pre-dental students, first took two terms of work that was identical for all types of officer candidates. During the second 16-week term, the trainees were screened to one of 19 upper-level curricula which were intended to prepare the trainee for the type of duty he would later be assigned. The advanced curricula varied in length from two terms for deck candidates to six terms for engineering and pre-theological students.

The V-12 Curricula

Naval Subjects: Indoctrination. Only two courses in Naval subjects were included in the V-12 curricula, except courses in the NROTC. A two-term course in Naval Organization, meeting one hour a week, was required of all regular students during their first two terms, and all except supply, pre-medical, pre-dental, and pre-chaplain students took a three-hour course in Naval History and Elementary Strategy with their more advanced courses. The course in Naval Organization was usually taught by one of the Naval officers attached to the unit and covered Navy Customs and Courtesy, Navy Law, Intelligence, Naval Communications, Naval Personnel, and Navy Organization, ashore and afloat. The course served an important function in introducing trainees, particularly those without previous Naval experience, to certain basic knowledge essential to all officer candidates. The quality of the course varied even more widely than is commonly true of courses taught by different instructors because most of the officers had had very little Navy experience and there was very little material at hand to help them in their initial efforts. Anxious as

Middlebury College, Navy V-12 Unit, 1943-44. First Row L to R: Metcalf, Urban, Sokal, Mulany, Rockstroh, Taylor Company Comm., Marsh Platoon leader, King, Shuster, Tobias, Hoover, King. Second Row L to R: Marotle, Lennart, Pinchitti, Speck, Shockely, Moseley, Sauter, Stambaugh, Wood, Spu, Mazzoil, Thaler. Third Row L to R: Strauch, Koester, Plumb, Pilcher, Trivinan, Scanlon, Whittimore, Morgan, Thrickel, Kelley, Becker, Wosinsku. Fourth Row L to R: Schaffer, Hutchins, Stidham, Williams, Kinker, Rork, Woenker, Styskens, Short, Jennikar, Irwin, Waters, Westerdale, Royson. (Courtesy of K.H. Irwin)

Commnencement Day, Feb. 23, 1944. Navy V-12 Unit, Wesleyan University, Middletown, Connecticut. Honor Guests and Speakers L to R: Cmdr. William S. Thompson, Director of Training, BuPers; Victor L. Butterfield, University President; Gov. Raymond E. Baldwin, Connecticut; Cmdr. Henry C. Herge (then a Lt.), USNR, Commanding Officer; Lt. K. Justus, Chaplin, U.S. Navy; Lt. Gerald W. Anderson, Executive Officer, USNR; Ens. Elizabeth Barrickman, Disbursing Officer. (Courtesy of H.C. Herge)

most trainees were to learn more about the service of which they had become a part, the trainees could have made very good use of an intelligently written book had it been available in sufficient numbers.

The only other Naval subject taught in the V-12 Program was the one on Naval History and Elementary Strategy, which presented no unusual problems except possibly for the civilian history instructor who suddenly found the course thrust upon him. Sound historical material in the form of specialized histories, biographies, and textbooks was available in quantities sufficient to meet most needs, and Brodie's *A Layman's Guide to Naval Strategy* proved to be a timely aid in teaching certain elementary and basic principles of that art.

Pre-Professional and Professional Curricula
(Medical, Dental and Theological)

Problems and Remedies. The medical, dental, and theological curricula gave rise to special administrative problems caused primarily by the fact that the training was professional in nature.

By an agreement between the Surgeons General of the Army and Navy and the War Manpower Commission (WMC), the available space in the professional schools of the country was allocated as follows: the Army was to use 55% of the space in medical schools, the Navy 25%, and 20% was to be left for the use of civilian students; of the dental space 35% was assigned to the Army,

20% to the Navy, and 45% was left for civilian use. It was further agreed that if the entire space available to civilians was not used, the Navy would be entitled to one-third of whatever space was in excess, thus permitting a maximum total of 31 2/3% of the medical space. To prevent undue expansion of the medical program, the Procurement-Training Committee ruled that if civilian students enlisted in the Navy, they and any ensigns H-V (P) on inactive duty in the professional schools were to be included in the Navy quota, which was to never exceed the 25% originally contracted for, plus a maximum of 6 2/3% of any unused civilian quota.

Considerable pressure was exerted from time to time by the WMC to have the Navy train men for civilian uses. Each such effort was in vain as the Navy took the position that responsibility for meeting civilian needs, however justifiable, did not properly belong to the armed services. One of these efforts concerned the medical program. In April 1944, the Army announced a reduction in its Specialized Training Program, which meant that 2,000 spaces would go unfilled in medical schools and 1,000 in dental. The chairman of the WMC was immediately concerned as the country faced a serious shortage of physicians that would be felt for years after the war if the flow of medical students was not continued. The situation was further aggravated by the decision of the director of Selective Service, due to the great demand for physically qualified young men in the services, to do away with the deferment of pre-medical and pre-dental students unless they had matriculated in a professional school by July 1, 1944. A meeting of the Surgeons General of the Army, Navy, and Public Health Service was immediately held with representatives of WMC. It was their unanimous opinion that the position taken by Selective Service would "result in a situation which would be disadvantageous and even dangerous to the armed forces and the civilian population." It was agreed that the matter was of serious enough import to be taken to the President, if necessary, to obtain action. They underscored their feelings in the following words:

"We all wish to go on record to the effect that should later developments such as an epidemic of great magnitude, war casualties of unanticipated numbers or any other unexpected demand upon medical manpower (occur)...we shall not be in the position of not having done everything in our power to have prevented this manpower production interruption. We don't believe that the places vacated in medical schools by the change in Army plans can be filled with women, 4-Fs, and men discharged from the armed forces. We feel that this is perhaps the most serious problem on which we have had to take a position, and for that reason we feel it is our duty to take an unequivocal position."

Despite the force of this argument, the director of Selective Service refused to change his policy, with the result that hundreds of young men who were ready to enter medical schools in the fall of 1944 were drafted. In a vain attempt to forestall the section of Selective Service, the chairman of WMC, therefore, appealed to the War and Navy Departments to place sufficient pre-professional undergraduates in an inactive duty status to maintain the flow to medical schools. The numbers were to be kept as low as possible, and the machinery for allocating the students to the professional schools could be handled, he suggested, through the Procurement and Assignment Service which provided medical officers to the armed services. Not more than 4,000 pre-meds and 2,000 pre-dents, including men both under and over 18, would be involved. In keeping with its policy on similar occasions, the Navy rejected the proposal on the ground it was not the Navy's mission to train doctors and dentists for the civil population.

During the operation of the V-12 Program, approximately 4,600 medical and dental students completed their professional training in addition to another 5,000 medical, and nearly 1,400 dental students who received a part of their professional education during that time. These figures do not include the thousands enrolled in the pre-professional curricula on the undergraduate level. At the end of the war a considerable proportion of the Navy's Medical and Dental Corps, approximately 23% of the medical officers and 20% of the dental, had completed their education in the V-12 Program. Thus, the program had accomplished much to overcome the serious shortage of officers in those fields, and in another year the proportions would have materially increased.

Theological Curriculum

Pre-theological and theological students were included in the V-12 Program on the basis of denominational quotas which were at no time very large, the total number being approximately 400. Pre-theological students were permitted eight terms of college work during which they carried some electives as well as the prescribed courses in the liberal arts and, on the professional level, theological students received six terms of work. Early in their training, students who wanted to become chaplains applied to the Bureau of Naval Personnel for such permis-

sion. If their academic record and physical and officer-like qualities qualified them for consideration, the applications were then sent to the Chaplaincy Commissions of the various denominations participating in the program. The Chaplaincy Commission of the applicant's denomination had the responsibility of ascertaining, from his synod, conference, presbytery, diocese, congregation, et cetera, whether or not he was a *bona fide* candidate for the ministry and of certifying the fact to the Navy. At any point in his training, the candidate could be dropped from the program for any of the reasons applying to other officer candidates and, in addition, he was subject to separation if his denomination withdrew its ecclesiastical endorsement.

Only 29 former V-12 students went on active duty as Naval chaplains before the Japanese surrendered, but by the time the theological portion of the V-12 Program closed in March 1946, another 45 had been commissioned in the Chaplains' Corps. It is, therefore, evident that the Chaplains' Corps did not profit materially from the men trained in V-12; however, had the war lasted for another year, the number of chaplains furnished by the program would have been much larger and would have been an important source, inasmuch as the procurement of chaplains from civil life had nearly ceased. Furthermore, there is again evidence to indicate that the program had real value in developing good public relations with the various denominations by showing a concern on the part of the Navy for the spiritual welfare of its men.

Auxiliary Enterprises[9]

In addition to the main V-12 Program, there were four auxiliary programs, small in size, administered by the College Training Section in accordance with much the same policies that controlled the larger program. The V-5 and V-7 academic refresher programs were college programs designed to enable V-5 and V-7 candidates from the fleet who had already had some college education to enter Pre-Flight Schools and Reserve Midshipmen's Schools with a preparation more nearly equal to that of the men coming from the V-12 Program. The other two programs, the Pre-Midshipmen's School and the Pre-V-12 School, were joint efforts to solve certain administrative problems which the V-12 Program had itself engendered.

The Pre-Midshipmen's School came into being because of the impossibility of sending on to Midshipmen's Schools all the graduates of the V-12 Program immediately after the completion of their undergraduate training. In order to make possible trans-

fers from one V-12 unit to another, the Navy had required all participating institutions, with minor exception, to go on a three-term calendar. The opening of classes in the Midshipmen's Schools did not wholly coincide, however, with the graduation dates in the V-12 Program. Consequently, interim duty had to be provided thousands of V-12 trainees awaiting transfer to Midshipmen's Schools, and it was decided to establish a Pre-Midshipmen's School in Norfolk to serve that purpose. In reality, the school was to be a receiving station or pool and, at first, little thought was given to the question of training during a trainee's stay there.

Because of the crowded conditions in the Norfolk area, the Pre-Midshipmen's School was moved to Asbury Park, New Jersey, on March 1, 1944, and a year later, to Princeton University. While the program never proved wholly satisfactory, it served a very real need and the only alternative to such a "pool" of trainees would have been to pull some trainees out of the middle of their last term of V-12 training so as to maintain an even flow to the Midshipmen's Schools - a policy to which all those sections in the Bureau responsible for the V-12 Program were unalterably opposed.

Had the Pre-Midshipmen's School operated by itself, it would have been extremely busy the first two months of each four-month term and then have practically no trainees on board the last two months. This might have necessitated disbanding all but a skeleton force of Ship's Company in the middle of each term and reassembling a new one at the beginning of the subsequent term. Fortunately, the Pre-Midshipmen's School was coupled with the need for Pre-V-12 training and made possible continuous operation of the school. During the first two months of a term, it was a Pre-Midshipmen's School and during the last two months, it became a Pre-V-12 School.

Experience during the first term of the V-12 Program, with enlisted men drawn from the fleet and shore establishments, indicated that a large proportion of them had difficulty in adjusting themselves immediately to the prescribed college work. Several factors were responsible: (1) Many of the men had been away from school for many months, with the result that they had forgotten much of the subject matter of the courses which they had passed in good standing; this was especially true in the field of mathematics and physics; (2) difficulty in readjusting their habits to the routine of effective study plagued many trainees; (3) some men had been selected under misinformation.

These academic refresher units were governed insofar as possible by the same regulations and directives that were used in

U.S. Naval Reserve Midshipmen's School, Twenty-second Class, Jan. 18, 1945. (Courtesy of William H. Griffy)

the V-12 Program, and the same basic manual for the operation of the larger program was applied to the two smaller ones. The men were given an examination upon arrival at their respective units, and each college determined whether the man needed eight, 16, or 24 weeks of refresher work, depending upon his preparation and the program in which he was enrolled. Thus, the colleges were given programs that they could well handle and that used the facilities abandoned by the Flight Preparatory Schools, and training that the Navy needed was provided.

Endnotes:

[1] *Navy Department, "United States Naval Administration in World War II: The College Training Program," Vol. IV (MS on file in Office of Naval History, U.S. Navy Department), p. 118.*

[2] *Ibid., pp. 68-69.*

[3] *Ibid., pp. 5-12.*

[4] *See Appendix A - Transitional Programs*

[5] *Ibid., pp. 70-76.*

[6] *Excerpted from Navy Department, Bureau of Naval Personnel, Training Activity, "U.S. Naval Administration in World War II; History of Line Officer Train-*

ing," Vol. VI (MS on file in Office of Naval History, U.S. Naval Department), pp. 1-3.

[7] *Navy Department, "The College Training Program," pp. 77-107.*

[8] *See Navy V-12 Bulletin, No. 101, pp. 35-36.*

[9] *Navy Department, "The College Training Program," p. 129-33.*

Chapter III

New But Much Different

Long before the Navy made public announcement of its intent to design a college training program for the preparation of officer candidates, it already had in focus the serious dilemmas created by the U.S. Army in the operation of its Student Army Training Corps (SATC) of World War I:

In the fall of 1918, just before the academic year began, the draft age was lowered to 18, the source of officer personnel was in short supply - unless young men were permitted to enlist but remain in college on an inactive duty status at their own expense.

All manpower procurement in World War I was under the War Department. There were 150,000 college students subject to conscription. Plans had already been formed to establish the SATC at more than 660 colleges, but in the matter of administration and curriculum planning, there were evidences of poor judgment and execution.

SATC units were still being established in October, hardly a month prior to the signing of the Armistice; yet, trainees were all demobilized by Christmas 1918.[1]

In the Navy's incipient planning phase, a basic principle was firmly established by officers in the Training Division of BuPers that the Navy V-12 College Training Program should be so designed as to present the least possible interference with established procedures, courses, and campus traditions. The sincerity of this intent was expressed by the Chief of Naval Personnel who, before the inception of V-12, highlighted its purpose at the first nationwide Columbia Conference in the spring of 1943 and reported upon earlier.[2]

Principles for Utilization

When the chairman of the War Manpower Commission (WMC) issued the principles by which non-federal institutions of higher education were to be selected, he simultaneously announced the establishment of the Joint Committee for the Selection of Institutions, its method for selecting colleges for contract training, and the procedures for choosing membership from within the War Department (3), the Navy Department (3), and the WMC (3). The mandate given this committee was to determine the availability of administrative and instructional facilities and to allocate their use. All committee decisions were to be unanimous. The WMC chairman, however, reserved the right to make decisions in the event of unanimity. The WMC General Order No. 2 contained these principles for selection:

1. ...selection shall be made on the basis, not only of administrative and instructional facilities, but also staff, library, laboratories, equipment, housing, messing, and recreational facilities.

2. Based upon scientific needs of the armed forces, selection should take into consideration both large and small institutions as well as their geographical distribution.

3. Consideration shall be given to ROTC and to Army and Navy programs already in progress as well as to institutions for the training of personnel for essential civilian activities.

4. Information gathered through personal inspections by representatives of the armed forces shall be considered in addition to data provided by the U.S. Office of Education and the American Council on Education.

5. Particular attention shall be given to institutions that are equipped to give, most effectively, instruction in certain specialized areas, e.g., engineering, medical, pre-medical, and other needed fields, as well as in the liberal arts.

6. Use shall be made of as many liberal arts and non-technical institutions as possible for basic training and to the limited capacities of colleges of engineering, medical, and other technical institutions in relation to anticipated needs by essential civilian activities and by the Army and Navy.

7. When and if additional institutions are needed, the extent to which the required additional facilities and staff can be obtained shall be considered.

In the winter of 1943, after the process for utilization of colleges by the military had become fully established, President George F. Zook of the American Council on Education declared:

"It might be assumed that two years after Pearl Harbor the colleges and universities, after many delays in arrangements with the War and Navy Departments, were at last fully immersed in the war effort.

Such an assumption would be considerably short of the truth. In the first place, large numbers of institutions, including nearly all teachers' colleges, colleges for women, small liberal arts colleges, colleges for Negroes, and junior colleges, had no contracts for Army and Navy units."[3]

According to the *Educational Directory* of the U.S. Office of Education, published for 1943-44, there were 1,702 accredited institutions of higher education in the nation, 690 of which were senior colleges or universities, 261 professional schools, 192 teachers' colleges, 20 state normal schools, 433 public and private junior colleges, and 106 Negro institutions of all types.

The total extent to which colleges and universities were utilized during the war (see table) was disappointing to many administrators in view of the emphasis placed upon our being engaged in a "total war" and upon the need for trained personnel.

Numbers of Institutions Cleared for Use by the Armed Services

Cleared for Use by:	Number of Institutions
Army (exclusively)	122
Air Corps (exclusively)	155
Navy (exclusively)	148
Army and Air Corps (jointly)	87
Air Corps and Navy (jointly)	13
Army and Navy (jointly)	96*
Army, Navy and Air Corps (jointly)	42
Total:	663

*Of the 96 schools used jointly by the Army and the Navy, 83 were medical and dental schools in which contracts were negotiated on a tuition basis.

It became obvious that President Roosevelt's Executive Order of October 15, 1942, that "...an immediate study be made as to the highest utilization of American colleges..." had not been fulfilled; for, at the peak of wartime training, less than 300 institutions of higher learning, exclusive of duplication, were under contract to the armed services.

About 70 percent of the approved institutions of the nation were excluded from any financial benefits derived from military contracts. The inability of many degree-granting schools to house and to feed large numbers of students was the primary factor in their not being used. For them, the only other source of financial assistance during the war came through defense contracts for research. But note, only 95 schools were used, with eight prestige institutions receiving 90 percent of these contracts, and these were colleges already "cleared for use" for college training programs.

No doubt, errors were made in the non-selection process of colleges, but in defense of the actions taken by the Joint Committee, it was their maintenance, insofar as possible, with the committee's mission of the American system of higher education (Principle No. 5).

The most outspoken critic of the selection of schools was voiced by President Carter Davidson, who wrote:

"It must be admitted that educational quality is not quite so evenly distributed, and the older colleges of the East have some financial edge, but the democratic method demanded even-handed justice. It was, of course, a great help to be on the approved list of the Association of American Universities, for that was a hallmark of quality that practically assured selection. It is hard for presidents of unapproved colleges to understand why none of the independent (not tax-supported) institutions in South Dakota were approved, and why every state teachers' college in Texas has a program. Were the Congressional telephones kept busy? It is not only in first aid to the injured that we need to study our pressure points.[4]

In general, the WMC principles were applied objectively and fairly; criticisms arose and political pressures, indeed, were made when certain small institutions learned they had been chosen for *no* program.

A principle of "saving the colleges" was not one which came into sharp focus as the Joint Committee undertook its mission, the selection of institutions, but it obviously did influence the number contracted and the procedure by which the Navy reduced its V-12 Program toward the end of 1944 and in early 1945. In the V-12 Program, there were 131 undergraduate colleges, not one of which was dropped (except by its own request) until after V-J Day, when the V-12 Program was merged with NROTC.

Education vs. Training

The total effort of the Army and Navy during World War II in training enlistees, draftees, technical specialists and officer candidates is now regarded as the most gigantic and effective human endeavor ever undertaken by a nation in active warfare.

Training programs were conceived quickly following "the period of vacillation" herein described. With the inception of several programs, there was also widespread criticism over the decline of the liberal arts during the war period.

Today, one wonders whether the attack was justified inasmuch as the armed forces, from the start, turned to the colleges and universities for assistance in designing programs and formulating prescribed courses. Consultants who served on planning committees were carefully selected representatives of the Commission on Liberal Education of the Association of American Colleges and "hand-picked" civilians from leading member institutions in the ACE, who were recognized proponents of the liberal arts.

The education consultants soon learned that the military emphasized mathematics, the sciences and technology at the expense of the social sciences and the humanities. So, the civilian educators ran into difficulty with their military counterparts. Accordingly, they protested the consistent exclusion of those courses an adult needs in order to develop human values, to think independently, to express himself clearly and logically, and to cope with the major issues of life. The civilians argued that liberal arts courses were just as essential in time of war as they are during peace. Training they maintained in vocational-technical education could produce mere automations. Therefore, they were opposed to training programs that were exclusively technical, scientific, or vocational.

The military consultants saw some merit in going the half mile and they agreed that all curricula for both Army and Navy College Training Programs should contain some liberal arts courses in the freshman year.

But, the military remained resolute in their argument that in a mechanized war the true test of a training program was the end product - the graduate - who is prepared to fight; not whether he is possessed with intellectual freedom. Therefore, during the emergency, all courses and programs that were to be designed had to eliminate irrelevant subject matter and be taught in the shortest possible time. They viewed training in terms of learning to fly an airplane, to plot accurately a course by means of navigation, to decipher coded messages with exactitude, to recognize on sight an enemy plane, to repair a diesel engine, and to maintain a motor in top performance.

During a tense period of interchange of ideas, but without any apparent resolution of issues, one individual began to be heard more and more clearly. It was Captain Arthur S. Adams,[5] a one-time Naval Academy graduate, who had been recalled to active duty with an important mission.

In focus, BuPers was addressing (a) ways to launch a new college training program for officer candidates and, (b) steps to absorb the several pre-war "V" college programs into the new design.

On Captain Adams' arrival, he found the military consultants were leaning toward planning the V-12 Program, with foundation courses in general education, containing some common elements associated with the liberal arts.

Academic representatives of the 131 contract colleges continued to question whether the V-1, V-5, and V-7 Navy Student Programs would continue, or if they could be absorbed gracefully into the new V-12 Program but, in short order, the change did occur smoothly and with an absence of criticism.

Another major hurdle in the nationwide initiation of the Navy V-12 Program was institutional contracting. Could it be achieved without major problems? The reader may formulate a judgment by reading the chapter dealing with the financial aspects of the college training program.

It is, nevertheless, important to note that the military took every precaution during the negotiation period to shy clear of the very embarrassing dilemma generated during World War I when the U.S. Army, in contracting with degree-granting institutions nationwide, was dictatorial in drafting contracts.

Thus, all contract negotiators for the Army and Navy were given clear instructions that all contracts during World War II were to be, in every sense, a truly "cooperative venture" in training military personnel. Fortunately, this area of concern was accomplished with a minimum of complaint.

Then, at a different level, there arose the question, regionally and nationally, what happens on campus when you mix civilian students, many categorized "4-F" by local draft boards, with students in Navy uniform? Would there be negative interpersonal problems generated? Similarly, would there be a hiatus created between civilian instructors and Naval officers?

Here, too, Navy foresight "calmed the waters." The reader may recall the words of Admiral Randall Jacobs and of Captain Arthur Adams during the Columbia University Conference when they announced, in reassuring terms, that the Navy V-12 Program, as designed, would in every sense be "a college program." As a result, college administrators returned home reassured.

The Columbia Conference assurances proved true when V-12 trainees on campuses nationwide learned that they were, indeed, "college students" who were fully eligible for membership in honor societies, Greek letter fraternities, social clubs, honor rolls and all extracurricular activities.

So, once again, the trepidations of college administrators melted away.

At this remote time, there is still some question whether the Navy required each trainee to carry an excessive load in that he had a minimum of 17 academic credits of course work, plus physical training, his routine military duties, and drill.

The academic program of every trainee was geared to consume 50-60 hours of concentrated effort weekly, wihtout any respite, except for a few days each of the three calendar year semesters.

Wartime calls for acceleration! Thus, the V-12 calendar evolved that was much more intensive than the normal college calendar before the war.

Once the V-12 Program became fully operative nationwide, question was raised whether the tempo for trainees was perhaps too demanding. Thus, the attrition rate of trainees was in constant focus in BuPers inasmuch as there were two primary factors known to unit commanding officers who were cognizant that many trainees had come directly into the program from the fleet, many of whom had been out of school for two to three years. These young seamen had been selected and recommended because of marked officer-like qualities and potential ability to do college work.

But, on many college campuses they were found to be weak in such courses as mathematics and the sciences. Many had difficulty in maintaining passing grades. Fortunately, on many campuses this problem was solved by dedicated faculty members who were imbued with strong patriotism. Faculty members on most campuses desired personally to contribute to the war effort. They gave freely of their time in coaching these trainees individually and in groups. This kind of dedication salvaged many who continued in good standing in the program once over the hurdle.

Even now, it is fitting that still another group of dedicated faculty members be accorded accolades. Their war effort contributions were similar in that they were responsible also for salvaging other V-12 trainees whose problem was periodic "physical fatigue."

Physical fatigue can impair one's academic status if unabated. Even periodically fatigue can keep a trainee from being abreast of his daily or weekly scholastic requirements. Constant fatigue would be more than he could cope with.

These trainees were not the "salty" fleet trainees; they were (1) trainees who had come from those secondary schools in which their preparation in mathematics and/or in the sciences was weak or, (2) trainees who, in their arduous routine, also acquired pronounced physical exhaustion - a condition that many faculty members observed during class sessions. There was an observable tendency among fatigued trainees "to doze" during daily lecture presentations.

Subjectively, some professors wondered whether subject matter content was dull or boring. The question on some campuses became a discussion topic in department meetings, even in faculty sessions.

Was the problem attributable to the heavy prescribed physical training requirements? Was the physical fitness program too demanding for some who required less pressure?

It was common knowledge soon that Gene Tunney, a former heavyweight champion, was in charge of the Physical Training Section of the BuPers Training Division. Commander Tunney's advocacy of strong body building exercises was reflected in the V-12 curriculum. As a result, the physical training staffs in many contract units were perhaps unrestrained in their personal enthusiasm for calisthenics, combative sports, drill, drill, and more drill. The typical physical education CPOs on many campuses radiated Commander Tunney's philosophy that there is a marked correlation between athletic achievement and leadership ability.

Whatever the causes of the physical exhaustion epidemics among trainees on certain campuses, faculty members came to their rescue with more individual coaching, generally during the evening study hour when the professor would normally be at home with his family.

This remarkable spirit of volunteer support was universal during the war. By working together cooperatively, trainees, faculty members and military staff members, it gave rise to an effective and remarkable operation resulting in one of this nation's most fruitful officer training programs in U.S. history!

Endnotes:
[1] *Samuel P. Capen, "The Experiences of Higher Education in 1917-18," Higher Education and the War, American Council on Education, Washington, D.C., 1943.*
[2] *Navy Department, Bureau of Naval Personnel, Conference on the Navy V-12 Program at Columbia University, May 14-16, 1943, NavPers 15012 (1943), p.4.*
[3] *ACE Bulletin, "Higher Education and National Defense," December 1943.*
[4] *Davidson, Carter, "Trial and Error in Wartime Programs," The Educational Record, XXIV, No. 3 (July 1943) p. 289.*
[5] *Elsewhere in this document, Captain Adams is given an appropriate tribute.*

Chapter IV

Mobilizing For Total War

Financial Aspects of Training In The Contract Colleges

When the Army and the Navy recognized a definite need for specialized training programs for part of their personnel and turned to the colleges and universities of the nation as the best medium for providing this training, the question of "compensation for service rendered" became important. As early as 1939, when the President declared a limited "national emergency," a hue and cry were heard. It became evident that Selective Service would eventually be invoked and that nearly all college students would fall within draft age groups. Thus, a drastic reduction in student enrollments was envisioned and, consequently, a decline in revenues. Nevertheless, it was obvious that not all of the approximately 1,700 approved colleges and universities in the nation could be made use of by the armed services.

But saving the schools from impoverishment was not the reason for designing the college training programs. Nor was it designed as an experiment in education *per se*. It was designed specifically as an officer procurement program to supplement the output of officer-candidate schools; as a program to continue the training at college level of young men who, after a period of seasoning and special training, would become valuable assets to the nation as officers in the armed services. The Army and the Navy were now calling upon civilian educational resources to build the supply of trained men, just as they were calling upon civilian industries to build up the supply of material.

Contract Problems and Relationship

Considerable credit belongs to the Navy for pioneering in the training of large groups of specialists at colleges and universities. Many of its experiences proved to be of great value in the events that led to the formulation of contract principles and to their subsequent adoption by the Army and Navy.

On February 13-15, 1943, a conference was convened in Chicago under the auspices of the ASTP Division. The purpose was two-fold: (1) to develop a suitable contract form and, (2) to develop uniform policies and procedures for publication in a manual to be used in negotiating college contracts. Present were representatives of the Navy and Army Air Force.

Of particular significance at this conference was the presence of 18 business managers and comptrollers of leading educational institutions who were invited for the purpose of offering questions and criticisms pertaining to the proposed contract provisions.

On March 1, the Negotiation Manual, as revised, was officially approved and circulated to interested governmental agencies.

Officials of both the Army and Navy were well aware of the fact that unity was essential. Of the 663 colleges cleared for use, 425 were used exclusively by one or the other services. Thus, the remaining 238, less the 83 schools used jointly for medical and dental training, left a balance of 155 schools in which contracts could be compared to insure uniformity.

Army-Navy Board for Training Unit Contracts

Cognizant of this need for uniformity between the services which were to utilize the colleges and university faculties and facilities, the Under Secretaries of War and Navy, on March 25, 1943, issued a joint memorandum establishing the Joint Army-Navy Board for Training Unit Contracts. The mission of this board was:

a. To approve one or more contract forms and to determine the extent to which the same shall be observed by the two departments;

b. To develop standards to be applied in

Navy V-12 Orchestra 1943-44. Apprentice Seaman Wally Muelder, Director. Illinois State University, Normal, Illinois. (Courtesy of M.F. Burrill)

Middlebury College, Navy V-12 Basketball Squad 1943-44. Front Row L to R: Card, Hutchins, Bobotas, Crescenti, Skuy, Dan Schaffer, and Brennan. Back Row L to R: Ens. Dean H. Hanley, Athletic Officer, Chief Acropolis Asst. Coach, Zuance, Tobias, Fisher, Irwin, Cooke, Salisbury Many, "Doc" Farrell, Trainer, and Coash Brown. Absent from picture: John Dillon. (Courtesy of Joan I. Fishel/K.H. Irwin.)

determining the rates of payment for facilities and services;

c. To keep informed as to the use made of such contract forms and such standards and as to the rates established in negotiations for contracts; and

d. To forestall or decide any question which might arise in the course of negotiations with respect to the application of such standards, and to take other steps designed to accomplish the above purpose.

Of relatively greater importance, as subsequent events confirmed, than even the establishment of the Joint Army-Navy Board was the appointment of its initial members.[2] It is generally conceded that no man is indispensable, but if there are exceptions to every rule, then certainly the board chairman was the exception in this instance. It is doubtful that anyone could have duplicated the ability, energy, and integrity displayed by Robert B. Stewart in guiding the destiny of the three-year enterprise involving the expenditure of well above $300,000,000.

Fully aware of its responsibilities, both to the government and to the contracting schools, the Joint Army-Navy Board immediately set about to adopt, formally, a standard form of contract and coordinate negotiating policies which would apply equally to each service and at the same time insure that the terms of such contracts were fair and uniform. As has already been stated, prior to the establishment of this Joint Board and, in fact, as far back as late 1942, both the Army and Navy operated independently in promulgating policy to govern college training programs. Fortunately, these independent operations provided the nucleus for the creation of the joint enterprise. Both agencies had developed valuable ideas that were to prove extremely beneficial in the coordination that followed.

At the time of the establishment of the Joint Board, the Air Corps had written 150 or more contracts with liberal use of so-called activating expenses, that is, expenses allowed for remodeling buildings and dormitories to accommodate a training unit. This use eventually proved extravagant and both the Navy and Army were forced to face the fact that they were following a too liberal precedent. Thus, the Army's contract, in attempting to increase the efficiency of operations, provided that certain expenditures be made by the college.

The colleges were, of course, of the opinion that it was unfair to ask them to spend money for expansion from which the government would receive benefit and the college suffer a loss. Now that the Joint Board was definitely an officially established organization, its prime and immediate concern was a revision of the *Negotiation Manual* to encompass the policies to be applied to these questions by all services.

On May 10, 1943, the so-called "ten commandments" were adopted and approved by both Under Secretaries. These principles of contract which were designed to protect fully the interests of the government and at the same time to consider most carefully the interests of colleges and universities, were as follows:

Principles of Contract

1. The definite purpose of the program is to serve the Army and the Navy. Normal functioning of the program will benefit the college by providing a paid-for utilization of its facilities and staff, but such benefit is clearly incidental to Army-Navy purposes.

2. To the extent the plant and facilities of any college or university are used specifically for and as prescribed by the Army and the Navy, the college shall be left in no worse position by reason of such use. This does not mean that *all* of the institution will be maintained by the program, nor that total over-all costs will be pro-rated on the basis of the ratio of service trainees and civilian students.

3. The Army and the Navy, so far as possible, shall make capacity use of those facilities of the institution which are used to meet Army-Navy requirements. Activities of the program shall be concentrated as much as possible in the least amount of plant.

4. The college will not be reimbursed for any instructional or other regular operating equipment which it must acquire to provide for the training unit it has accepted, except that an allowance for depreciation will be included under costs on the basis of the uniform contract provisions.

5. Activating expense shall be restricted to (1) specialized equipment not usable by the college in its normal operations and, (2) plant alterations requested by duly authorized and competent authority for the Army or Navy to meet their particular requirements.

6. The basis of the uniform contract terms is known as "budgeted cost basis," i.e. the payments agreed upon are based upon a budget of expected costs predicated upon current (latest fiscal report) facts of operation, reviewed at sufficiently frequent intervals as to permit such adjustments as will correct differences between estimated costs and actual costs.

7. Whenever the service or facilities are provided particularly for or as prescribed by the Army-Navy (for example, Ground School, A-12, V-12), the college shall be paid for its costs on the basis of the uniform contract provisions, and a fee for "use of facilities" based upon pre-war book values or cost of buildings, not exceeding 50 cents per cubic foot for the best facilities used on a capacity basis. No payment other than maintanance will be made for use of land.

8. Whenever a trainee pursues a course of study available to all students and not involving special requirements by the Army-Navy (for example, medical, ROTC), and the circumstances do not justify a cost analysis, the college may be paid the same fee for such services as it regularly is collecting from civilian students, *except* that in publicly supported institutions, the non-resident fees shall be paid for all Army-Navy trainees.

9. The college shall keep its books of account in its usual manner, except that extra records may be required to enable proper reporting upon the facts of actual financial operations. Statements of actual costs, when submitted, shall be sworn to by the chief financial officer of the institution in such manner as requested. The college shall fur-

nish proof of the facts submitted if requested to do so by the Army-Navy authorities.

10. The college shall be the prime contractor for all space and facilities for use at the college, except that where commercial property such as a hotel is to be used, the lease shall be subject to approval of the government and may be made directly by the government if required by special circumstances.

Upon the adoption of these principles by the Joint Board, a most decisive and beneficial conference relating to contract work was held in Omaha, Nebraska, on May 28 and 29, 1943. Broad discussions encompassing every phase of the contract program were held and policies and procedures were definitely crystallized; any skepticism entertained by either of the services of the colleges gave way to an almost complacent attitude. The Joint Army-Navy Board, however, not satisfied that the road ahead would be entirely smooth and realizing that questions involving the interpretation of principles and methods might cause considerable unrest and dissatisfaction, immediately made plans to increase its membership.

In view of the complexities and importance of the problems involved, it appeared advisable to expand the size of the Joint Board and thus obtain the benefit of the experiences and judgment of additional members and make the board more fully representative of the various educational institutions. On August 7, 1943, upon recommendation of the board, the Under Secretaries expanded it to include three additional members.[3] It continued in operation until placed on inactive status December 1, 1945, subject to call by either Under Secretary in the event of controversy or appeal by a contracting institution. In essence, the Joint Board acted as a policy-making group and as a Board of Arbitration to decide controversies between the colleges and the representatives of the Army and Navy.

The Contract Form

The contract form is divided into two major subdivisions, those dealing with the operational articles and those concerned with normal mandatory contract provisions. The main compensatory articles were entitled as follows: (1) activation, (2) use of facilities, (3) instruction, (4) medical services, (5) subsistence and, (6) maintenance and operation.

Working data forms were included and annexed to the contract as a means of implementing and expanding the provisions of the basic articles. These working data forms provided for complete breakdowns of the rates chargeable under each article and provided a more complete analysis of cost fig-

	Type of	Number of	Total	Average Cost
Service	Training	Schools	Expenditures	per School
Army	ASTP	195	$ 3,442,950.39	$17,656.15
Navy	V-12	131	849,338.00	6,483.50
			3,150,662.00	24,050.85
Air Corps	Aircrew	147	3,299,709.94	22,447.00
	Total:	473	$10,742,660.33	$22,711.54

Table 1

Costs of Activating College Training Units

*Approximation made by the Navy Department representing purchases by the Navy of housing and administrative equipment for shipment to institutions.

ures. Each article of contract, except activation, was considered on the basis of "budgeted costs," that is, the negotiating officer would meet with school officials, determine an approximation of costs under the different articles, and write in such rates as were agreed upon.

These rates were subject to revision at any time upon the request of either the college or the government. For example, if the rate for instruction appeared to be excessive as compared to rates charged at other institutions, a request could be made for a review of the items leading to the establishment of that particular contract figure. If the rates, based on actual experience, were then found to be either too high or too low, an agreement could be negotiated amending such articles and providing for either a retroactive or subsequent revision as of a specified date.

At first glance, perhaps the entire situation appears arbitrary. Why should the school be required to remodel its facilities only to have the government retain the title to buildings so remodeled? It was not the intent of the government to retain title or interest in any institution, nor was it intended that the added property should become a fixture or realty by reason of having been installed, but rather that such fixtures and property would be removed or otherwise disposed of upon the expiration or termination of the contract. It was the intent of the government and the contractor to be able to agree on a fair arrangement for its use by the school at such prices as would reflect a great saving to the institution.

The contractor, in effect, under Article XIII of the contract, was allowed to purchase any item of equipment utilized in this program and the government was not obligated to remove any of its property or restore the premises upon expiration or termination of the contract. There was little

objection to making alterations which would prove to be a decided benefit to the institution and, as later events proved, colleges were able to purchase items of equipment from the government for as little as one-tenth of the original cost. In fact, in some instances property was abandoned and allowed to remain with the institution since it was not considered profitable for the government to remove it and thus, the value of many buildings was enhanced with actually no cost to the institution.

Table 1 is a resume of the expenditures made by the services in their major programs to cover activating or commissioning costs. (See Table 1.)

The extent to which the services benefited by the expenditures for activation or commissioning varied in proportion to the length of time each unit was utilized. The Army Air Forces Aircrew Training Program was of relatively short duration (17 months only), whereas certain Army and Navy units continued operations three full years, the last two years of which saw a utilization of from 20 to 25 percent of the schools initially activated. Thus, to reduce the figures in Table 1 to terms of a per-man rate based on either initial activation or extent of use would be of little consequence. It is interesting to note, however, that certain schools were able to accommodate training units without one cent of activating cost, whereas others required an expenditure of as much as $197 per man to activate a unit of 350 men.

From the experience gained, definite implications can be drawn in the event that the government should require the use of colleges and universities under similar circumstances. From the taxpayers' point of view, perhaps too much money was spent in preparing colleges for service use; but the lessons learned, particularly after the Joint Board came into operation, definitely prove that colleges can be made ready to accommodate large numbers at a minimum expense.

From the point of view of the college, it has become apparent that any proposed alterations or remodeling should be agreed upon completely at the time the contract is first negotiated in order to forestall any unwarranted expenditure resulting from the college's desire to make its facilities more attractive for government use. In 1943, many colleges acted in good faith when proceeding with improvements in accordance with their conception of what the Army or Navy required, without the specific approval of competent authorized personnel of the services. The services could not penalize them for such action, however, and they were properly compensated. Now, based on these experiences, the matter can be approached with a clear understanding of the problem and a mutual desire to cooperate fairly and economically.

Classrooms and laboratory facilities in buildings used exclusively by the services were computed on the basis of full valuation of the buildings. Where the services used only portions of a building and where civilian students utilized other portions, calculations were based on a percentage use by the service, times the value of the building, times the percentage of use during four hours of a normal day. Four hours a day was considered the average daily use of a classroom, but if the services utilized certain rooms more than four hours a day, the basis of calculating the percentage was to be changed accordingly. As an alternative means, where percentages could not reasonably be applied, it was agreed that 1,200 cubic feet of space per trainee would be normally required for instruction; hence, the number of trainees in attendance, multiplied by the 1,200 cubic feet, times the value per cubic foot (not to exceed the 50 cent limitation), times four percent, was to equal the total annual cost of facilities charged. Here again, individual circumstances largely dictated the determination of an adequate rate.

Auditoriums, libraries, student union buildings, medical service facilities, administrative space, and athletic facilities were required for use by the services. In these instances, each utility was calculated on the basis of the nature of its use rather than on a percentage, since the four percent factor was intended in comtemplation of capacity use. For example, payment for use of auditoriums or libraries being used only occasionally for meetings or lectures could not reasonably be calculated on the same basis as when they were used regularly as study halls or classrooms.

With the possible exception of computing costs for instruction, the matter of use-of-facilities computations was the most constant source of contention and discussion. In some cases, fraternity houses situated on privately owned land had to be utilized, necessitating a different basis of consideration. In still other instances where campus facilities were not available, nearby hotels and other commercial property were used, and such action required special sanction by the service headquarters in order to be negotiated on lease or contract.

Subsequent changes in the manner of computing instructional costs were later to affect the extremely complicated negotiation for use of facilities. The figures in Table 2 indicate the rates paid by the services for facilities. (See Table 2.)

In spite of the difficulties experienced in this connection, the rates paid to institutions for use were eventually to prove extremely economical for the government; at the same time, they allowed the contractor to meet normal expenses and covered building depreciations that occurred during the war period. That the colleges were, for the most part, satisfied with the treatment accorded by the services in connection with payments for use is indicated by the fact that only eight requests were received by the Joint Army and Navy Board pertaining to payment of restoration of net income rather than normal four percent use of facilities allowance. Three of these protests were made on the basis that four percent was inadequate.

It is quite naturally implied that on the basis of only eight complaints as related to the total number of institutions utilized, the schools were dealt with fairly and were satisfied with the end results. In the event, however, of a subsequent need for the services to repeat such a program, it is considered most desirable to attempt to simplify procedures for determining specific use of the general plant in order to align more closely the determination of rates based on more definite and serviceable records. The following quotation summarizes the general attitude of the college officials:

"And when peace comes and we settle down with a 100% civilian enrollment, we shall discover that Doane has gained much from its experience with the V-12 Program. In the first place, it has become a more efficient college; the necessities of the case have demanded more strict financial accounting, closer analysis of costs, and more serviceable records. The college runs in a more ordered fashion than in the past, as would inevitably be the case with an enlarged faculty and the complex problems of military and civilian establishments. For the time being, at least, some of the old free and easy ways have had to be sacrificed.

In the second place, the faculty has become more efficient, having learned many lessons during this past year. The faculty also is more aware of its responsibility; many of its members realize that they have a share of responsibility for the present state of the world.

In the third place, the college has a better physical plant. Nearly every classroom is better lighted than before. Laboratories have been increased in number and hundreds of dollars worth of new laboratory equipment has been purchased. The library has purchased more books than in the average year; fluorescent lights have been installed in reading rooms, and additional stacks have been built."[5]

Instruction

Computation of rates applicable to instruction was generally the most difficult portion of the training contract. The methods utilized in computing instructional costs appeared to be in a state of constant dispute. Here again, the differences apparent between the services made difficult the realization of the desire that rates be made uniform.

It was the Navy's normal practice merely to take students at the particular school where they happened to be enrolled and allow them to continue the same courses that they had been taking. Unlike the Army, the Navy trainees were separated into classes and intermingled with civilian students.

Service	Man Months	Weighted Average per Man Month	Total Cost
		Table 2	
		Payments For Use of College Facilities	
Army ASTP	1,165,082	$3.81	$ 4,438,962.42
Navy V-12	1,626,738	3.31	5,384,502.78
Air Corps	857,542	3.48	2,984,246.16
Total:	3,649,362	$3.51	$12,807,711.36

Table 3

Payments For Instruction

Service	Man Months	Weighted Average per Man Month	Total Cost
Army ASTP	1,165,082	$27.01	$ 31,468,864.82
Navy V-12	1,626,738	35.60	57,911,872.80
Air Corps	857,542	17.03	14,603,940.26
Total:	3,649,362	$28.48	$103,984,677.88

Also, the Navy's normal academic term of instruction consisted of four months, whereas the Army curriculum specified a 13-week term. Payment was first made by the Navy on a "token" basis, that is to say, a nominal payment was made. This by no means insignificant payment was not necessarily intended to equal the cost of final payment. It was done with the express purpose in mind of merely estimating a rate so that the school, in presenting invoices, would be able to receive compensation without a long waiting period. Thus, by the end of the first month, the school would know how many students it had and what instruction requirements were necessary. Then, upon the determination of actual costs, adjustment could be made against the token payment already established, whereby the Navy would either receive a refund or would make up costs extending beyond the token payment.

The Army's much more definitive courses of study necessitated a reshuffling of instructors and classroom space. According to the Army's method of considering instructional costs, operation became more complicated. The ratio of teaching faculty and the average class hours each week were specified. For example, if trainees were to have 26 contact hours a week, the approximate number of instructors needed to teach a group of 750 trainees was computed as follows: 750 trainees, divided by 30 (the average size of classes), equals 25 class sections, multiplied by 26 contact hours, divided by 20 teaching class hours, equals 32.5, or 33 teachers. Therefore, if a full-time faculty were used, from 32-35 members should provide complete instruction for a training unit of 750 trainees. If classrooms or laboratories required a different average class size, the school would require more or less instructors as the case might be.

Without going further into the Army's complex system of formulae, proportions, equations, and graphical representations, it can be plainly seen that here are factors which would tax the abilities of the most competent negotiators. Each factor was meticulously specified. Direct teaching salaries, salaries and wages of supervisors, including share of cost of the office of the dean, maintenance and repair, equipment directly used in instruction, depreciation, general university administrative overhead, and textbooks, were considered in an attempt to establish instructional rates.

Since salaries paid to instructional personnel constituted the major portion of costs in furnishing instruction, a close analysis was imperative. Because of the manpower shortage, service officials believed that faculties ought to handle more than their 1937 teaching loads. The question arose concerning what size a class ought to be for maximum results. This situation, together with the number of hours a teacher was required to instruct, presented not only difficulties from the business officers' standpoint, but also from that of academic officials as well.

In consequence, it became advisable to review the method of fixing contractual obligations under the several programs.

Thus, the contract pertaining to instruction was modified to provide for a new rate per man at a flat rate for the entire term as contrasted with the monthly payment under contracts then current. Sufficient information was not available to enable the contracting officers and the schools alike to arrive at fair and equitable rates. Table 3 is a comparison of the costs based on the experiences of all services. (See Table 3.)

It should be borne in mind that the apparent differences in man-month rates are not due to the fact that certain services were shrewder in negotiations, but rather that different levels of academic work and the manner in which instructions were conducted, account for these variations.

Since instruction represents the most important item to be considered in these contract negotiations and was the item which had heretofore caused the greatest difficulty between the services and the contracting institutions, the implications to be drawn are of utmost importance. Primarily,

the complicated formula for prescribing the ratio of teaching personnel to class sizes should be discarded. Then, unless a specific course of study is requested or prescribed for service trainees, the institution could maintain complete autonomy in providing the number of instructors as is its general practice under normal operations. If a school elects to conduct a mass-teaching type of schedule or prefers to conduct instruction with minimum class sizes, such action should remain a matter of individual discretion. In either event, payments by the services would be made in relation to actual costs experienced by the institution.

In considering the items that make up the rate for instruction, it is believed that a simpler approach will prove infinitely more economical. In lieu of the time-consuming analysis of the various items that were heretofore considered, and in order to eliminate the meticulous appraisal of the cost of classroom space, instructional equipment, depreciation, etc., the wisdom of adopting an alternative means becomes apparent. Since instructional salaries constitute the major portion of cost in furnishing instruction, it is suggested that a percentage of such salaries be allowed as overhead to include all other miscellaneous items directly connected with the instructional operation. A rate so agreed upon should by no means be considered a fixed rate, but should be adjusted by mutual agreement from time to time because of the variable factors that cannot be reasonably predetermined. Occasionally, small gains might accrue to an institution, but such gain is in no way comparable to the tremendous cost involved in initial negotiations and subsequent continuous audit of cost records.

In most cases, the initial establishment of a rate for instruction could be based upon the most recent, or perhaps the previous years' experience. It is believed that such a system would provide a satisfactory starting point upon which prospective and retroactive adjustments could be made monthly or semestrally in order to reflect current cost fluctuations.

Medical Services

Normally, the care and treatment provided for members of the armed services is far in excess of that ordinarily received by civilians. This, of course, is because physical standards maintained by the services are different, and the possibilities of disease and infection that regimentation inflicts are greater.

The medical services to be provided by the contractor were simply and briefly set forth in working data sheets attached to the contracts. They merely stated that the contractor agreed to furnish the following services for all military personnel assigned or attached to the training unit: dispensary services, hospital service, full hospital care and bed service, laboratory service, ambulance service, and the keeping of records.

Normal college facilities ordinarily provided for some form of health service on the campus. Thus, little thought was given to utilization of established Army or Navy facilities. Here again, Army and Navy methods were different. The Navy provided resident Naval medical officers who would hold sick call and conduct routine physical examinations and give required inoculations. The Army placed the entire responsibility upon the school, which provided the necessary medical personnel. The question might be raised as to why, in some cases, adjacent Army facilities were not utilized, thus effecting a saving of the rates paid to schools. There is no reasonable answer except that a precedent had been established and the costs involved were considered too inconsequential to warrant a thorough investigation of the matter.

Competent medical authorities from the Army made inspections of college medical facilities and were able to determine a fair figure on the basis of a per-man cost for such services. An average of from $2.00 to $3.00 per man per month was considered equitable for the most complete facilities. It should be remembered that medical services as well as all other articles of contract, except activation, were subject to a revision of rates based on actual costs. Table 4 indicates the rates that were paid for these services. (See Table 4.)

Based on the experience gained regarding medical service facilities in the event that future needs should arise, it is believed that conveniently available facilities maintained by the services should be utilized to the maximum. This doee not mean that services rendered by the institutions were not entirely satisfactory, but rather that medical service facilities and the rendering of service by the school did not operate to the financial advantage of the institution nor were the facilities established at the school of any intrinsic value in the course of their normal peacetime operation.

Subsistence

Subsistence, which cost more than any other item of contract - in fact, almost as much as the entire balance of the contract combined - was a comparatively new experience in many instances. A good many schools which had had no such previous experience were subsequently reluctant to commit themselves as to what might be a reasonable figure in the preparation of a subsistence budget. Both the Air Force and the Navy, prior to the V-12 and ASTP ventures, had had some experience with such costs. These experiences again operated to the distinct advantage of the Army in that it profited from them. This item happens to be one which was given little consideration in the sense of being figured to the last penny, yet, by simple mathematics, one can realize that for 300,000 men in the service program, one or two cents a day might result in a tremendous leverage at the end of the year.

This particular section of the contract, plus the cost of instruction, represents about 75 percent of the entire contract costs and comprises the area in which the school can operate profitably or lose money to a serious degree during normal operations. The possibility of extremely wide variation between different schools was taken into consideration and also the possibility that considerable savings could be passed along to the training units as to ways in which messing might be handled when applied to the contracts. All the factors were set forth in a clear and concise manner. The contract articles stated:

...the contractor will provide and serve three meals per day to the trainees at the training unit but not to exceed (a specified number) at any one time. The hours during which meals are served shall be specified by the commanding officer of the training unit and the quality, quantity, and type of food and the purchase, preparation, and serving thereof will comply with service standards. The government will pay the contractor compensation for subsistence at the rate of blank dollars per day for the number of men reported by the commanding officer to the contractor to be on rations.

Spaces were provided in working data forms for a completely analytical estimate based on salaries and wages, current repairs and maintenance, utilities, telephone and telegraph, insurance, operating supplies, laundry and dry cleaning, small equipment replacement, depreciation, administrative expense, and cost of raw food. A practical means of establishing raw food cost, which normally approximated about 68 percent of the cost of messing, was accomplished by analyzing typical menus in collaboration with the institution's director of dining halls.

Fortunately, the item of subsistence, while representing almost half of the cost of the entire contract, presented few negotiating problems. In contrast to academic procedures, the service negotiators could take the initiative, having had considerably more experience in the field of group feeding techniques. Recorded experiences indicated the small range of from 66.5 to 74.3 percent in costs of raw food, from 16 to 29 percent for labor, and from 10 to 15 percent for utilities, depreciation, and general administrative expense. Thus, a general formula could be utilized in the preparation of a messing budget. This formula, while not constituting a hard-and-fast rule, was eventually to prove effective in establishing uniform rates. Standards required by certain commanding officers were found to vary widely, even in similar locations, and consequently, the cost of raw food would vary even within the same area. Costs of mess management and labor were very often found to be out of proportion under almost identical situations. Thus, the college officials and the service negotiators could review all the items of messing cost to the end that maximum efficiency could subsequently operate to mutual advantage. Once these budgets were reasonably established, the changing conditions could readily be adjusted at the end of each 90-day period, at which time definite audits were made.

Some messes were conducted only slightly within service standards, whereas others were more liberal in prescribing the quantity and quality of the food served. Efficiency in purchasing and preparation also had considerable bearing on final costs. However, considering the human element, the end results were surprisingly uniform. By and large, food costs varied, depending upon the geographic areas. The West Coast

Table 4

Payments For Medical Services

Service	Man Months	Weighted Average per Man Month	Total Cost
Army ASTP	1,165,082	$1.88	$2,190,354.16
Navy V-12	1,626,738	1.42	2,309,967.96
Air Corps	857,542	1.77	1,517,849.34
Total:	3,649,362	$1.65	$6,018,211.46

and the Northeast were generally high as compared to the South and Southwest. Table 5 indicates the variations in cost of subsistence in relation to geographic areas. (See Table 5.)

In cases where messing was a completely new venture for the institution, valuable assistance was offered by the Quartermaster Corps of the Army or by the Navy Messing Officer of the Naval district who, in practically all cases, exercised a guiding hand in the matter of service standards, menu construction, methods of buying, and sanitation. Cooperation by the Army sometimes extended to making available food supplies from government-issue stores to both Army and Navy units.

Occasionally, problems were created with regard to adequate facilities. Mess halls were frequently located in the dormitories or adjacent buildings. In such cases, the use charges, having already been computed for the building as a whole, were not separately estimated for the dining room and kitchen facilities.

It has been pointed out that the messing rate was on a per capita basis, whereas other contract rates were on the monthly or term basis. The reason for this is that the cost varies with the number of individuals, and even minor variations during a longer period than one day result in a tremendous change, both in the preparation of raw foods and the daily cost of operation. In certain schools, facilities were not available. Therefore, private caterers or restaurant suppliers were subcontracted. This arrangement presented a problem since the element of profit was obviously essential to such enterprises, proving difficult from the standpoint of auditing and adjusting rates in proportion to the costs of services being rendered. (See Table 6.)

The implication to be drawn in connection with feeding furnished by colleges and universities is certainly clear and decisive. The manner in which messing was handled is noteworthy and impels commendation to the contracting institutions. It was an entirely new venture for some schools, and experiences, while definitely minimizing costs for subsistence to service students, provided the school with a thorough knowledge and experience in feeding techniques to the end that their peacetime operation will be conducted on a more profitable basis. From the point of view of the services, they have only to make comparison between the $1.80 a day paid as commutation in lieu of ration (as in the case of medical students) as against the average rate obtained under the contract. It is further of considerable importance that since the cost of messing represents the major portion of college training, schools which cannot provide messing

Table 5

Payments For Subsistence Per Man Day:
High, Low, and Average, by Areas

| | High Cost | | Low Cost | | Average* | |
| | Raw | | Raw | | Raw | |
Area	Food	Total	Food	Total	Food	Total
New England	$ .85	$1.25	$ .70	$1.00	$ .793	$1.200
New York-PA	.85	1.30	.64	1.05	.777	1.144
Southeast	.84	1.23	.59	.93	.725	1.068
North Central	.84	1.16	.59	.91	.722	1.055
Southwest	.78	1.10	.62	.92	.695	1.038
West Coast	.88	1.38	.58	.93	.754	1.203

*Average based on review of contract rates by areas, not on arithmetic average of high and low.

Table 6

Payments For Subsistence

Service	Man Months	Weighted Average per Man Month	Total Cost
Army ASTP	1,165,082	$33.80	$ 39,379,771.60
Navy V-12	1,626,738	33.60	54,658,396.80
Air Corps	857,542	30.70	26,326,539.40
Total:	3,649,362	$32.91	$120,364,707.80

services should not be considered as potential training establishments, thus eliminating undesirable subcontracting.

Maintenance and Operation

The item maintenance and operation was of great importance from the standpoint of the school in arriving at an understanding of the maintenance involved in proportion to the facilities actually utilized by the services. Here again, the principles of contract were of major importance in arriving at sums to be paid for maintenance and operational costs.

Based on this theory, the contractor was required to provide light, heat, water, power, janitorial services, and all other supplies for the operation and maintenance of the property and facilities used by the training unit, for a capacity of a designated number of men and in accordance with the requirements of the services. For the performance of these services, a contract rate was established on the basis of so many dollars a month, commencing with a specifically designated period. These estimates were determined on an annual basis and the annual sum was reduced to a monthly amount to be stated in the contract.

To the greatest extent possible, the published financial reports were considered in determining cost estimates. However, elements of year-around operation and more intensive use were also carefully considered. The working data forms annexed to the contract were inclusive and covered practically all the factors involved in adequately arriving at budgeted cost figures. This article, representing the third highest item of cost, required careful study and contracting officers were cautioned to exercise meticulous care in properly setting forth costs applicable to each facility utilized. As usual, questions arose with relation to special circumstances peculiar to an individual institution.

The care of the campus, laundry and dry cleaning, taxes, libraries, athletic facilities, normal and irregular repairs and maintenance, and insurance presented individual problems, yet all items of maintenance and

Table 7

Payments For Maintenance And Operation

Service	Man Months	Weighted Average per Man Month	Total Cost
Army ASTP	1,165,082	$ 9.00	$20,485,738.00
Navy V-12	1,626,738	6.29	10,232,182.02
Air Corps	857,542	8.55	7,331,984.10
Total:	3,649,362	$23.84	$28,048,904.12

operation were reasonably and honestly considered to the end that mutual satisfaction was ultimately obtained. (See Table 7.)

As the programs progressed, the experiences gained by the services were of considerable value in subsequent reviews of costs. The experiences were of similar significance in that they enabled comparative analyses to be made, thus providing reviewing authorities with definite evaluation that could be conveniently utilized in the establishment of more definite policies.

As has been the experience in computing other articles of the contract, maintenance and operation required close scrutiny of past records of the institution. The initial establishment of rates was extremely difficult in itself, and so, subsequent revisions and adjustments of these rates were not only beneficial to the service in determining more uniform rates, but were even more beneficial to the schools in enabling them to realize the importance of adequate and thorough financial records.

It is considered a conservative estimate that not more than 10 percent of the colleges had records as complete as the services required. In some cases, in fact, the status of financial records was substandard.

Revision of Rates

The Articles of Contract known as "Revision of Rates" was perhaps the most important and conclusive article of the contract. To know this article completely was to understand contractural relationships between the services and the colleges. The contract form was extremely simple; yet, it had internal latitudes and flexibilities within the Revision of Rates Article, which were bound to accomplish ultimate success in a wide variety of situations.

In essence, the contract was negotiated on a budgeted cost basis. Foundation budgets or estimates were inserted into the contract under various heads; estimates were then modified periodically during the ac-

tual course of operation of the contract to conform as nearly as possible with true costs. Thus, Article VII, "Revision of Rates," implemented the action resulting from the determination of facts in relation to the estimates. This Article stated in effect that Article III, Instruction; IV, Medical Services; V, Subsistence; and VI, Maintenance and Operation might each be revised from time to time.

The demand for revision might be made either by the contractor (college) or by the government within a period of 30 days after the expiration of each three months, and any rates so revised might be made retroactive to any period covered by the contract and would continue in effect until a subsequent revision was made in accordance with future demands of either party. In the event contracting officers and college officials were unable to agree on the revised rates, the Secretary of War or the Secretary of the Navy was to have full authority to determine the revised rates and the effective dates thereof. The agreement resulting from revision was an agreement, not only as to the revised rates, but as to the date at which the revised rates should take effect. Thus, theoretically, adjustments could be made back to the original date of the contract and, moreover, revised backward as often as was necessary.

The agreements to review were stated in the form of a supplemental agreement to the existing contract. It was not necessary to revise all articles of a contract and, for example, if both parties were satisfied with the rates set forth under everything except the messing cost, supplemental agreements could be drawn to the contract providing for that one revision only.

This article also contained an important connection with Article XV entitled, "Records," which dealt with financial statements required of the contracting institution. It provided that the contractor might fix the time at which the institution should furnish such statements. Here again, is an example of how some of the demands of government, while perhaps appearing arbitrary, aided the

institutions in accumulating and maintaining proper records.

Termination and Property Disposal

At the beginning of the college programs it was anticipated that institutional facilities would be utilized for a comparatively lengthy period, since all indications pointed to a long conflict. Moreover, the successes of our military forces and the urgent need for more effective concentration of military strength on all battlefronts resulted in a drastic reduction in the ASTP Program before the operation had been in effect for a year. The Army reduced its enrollment in colleges from approximately 145,000 to about 35,000 students, thus obviating the need of using many institutions. Meanwhile, the Navy's enrollments remained fairly constant while the Air Corps experienced a drastic curtailment of its training program which resulted in the complete termination of air crew training by July 1944. Machinery had to be set in motion immediately to consider the problems of termination and to make adequate provisions for the disposal of property which had, in many instances, only recently been acquired.

While the original contract contained an article relative to termination, such provisions contained the mandatory requirements covering the rights and obligations of both parties and a statement to the effect that, in the event of such termination, the contractor and the government would put forth their best efforts to mitigate any losses or commitments in connection with such termination. Likewise, Article XIII provided generalization with regard to the intent of both parties and also to the disposition of property acquired under the terms of the contract. It also provided an option to purchase such property as may have been utilized.

Now that the college programs were in a process of rapid decline, special policies and provisions had to be drafted to effect, expeditiously, terminations and disposals of property. Specific rules were drawn and approved by the Joint Army-Navy Board on March 7, 1944, and a standard form of supplemental agreement was provided in order to maintain uniformity in adjusting rates of compensation and to adhere to established government policy with regard to property disposition.

The intricacies of termination procedures and the fact that termination came about suddenly and unexpectedly, demanded an exact reversal just at the time the armed services' programs had attained their maximum operation efficiency.

Details related to the creation of the termination policies were just as involved as were the initial creation of contract policies and procedures.

Marine V-12, Duke University Campus, 1943. Standing (front) L to R: Maj. Walter G. Cooper and Warrant Officer Blanchard. (Courtesy of Walter G. Cooper)

The considerations and discussions of the many factors considered in establishing termination policies were too voluminous for complete review here, but it can be safely stated that such policies were fairly and honestly conceived and were carried out in keeping with the spirit of the Principles of Contract. Many institutions profited considerably by reason of the alterations and the improvements made in the life of the contracts.

Professional Training Programs

Up to this point, mention made of the medical, dental and veterinary programs has been purposely avoided. This training, while constituting an important phase of the AST and V-12 Programs, was on an entirely different contractural basis.

In the early months of 1943, when the college training programs were in the process of development, the matter of a basis of payment for professional training was approached with considerable caution and hesitation. It was clear at the outset that the budgeted cost basis of contracting could have no relation to medical instruction. Except for accelerating a continuous instructional program, there was no change in normal operation or academic requirements. Hence, the only difference between a service trainee and a civilian student was the uniform and the fact that the government was paying for the service trainee, and the parent for the civilian. Accordingly, the basis of normal tuition and fees was adopted with slight modifications as to the fees that were properly chargeable to either the student or the government. In order to equalize the normally high rates charged by privately endowed schools, it was further decided that in state-supported institutions the nonresident fee would be

Wesleyan University Professor Neumann lecturing History II Class on the background of World War II. U.S. Navy V-12 Unit, Wesleyan University, Middletown, Connecticut. (Courtesy of H.C. Herge)

paid, regardless of the residence of a trainee.

Many schools were greatly concerned over the housing and messing of medical students since few schools were equipped with the necessary facilities. In view of this fact, and to avoid the least possible interference from military housekeeping duties, the Navy was definite in its decision to place all students on a commutation allowance. The Army concurred. The only exception was that where facilities were manifestly available, all such facilities would be utilized. During the entire program, only about 20 percent of the medical students were quartered and fed under contract.

Considerable discussion centered on the problem of furnishing the necessary textbooks and instruments required for these courses. This question resolved itself into a final conclusion that the government would provide all textbooks and instruments and would retain ownership of them. Subsequent developments enabled the students to purchase the textbooks they desired to retain, thus aiding the government in disposing of tremendous quantities of property.

Summary of Statistics

Statistics of the estimated aggregate costs of the major college training programs are condensed in Table 8. This table is by no means representative of all training conducted at colleges and universities.

The Navy Department, in the absence of Naval training stations, utilized approximately 90 colleges for special training other than V-12 in technical and administrative fields peculiar to that service. The contractural relations for this type of training, while following a pattern similar to that used for V-12 training, differed in a great many instances in that facilities and faculties were utilized in varying amounts. Training in specialized fields often required the installation of special equipment and, in effect, created a Naval installation on the college campus with little or no relation to the school itself, except for the utilization of its facilities.

The data presented herein is not final cost figures of all the programs. They are approximations only. (See Table 8.)

Chapter V

How Blacks Became Naval Officers

During World War II, President Franklin D. Roosevelt constantly used his high office in pressuring the armed forces to enable blacks to achieve status, especially in the Navy. His efforts, together with Mrs. Eleanor Roosevelt, who shared her husband's concerns, is a well-established fact.

The story of how segregation in the Navy, a long tradition, was changed with great difficulty is the theme of this chapter.

As World War II loomed over the United States in the early 1940s, Secretary of the Navy, Frank Knox, believed in the desirability of continued segregation within the Department of the Navy. Knox depended

for advice on the senior admirals, a group that wanted to uphold the tradition of a segregated service.[1]

The Washington Press Corps was apprised that the President appointed Mr. Knox to this Cabinet position largely for political reasons, even though he was an avowed segregationist.

Early in the war, the National Association for the Advancement of Colored People (NAACP) began its long campaign for a change in the armed forces' personnel policies and practices.

To comprehend the Navy's resistance to any change, historians reviewed the situation before World War I, when the assignment of blacks was strictly as "messmen," and the reflected segregationists' influence of society at large.[2]

Similarly, it was of interest to writers to learn that just before World War II black enlistments in the Navy were negligible inasmuch as large numbers of Philippines in the decade of the 1930s were recruited to fill all food service vacancies aboard Naval vessels.

This practice changed in September 1940, with the enactment of the first Compulsory Military Service Act and the ensuing manpower shortage just at the time the Navy was embarked on its long-range program of growth in the number of ships, shore installations and personnel.

The draft was the major factor for change in the Navy's personnel policies. No longer was it possible for the Navy to operate solely with volunteer enlistments. As the initial alternative, the Navy instituted a "tactic" in the assignment of blacks to segregated training centers. Subsequently, upon their training completion, blacks were assigned to a wide variety of Naval duties, such as dockside ship loading crews, to auxiliary ships in unrestricted waters, to anti-submarine patrols, and then to Seabee units, but no longer aboard ships as "messmen."

While all this was transpiring, dedicated Naval officers in BuPers were engaged in designing an inspiring new officer training program called "V-12," in anticipation of an acute shortage.

Institutions of higher education nationwide were to play a major role in the war effort quite unlike any program in history. Information began to trickle out as contracting officers held regional planning sessions with college and university administrators. Young men learned more about the V-12 National Qualifying Examination.

Soon newspapers, the radio and other media carried announcements that attracted the attention of young high school graduates nationwide. No mention was ever made about race, creed or color, rather that the program would accept the country's best qualified young men on a broad, democratic basis without regard to financial resources, and thus permit the Navy to induct and train young men of superior ability for officers and specialists.

The National Qualifying Examination date, as widely announced, was April 2, 1943, when over 300,000 took the test. How many of these test-takers were black no one knows, inasmuch as that matter is still in question today.

The most intriguing question that will forever darken the pages of Naval history is whether "some" blacks actually did take the Qualifying Examination on that appointed day.[3]

The dispute arose because the application form contained no instruction that the test-taker was to record his race.

The question, therefore, reverted to the Bureau of Personnel, which in turn routed a memorandum, via chain of command, to the Bureau chief, Admiral Randall Jacobs. He, in turn, passed the memorandum higher to the secretray, Frank Knox, for official clarification.

Secretary Knox, true to form, made the following very ambiguous reply on April 3rd, *one day following* the national examination:

The Navy College Training Program admits all students selected for this program, including Negroes, to the possibility of becoming officers in the Navy, and the examinations offer the first step toward this end.[4]

Secretary Knox's ambiguous statement muddied the water, especially for Naval officers in the Training Section of BuPers: How could a bright black student take the Qualifying Examination when the scheduled date has passed?

By this time, the matter had become so controversial it precipitated still another "memorandum" from within the Training Section of BuPers that moved on up the ladder until Secretary Knox brought it to the attention of the Commander-in-Chief, President Roosevelt, himself.

Without any fan-fare whatever, the President returned the memorandum with this terse comment: "Of course, Negroes will be tested." *F.D.R.*[5]

President Roosevelt, the adroit politician, was attuned constantly to problems in the military, especially in the Navy, and its need to provide greater opportunities for blacks. His efforts, along with others, came to fruition on April 7, 1942, when the Navy finally agreed to provide for blacks a process by which they might acquire general service ratings, on June 1st.

Another individual in the administration, who shared fully the President's empathy for blacks in the Navy, was Adlai Stevenson who, at that time, served as the Assistant Secretary of the Navy. Apparently, he could no longer remain silent. The following is the communication that Mr. Stevenson prepared and submitted to his superior, the Secretary of the Navy, Mr. Frank Knox:[6]

"I feel very emphatically that we should commission a few negroes. We now have more than 60,000 already in the Navy and are accepting 12,000 per month. Obviously, this cannot go on indefinitely without making some officers or trying to explain why we don't. Moreover, there are 12 negroes in the V-12 Program and the first will be eligible for a commission in March 1944.

Ultimately, there will be negro officers in the Navy. It seems to be wise to do something about it now. After all, the training program has been in effect for a year and a half and one reason we have not had the best of the race is the suspicion of discrimination in the Navy. In addition, the pres-

sure will mount, both among the negroes and in the government as well. The Coast Guard has already commissioned two who qualified in all respects for their commissions.

I specifically recommend the following: (1) Commission 10 or 12 negroes selected from top-notch civilians just as we procure white officers, and a few from the ranks. They should probably be assigned to training and administrative duties with the negro program. (2) Review the rating groups from which negroes are excluded. Perhaps additional class of service could profitably be made available to them.

I don't believe we can or should postpone commissioning some negroes much longer. If and when it is done, it should not be accompanied by any special publicity, but rather treated as a matter of course. The news will get out soon enough." *Adlai Stevenson, September 29, 1943.*

With the V-12 Program well under way in the fall of 1943, the earliest any trainees might be commissioned as ensigns would be March 1944. The Assistant Secretary of the Navy, Adlai Stevenson,[7] inadvertently created "a tempest in a teapot." He urged the creation of an alternate route for the commissioning of blacks as Naval officers.[8]

Mr. Stevenson's prodding was well timed. Mr. Knox yielded! He agreed to the commissioning of the "Golden 13" as Naval officers: These are the men who came in time to be known as the "Golden 13" (who)....were thoroughly screened, and the FBI conducted detailed inquiries into their background....Another observation is that all of the candidates had proved their proficiency as enlisted leaders.

In January 1944, these black sailors began their training at Great Lakes (Naval Training Center)....All successfully completed their training....in March 1944.[9]

In addition to the "Golden 13" who were first commissioned ensigns, there were approximately 60 blacks, of whom six were women. The report also stated that there were 36 graduates of the V-12 Program, and a few officers were commissioned as staff specialists.[10]

Currently, during peacetime, the NROTC contract units in certain colleges and universities supplement the role of the Naval Academy in supplying some black officer graduates annually.

With an elapse of more than a half century since this nation entered World War II, it is fitting to pay high tribute to the myriad black men and women who served in the armed forces and overcame the traditional policy of segregation, with some rising to high levels of responsibility in the Navy.

Perhaps no black Naval officers illustrate better the frustrations and difficulties encountered in attaining officer status than the "Golden 13." Likewise, no one can express better that struggle to achieve rank status than the chairman of the Joint Chiefs-of-Staff:

"In March of 1944, America was in the grip of world war....two and a half million men were serving in the U.S. Navy uniforms that month. Over 3,000 brand new ensigns were commissioned in the first six months of 1944 alone.

The story of one small group of 12 of those new ensigns - plus one Navy warrant officer - might well have been forgotten in the sweep of world events that spring....

Nevertheless, those 13 men made history....those 13 men - the Golden 13 - in fact, helped to change the face of American military. On active duty in the U.S. Navy today are more then 3,000 black officers who walked through the door that the "Golden 13" opened in 1944." [11] *Colin L. Powell, General, U.S. Army; Chairman, Joint Chiefs of Staff, Washington, DC - September 1992.*

With an elapse of nearly five decades since the Navy's largest World War II officer training program terminated, a historian who was also an alumnus, decided to learn whether any trainees achieved prominence as leaders in the military or as civilians in American life. His findings proved his assumption - an amazing number became national leaders.[12]

Among his immediate findings was to know the number the V-12 Program produced as commissioned officers in the Navy and Marine Corps: 60,000.

A second surprise was to learn that many trainees who grew up in the Great Depression had the opportunity to go to college.

The biggest revelation came in his listing, by name, the long roster who, in postwar civilian life, became corporate executives, CEOs, doctors, lawyers, judges, bankers, prominent by-line newspaper reporters, television newsmen, educators, clergy, entertainers, sports officials, college and university presidents.

For those who remained in the armed service, the V-12 Program produced 38 admirals and 20 Marine generals. The list was extensive for those who remained in the active Reserve and were called back to active duty in Vietnam and Korea during the pivotal 40 years after World War II.

Last, but certainly not the least, was the item pertaining to the blacks who were enrolled in the V-12 Program nearly a year before there were any black officers in the Navy, one of whom later became a vice-admiral, Samuel L. Gravely Jr., the first black to command a warship and the first black to advance to the rank of Admiral.

Yes, indeed, the V-12 Program goes down in history as the "Open Sesame," the milepost for human relations in that the armed services policy today carries the authority of a directive, issued by the Chief of Naval Operations, dated December 14, 1970, and broadened to include under a title reissued as the "Department of Defense Equal Opportunity," dated June 3, 1976. (Herbert R. Northrup, p.1)

Endnotes:
[1] *Paul Stillwell, p. xix-xx.*
[2] *Herbert R. Northrup, p.11*
[3] *As one commanding officer of a Navy V-12 unit, the writer can testify that "some" blacks must have taken the V-12 Qualifying Examination as scheduled. Quite unexpectedly, I was the recipient of a communication from Mrs. Eleanor Roosevelt who graciously urged my cooperation. She informed me that a black trainee would arrive at the start of the new semester. I must have taken the appropriate course of action. Although the V-12 unit contained many trainees from the deep South, we encountered only smooth sailing with the black trainee on board. —Henry C. Herge.*
[4] *James G. Schneider, p.150*
[5] *Ibid., p. 151*
[6] *Paul Stillwell. Appendix A, pp. 279-80*
[7] *Mr. Stevenson, after the war, distinguished himself politically. He became governor of Illinois, twice as a candidate for the Presidency and as the Ambassador to the United Nations.*
[8] *Paul Stillwell, p. xxii*
[9] *Ibid., p. xxiv*
[10] *Ibid., p. 266*
[11] *Paul Stillwell. Foreword, p. vii-viii*
[12] *James G. Schneider, p. 341-355*

Navy V-12 Bulletins
The Program's "Bible"

The Bureau of Naval Personnel (BuPers), during World War II, was located on a hill high above the Pentagon, in what was popularly called "The Navy Annex." Other bureaus in the Navy Department were scattered in different federal buildings in the Nation's Capital.

BuPers was a complex of divisions. Each division was further subdivided into sections. So, it is appropriate to know that the Navy V-12 Program, from its inception and until its termination, was administered by personnel in the Officer Program unit in the Training Division.

The Navy's Officer Training Program was, in reality, a partnership between 131 contract colleges and universities and the Navy Department. The medium for communicating V-12 policies and procedures was the consistent issuance of the "Navy V-12 Bulletin," of which there were over 375 distributed in the duration of the program. The combination of all these "Bulletins" is a complete record of the Navy V-12 Program from alpha-to-omega, and they reveal how cleverly the U.S. Navy kept the civilian personnel in the contract institutions, as well as the Navy officer personnel at each training unit, clearly informed of any

essential changes in procedures. The commanding officer in each institution, in keeping with BuPers instructions, maintained a cumulative file of "Bulletins" and also supplied appropriate academic administrators with copies of every directive.

To facilitate the proper distribution of the "Bulletin," symbols were used for proper classification of each as issued:

A - Curriculum
B - Training Aids
C - Administration
D - Finance and Contracts
E - Procurement
F - Physical Training
G - Medical and Dental Education
H - Theological Education
I - Marine Corps
J - Naval ROTC
K - General Information on Entire Program
L - Supplies and Accounts
M - Naval Academy

Many "Bulletins" did not pertain to on-campus administration of the V-12 Program. For example, there were certain issues that were primarily the concern of the commanding officer who played an important role in hosting traveling deans of professional schools of medicine, dentistry, theology and engineering. The commanding officer was the key person in providing and interpreting vital information needed in the screening procedures.

At times, unit commanding officers needed interpretation of a particular "Bulletin" inasmuch as its clarity was in question. On these occasions, the director of training in the Naval District Headquarters was called upon for the appropriate interpretation.

As the number of "Bulletins" increased, the College Training Section of BuPers devised an index, which enabled the reader to cross-reference subjects as needed. Thus, the "Bulletins" served a useful role on many campuses when questions arose pertaining to Navy policy or procedure.

U.S. Navy Phraseology
Navy Terms and Definitions

As each Navy V-12 trainee came on board, he was directed to use Navy idioms in his oral conversations and written correspondence, both official and unofficial. Most trainees were quick to comprehend the new terminology which they privately labeled "ship linguistics," or "Navy lingo."*

The following is a list of the terms that were employed at all times by Navy personnel in substitution for the usual civilian expressions, while aboard ship:

Abaft - A relative term used to describe the location of one object in relation to another. Thus: The mainmast is abaft (farther aft than) the foremast.

Abandon - For all hands to leave - as abandon ship

Abeam - The bearing of an object 90 degrees from ahead

Abreast - Abeam of

Accommodation Ladder - Portable steps from gangway to water line

Adrift - Loose from the moorings

Aft - At, near, or toward the stern

Ahoy - A term used in hailing a vessel or boat

Alee - to the leeward side

Alive - Alert

All hands - The entire crew

Aloft - Above the upper deck

Alongside - Side to side

Amidships - In or towards the middle of a ship in regard to length or breadth

Anchor Lights - The riding lights required to be carried by vessels at anchor

Anchor's Aweigh - Said of the anchor when just clear of the bottom

Anchor Watch - A detail on deck at night, when at anchor, to safeguard the vessel

Arming - The tallow placed in the cavity at the bottom of a lead for the purpose of bringing up a sample of the bottom

Astern - The bearing of an object 180 degrees from ahead

Athwartships - At right angles to the fore-and-aft line

Avast - An order to stop or cease hauling

Aweigh - Said of the anchor when hoisted just clear of the bottom

Aye Aye, Sir - Reply to an officer's order signifying it is understood and will be obeyed

Yes, Sir - An affirmative answer made in reply to a senior's question

Very Well - A statement signifying assent made by a senior to a junior

Ballast Tanks - Double bottoms for carrying water ballast and capable of being flooded or pumped out at will

Barge - A large motor boat for the use of flag officers

Barnacle - Small shellfish found on the bottom of vessels

Batten Down - To make watertight, said of hatches and cargo

Battle Lights - Lights used for illumination during "darkened ship"

Beachcomber - The expression today connotes tramp of the sea, unreliable drifter

Beam - The broadest or central portion of a vessel

Bear A Hand - To hurry, "shake it up"

Belay - To make fast to a pin or cleat, to rescind an order

Belaying Pin - A wooden or iron pin fitting into a rail upon which to secure ropes

Below - Beneath the decks

Bitter End - The very end of a rope or the last link in an anchor chain

Bitts - Vertical wooden or iron posts projecting above the deck used for securing lines

Black Gang - Members of the Engineer's Force

Blinker Tube - A tube carrying a signal light capable of being operated to indicate dots and dashes; the tube directs the beam.

Blister - A bulge built into a man-of-war's side as a protection against torpedoes, capable of being flooded or pumped out

Block and Tackle - The apparatus consisting of pulleys and the necessary lines

Blue Jacket - An enlisted man in the Navy; "Gob" is not used.

Boat Boom - The boom swung out from the ship's side when at anchor and to which boats in the water secure

Boat Falls - The tackles used to hoist a boat to its davits

Boatswain's (bo'sn) Pipe - A small shrill whistle used by the boatswain's mate in passing a call or in piping the side

Boatswain's Chair - A piece of board on which a man working aloft is swung

Bollard - An upright wooden or iron post on dock to which hawsers may be secured

Boot - (slang) A newly enlisted recruit undergoing training in the Navy or Marine Corps

Bow - The forward part of a vessel's sides, or the forward end of the ship

Breaker - A small cask carried in ship's boats for drinking water

Breakout - To unstow

Bridge - The raised platform extending athwartships in the forward part of the ship and from which the ship is usually navigated

Brig - Where prisoners are confined on board men-of-war; the ship's prison

Bright Work - Polished metal work

Brow - A portable gangplank

Bulkhead - Partitions separating portions of the ship

Bull Ensign - (slang) Senior ensign on board

Bulwark - A light plating or wooden extension of the ship's side above the deck

Bumboat - A boat employed by civilians to carry provisions, vegetables, and small merchandise for sale to ships

Burdened Vessel - The vessel required by the rules of the road to keep clear, while the "privileged vessel" is required to hold course and speed

Burgee (bur'je) (g as in gem) - A swallow-tailed pennant or flag

Bust - To reduce in rate

Cabin - The captain's quarters

Cable - A rope or chain of great strength, generally used in reference to chain or rope bent to the anchor

Call Away - To summon a boat's crew by a boatswain's pipe or bugle

Camel - A wooden float placed between a vessel and a dock and acting as a fender

Capstan - The vertical barrel situated on the forecastle and geared to the anchor engine

Carry Away - To break or tear loose

Carry On - A command to resume work or continue that which was in progress

Casemate - A compartment in which a broadside gun is mounted

Cast Off - To let go

Caulking (corking) Off - Sleeping

Chafing Gear - A guard of canvas or rope around spars or rigging to prevent chafing (rubbing)

Chains - The station for the leadsman

Charley Noble - Galley smoke pipe

Check - To ease off gradually; to check a line means to keep a strain on it but to slack it as necessary to avoid parting.

Chock - A heavy wooden or metal fitting secured on deck or dock, with jaws to give a fair lead to lines or cables

Chockablock - Full; filled to the extreme limit

Christening A Ship - Ceremony at launching

Clamp Down - To sprinkle and swab down, as a deck in hot weather

Cleat - A fitting of wood or metal with horns, used for securing lines

Coaming - The raised collar about deck openings which prevents water from washing within them

Coil - To lay down rope in circular turns (see flemish down)

Colors - The national ensign; also the ceremony of raising or lowering the colors

Command - Direction by a superior to perform a certain act in a certain definite way; no discretion is allowed in the manner of execution.

Commission Pennant - The pennant flown at the main truck of a ship of war in command of a commissioned officer, not a flag officer

Companion Ladder - A ladder leading from one deck level to another

Companionway - A passage for communication purposes from one deck to another

Conning - Directing the steering by orders to the steersman

Coxswain (cox'n) - The enlisted man in charge of a boat and usually serving as steersman; his rate is boatswains mate third class.

Crossing The Line - Crossing the equator

Crow's Nest - The platform on the mast for the lookout

Cut-of-the-Jib - General appearance of a vessel, sometimes applied to a person

Day's Duty - A tour of duty lasting 24 hours

Dead Ahead - In line with the vessel's keel extended and ahead

Dead Reckoning - A navigator's reckoning with courses steered and distances run independent of sights or bearings

Deck, Main - The highest complete deck extending from stem to stern

Deck, Second - A complete deck next below the main deck

Deck, Third - A complete deck below the second deck

Departure - Before losing sight of land, it is the duty of the navigator to take a departure, which consists of fixing the position of the ship by observation of the best landmarks available. This position is the origin for dead reckoning.

Derelict - An abandoned vessel at sea

Dip - A position of a flag when hoisted part way of a hoist; to lower a flag part way then hoist it again

Ditty Bag - A bag used by sailors for stowing wearing apparel

Dock (see pier) - The water alongside a pier

Dog - A securing device for watertight doors, hatches and manholes

Dog Watch - One of the two-hour watches from 1600 to 2000; from 1600 to 1800 is the first dog and from 1800 to 2000 is the second dog.

Dolphin - A piling or a nest of piles off a wharf or beach or off the entrance to a dock for mooring purposes

Double Bottoms - Watertight subdivisions of a man-of-war next to the keel and between the outer and inner bottom

Double Up - To increase the vessel's securing lines

Douse - To lower; to let down, as to "douse" sail

Drift Lead - A lead of from 30 to 50 pounds dropped over the side when at anchor to give notice if the vessel drags

Dutch Courage - False courage due to liquor

Ease Off - To slack up

Easy - Carefully

End for End - To shift one end of a rope to the position occupied by the other

Ensign - The national flag; a junior officer in the U.S. Navy

Eyebrow - The metal lip over a port to carry the water to the side of the opening

Eyes - The foremost portion of the weather deck in the bow of a ship

Fake - A single turn of line when a line is coiled down

Fair Tide - A tide running in the same direction as the vessel

Fairway - An open channel

Fantail - The part of the stern of a vessel extending abaft the stern post; in light craft such as destroyers and cruisers, the after section of the main deck

Fathom - Six feet

Fathometer - An electrical sound apparatus for measuring automatically the depth of the water

Fenders - Canvas, wood, or rope used over the side to protect a vessel from chafing when alongside another vessel or pier

Fend Off - To shove off

Field Day - A day for general ship cleaning

Figurehead - A carved wooden figure carried on old sailing vessels under the bowsprit

First Lieutenant - The officer charged with the cleanliness of a ship, and who is in charge of the C&R equipment and ground tackle

Flag (see pennant and burgee) - A term used only with reference to a four-sided flag; namely, rectangular in shape

Flag Officer - A designation of those who have attained the rank of rear admiral and the grades senior thereto

Flash Plate - A steel plate to protect the deck from the anchor chains

Flemish Down - To coil a line flat down on deck, each fake outside the other, beginning in the middle and all close together

Flogging - An old form of punishment in the Navy

Flotsan (see jetsam) - Floating wreckage or goods

Fogy (slang) - An increase in pay due to length of service without an increase in rank; technically, longevity pay

Forecastle (focsl) - The topmost deck extending from the superstructure to the stem

Forepeak - The part of the vessel below decks at the stem

Fouled - Jammed, not clear

Frame - A term used to designate one of the transverse ribs that make up the skeleton of the ship

Franklin Life Buoy - A brass life buoy which automatically ignites two lighted arms upon striking the water

Frap - To pass a line around another line or set of lines, at the same time maintaining control of both ends of the frapping line; this is often used in connection with lowering lifeboats.

Freeboard - The distance from the waterline to the rail or covering board

Gaff - The spar from which the colors are usually flown when at sea

Galley - The ship's kitchen

Galley Yarn - A rumor

Gangway - (1) An opening in the bulwark or rail to give entrance to the ship; (2) An order to stand aside

Gear - A general name for all ropes, blocks, miscellaneous material, etc.

Gig - The captain's boat

Glass - A term used by mariners for barometer

Gob - An enlisted man; not considered in good taste and its use should be frowned upon

Grapnel - A small anchor with several arms used for dragging purposes

Gripe - A lashing, chain, or the like, used to secure small boats in the chocks and in sea position in the davits

Grog - Today, means any intoxicating drink

Ground Tackle (taykle) - A term used to cover all of the anchor gear

Guess Warp - A line from forward rove through a thimble at the outer end of a boat boom, used for securing a boat to the boom

Gunwale (gunnl) - The rail or edge of a boat

Guys - Side braces for booms

Hail - To address a vessel; to come from, as to hail from some port

Halliards - The lines by which flags are hoisted

Hand Lead - A lead of from seven to 14 pounds used with the hand lead line for ascertaining the depth of water in entering or leaving port

Handsomely - Carefully, not necessarily slowly

Handy Billy - A small portable force pump

Hash Mark - Slang expression for a diagonal strip on an enlisted man's sleeve to denote a previous enlistment

Hatch - A cover for a hatchway

Hatchway - An opening in a ship's deck for communication or for handling stores or cargo

Haul - To pull; a change of wind in the direction of the hands of a clock

Hawser - A large rope used for heavy work, such as towing

Hawsepipe - The pipes through which the hawser or chain leads to the anchor

Head - The ship's water closet

Heads Up - A warning to look out

Headway - Moving ahead

Heave - To throw; the rise and fall of a vessel in a seaway

Heave Away - An order to haul away or to heave around a capstan

Heave Round - To revolve the drum of a capstan, winch or windlass

Heave Short - To heave in until the vessel is riding nearly over her anchor

Heave the Lead - The operation of taking a sounding with the hand lead

Heave To - To deaden a vessel's headway by bracing some of the sails back; to stop a vessel under sail without taking in sail

Heel and Toe - Any watch stood four hours on and four hours off

Hog (see sag) - The strain of the ship in which the two ends of the vessel are lower than the middle

Hold - (1) The space below decks utilized for the stowage of ballast, cargo, and stores; (2) to hold a line means to make a line fast in such a manner that no slipping will occur (see check)

Holiday - A space unintentionally missed while painting or cleaning

Holy Joe - The chaplain

Holystone - A flat stone used to clean a vessel's deck

House - To stow or secure in a safe place

House An Awning - An awning is housed by hauling some of the stops down and securing them to the rail.

Hug - To keep close

Idlers - Members of the ship's company with no night watches

Inboard - Towards the fore-and-aft line of the ship

Inland Rules - The rules of the road enacted by Congress and governing the navigation of inland waters of the United States

Inshore - Towards the shore

International Rules - The rules of the road established by agreement between maritime nations and governing the navigation of the high seas

Irish Pennant - Any piece of line loose or adrift; considered very unseamanlike

Jack - Flag similar to the union of the national flag; flown at jackstaff

Jack-O'-The-Dust - The petty officer charged with the care of the provisions' store room; the "jack" in charge of the flour or "dust"

Jackstaff - The spar or staff from which the jack is flown

Jacob's Ladder - A ladder of rope with wooden steps

Jetsam (see flotsam) - Goods which sink when thrown overboard

Jigger - A light luff tackle used for various deck work

Jump Ship - To leave a ship without authority

Jury Anchor - A heavy weight used as an anchor

Jury Rig - A makeshift rig

Keel - The timber or bar forming the backbone of the vessel and running from the stem to the stern at the bottom of the ship

Keelhauling - Today, means a verbal reprimand

Ki Yi - A scrubbing brush

Knock Off - To stop, especially to stop work

Knot - One nautical mile per hour

Ladder - A metal, wooden or rope stairway

Land Fall - The first sighting of land at the end of a sea voyage

Land Ho! - The hail from the lookout when land is sighted

Lanyard - A line used to secure some object from becoming lost or going adrift

Larboard - A term used in earlier days to denote the present day port side

Lay - Preliminary order, e.g. lay below, lay aloft

League - Three nautical miles

Leeward (lou ard, accent on first syllable) - The direction away from the wind

Leeway - The drift of a vessel to leeward caused by wind or tide

Lend A Hand - Request for assistance

Lie Off - The practice of having a boat stand by, under control, but not made fast to anything; underway, under control, but with no way on

Lighter - A boat used in harbors for transporting merchandise

Lime Juicer-Limey - This expression refers only to British seamen.

List - The athwartships inclination of a vessel

Log - A book containing the official record of a ship's activities together with remarks concerning the state of the weather

Lower Away - An order to lower down

Lubber's Line - The vertical black line marked on the inner surface of the bowl of a compass indicating the compass heading of the ship

Lucky Bag - A locker or compartment for the stowage of loose articles of clothing found about the ship

Magazine - The ammunition stowage in a ship

Make Colors - Hoisting the ensign at 0800

Man the Boat - An order to embark

Marlinespike - A pointed iron instrument used in working with rope and wire

Martinet - A stickler for discipline

Mast - Morning court

Meal Pennant - The red pennant flown at the yardarm of men-of-war at anchor during meal hours

Messenger - A light line used to haul over a heavier line or cable

Mess Gear - Equipment for serving meals

Midshipman - A naval cadet

Monkey Fist - A knot worked into the end of a heaving line

Mooring - Securing to a dock or a buoy, or anchoring with two anchors

Morning Order Book - The book in which the executive officer writes his instructions for the next morning's ship's work

Muster - To assemble the crew and verify the absentees

Nun Buoy - A buoy with a conical top

found on the starboard hand on entering a channel, and painted red

Old Man - The captain of the ship

On Soundings - Said of a vessel when the depth of water can be measured by the lead, within the 100-fathom curve

Order - Direction by a superior to perform a certain act, leaving the method of execution to the individual

Outboard - Towards the sides of the vessel

Out of Trim - Not properly trimed or ballasted

Overboard - Outside of the ship

Overhaul - (1) To separate the blocks of a tackle; (2) to overtake; (3) to examine and repair

Overhead - A ceiling

Painter - The line by which the bow of a small boat is made fast

Paravanes - Fish-shaped metal bodies towed underway from the stem of men-of-war as a protection against mines

Part - To break; as, to part a line

Passageway - A corridor

Pass A Line - To reeve and secure a line

Pass the Word - To repeat an order or information to the crew

Pay Out - To slack out on a line made fast on board

Pelorus - An instrument for taking bearings

Pendant - A length of rope with a block or thimble at one end

Pier (see dock) - A structure built out into the water on pilings for use as a landing space

Pilot Rules - The rules supplementing the Inland Rules and established and published by the Board of Supervising Inspectors of steam vessels

Pipe Down - An order to keep quiet; an order dismissing the crew from an evolution

Pipe the Side - The ceremony at the gangway when the boatswain's pipe is blown when an official comes aboard or leaves a man-of-war

Pitch - The motion of the ship about the athwartship centerline

Pollywog (slang) - One who has not crossed the equator (see shellback)

Poop Deck - A partial deck at the stern over the main deck

Position Buoy - The spar towed by men-of-war in a fog as an aid to the next astern in column; also called a towing-spar or sea pig (slang)

Pratique - A limited quarantine; a permit by the port doctor for an incoming vessel being clear of communicable disease to have the liberty of the port

Prolonged Blast - A blast from four to six seconds' duration

Prow - The part of the bow above the water

Quarterdeck - A name applied to the part of the upper deck used for ceremonies

Quarters - Early morning muster formation for all hands aboard ship

Rank - The grade of a commissioned officer

Rating - The grade of an enlisted man; a petty officer

Reeve - To pass the end of a line through any lead such as a sheave or fair lead

Ride - To lie at anchor; to ride out, to weather safely a storm, whether at anchor or underway

Rise and Shine - A call to turn out of bunks and hammocks

Rocks and Shoals (slang) - Articles for the Government of the Navy

Roll - The motion of the ship about its fore and aft centerline

Ropeyarn Sunday - Usually Wednesday afternoon; no ship's work is scheduled at this time and the men are permitted to do their personal work.

Rough Log - The ship's log as written up in pencil by the quartermaster and the officer-of-the-deck

Round Turn - To take a turn around a bitt or bollard, to check a strain or weight; "To bring up with a round turn," is nautical phraseology for a "call-down" or reprimand.

Run Away - To seize a line and haul on the run; to desert

Run Down - To collide with a vessel head on

Running Lights - The lights required by law carried by a vessel when underway

Sag (see hog) - The waist of a ship having settled down below the level of the bow and the stern

Sail Ho! - The hail from a lookout to notify that a vessel has been sighted

Scope - The length of chain out

Scullery - The compartment for washing and stowing dishes

Scuttlebutt - (1) The drinking fountain; (2) a rumor

Sea Anchor - A float, usually of canvas stretched on a conical frame, dragged by a vessel with the large end toward the vessel, to keep the vessel from drifting or to keep her head to the wind

Sea Ladder - A rope ladder, usually with wooden steps, for use over the side

Sea Lawyer - A seaman who is prone to argue, especially against recognized authority; one who by his wit tries to avoid difficult duties

Sea Room - Far enough away from land for unrestricted maneuvers

Secure - To make fast; the completion of a drill or exercise aboard ship; knock off

Shellback (slang; see pollywog) - A seaman who has crossed the equator

Short Stay - When the scope of chain is slightly greater than the depth of water

Shot - A short length of chain, usually of 15 fathoms

Shove Off - To leave

Show A Leg - An order to make haste

Sick Bay - The ship's hospital

Side Boys - Enlisted men of the watch, in port or at anchor, attending the side to render honors

Silence - A command used in gunnery, meaning to stop immediately whatever is in progress; emergency order

Sing Out - To call out

Single Up - To come up double lines so that only single parts remain secured

Sister Ships - Ships built on the same design

Skivvies (slang) - Underwear

Sky Pilot - The chaplain

Slick - The smooth water between the wash and the wake; also, smooth water made by pouring oil on it

Small Stuff - Small cordage; it is designated by the number of threads or by special names based on its use.

Smart - Snappy; seamanlike; a smart ship is an efficient one.

Smoking Lamp - A real or mythical light aboard ship signifying whether or not smoking is permitted at any time; when the word is passed, "the smoking lamp is out," it means "knock off smoking."

Snub - To check suddenly

Speed Cones - Conical shapes used in formation to indicate engine speeds

Spitkit - A receptacle for cigarette butts and matches

Splinter Deck - The lower of two armored decks; a single armored deck is called the protective deck.

Square Away - To put in order

Squilgee (squeegee) - A deck dryer composed of a flat piece of wood shod with rubber and a handle

Station Bill - The posted bill showing stations of the crew at emergency drills

Steady - An order to hold a vessel on the course she is maintaining

Steerage Way - The slowest speed at which a vessel will steer

Stern - The after end of a ship

Stern Sheets - The space in a boat abaft the after thwart

Stopper - A short length of line or chain, secured at one end, and used in securing or checking a running line

Stow - To put in place

Striker - An apprentice or student for a special job

Strongback - A light spar lashed to a pair of boat davits and serving as a spreader for the davits

Sundowner - A martinet

Swab - A mop

Tackles (commonly taykles) - Commonly called a pulley or system of pulleys in civilian life

Taffrail Log - The log mounted on the taffrail and consisting of a rotator, a log line and a recording device

Tarpaulin - Heavy canvas used as a covering

Thwarts - The athwartship seats in a boat

Thwartships - At right angles to the fore-and-aft line

Tier - To stow cable in a chain locker

Tiller - Short piece of iron or wood fitted into the rudder head and by which the rudder is turned

Toggle - A wooden or metal pin slipped into a becket and used for securing gear

Tompion (tompkin) - The wooden plug placed in the muzzle of a gun to keep out dampness

Topsides - On the weather deck

Transom - Any seat that is built in officers' country, and is a permanent fixture, is, by usage, called a transom.

Trick - Steerman's period at the wheel

Trimming Tanks - Tanks used for ballast

Turn To - An order to commence ship's work

Two Blocks - Hoisted as high as possible

Unbend - To cast adrift or untie

Underway - Said of a vessel when not at anchor, nor made fast to the shore, nor aground

Unship - To take apart or to remove from its place

Up Anchor - The order to weigh the anchor and get underway

Veer - To slack off and allow to run out; said of a change of direction of wind

Very's Signals - Red, white, and green stars fired in a pistol and used in a special code for signalling purposes

Waist - The portion of the deck between the forecastle and quarterdeck

Wake - A vessel's track; behind

Wardroom - Commissioned officer's mess room and lounge

Warp - To move the ship by a line made fast to a pier or anchor

Wash - The waves pushed out by a vessel in moving through the water

Water Breaker - A small cask carried in ship's boats for drinking water

Waterway - The gutter at the sides of a ship's deck to carry off water through the scuppers

Way - The motion of a ship through the water

Weather - Exposed to the weather, as "weather deck"

Weather Eye - To keep a weather eye is to be on the alert.

Weather Side - The windward side

Weigh - To lift the anchor off the bottom

Well! - An order meaning sufficient

Well Deck - The part of the main deck between superstructures

Whaleboat - A sharp-ended pulling life boat

Lawrence College, V-12 Unit, Neenah Nodaway Yacht Club. They invited 100 members of the unit to participate in their races.

USS Saluda, IX-87, New London, Connecticut, 1944. (Courtesy of Herbert Humphrey)

Where Away - A call in answer to the report of a lookout of an object sighted

Wide Berth - At a considerable distance

Wildcat - A sprocket wheel on the capstan for taking the links of the chain cable

Winch - An engine secured on deck and fitted with drums and driven on a horizontal axis

Windsail (winsel) - A canvas trunk spread to admit air below decks

Windward (second "w" is not pronounced) - Toward the wind, or into the wind

Work (to work a ship) - To handle by means of engines and gear

Yard - A spar suspended horizontally from the mast

Yard-Arm Blinker - A signal light carried on the yard-arm of men-of-war and operated to indicate dots and dashes

Yaw - To steer wildly or out of line of the course

The above has been compiled from the following books: A.M. Knight, *Modern Seamanship; Reserve Officer's Manual, U.S. Navy; The Bluejackets' Manual; L.P. Lovette, Naval Customs, Traditions and Usage.*

At some V-12 Units, the campus lexicon of Naval terms and definitions was more extensive. Trainees were instructed to use this additional phraseology in their conversations and correspondence:

Addendum

Brig - A place for temporary confinement of offenders

Bunk - A built-in sleeping place aboard ship

Caulk - To caulk wooden decks or sides of ships; or (corking off) sleeping

Doldrums - A spell of listlessness, as before an examination; the oceanic area close to the equator

Duffle - Name of a seaman's personal belongings or the sea bag for storing them

Flagship - The college hall in which the offices of the unit's command is located

Locker - The closet assigned a trainee for his personal belongings

Mess Hall - The on-campus building in which the dining area is located

Ship - Each of the on-campus dormitory buildings

Skipper - The nickname for captain or officer in command

Skylark - To play in the rigging of a sailing ship; or to goof-off

Three-Mile Limit - The on-shore limit for trainees; or weekend leave

Watch (on watch) - The on-duty officers and crew members in the ship's company assigned the watch during particular intervals of a 24-hour day

Bibliography

During 1946-1948, 10 studies and the general staff report were published by the American Council on Education for the Commission on Implications of Armed Services Educational Programs, one of which dealt specifically with the Army Specialized Training Program (ASTP) and the Navy College Training Program V-12. This publication, as well as the staff document containing condensations of all 10 commission studies, were quoted in this history. By permission, those two volumes are:

Grace, Alonzo (ed.). *Educational Lessons from Wartime Training: General Report of the Commission.* Washington, DC: American Council on Education, 1947.

Herge, Henry C., et al., *Wartime College Training Programs of the Armed Services,* Washington, DC: American Council on Education, 1948.

In addition, the following references were used:

Association of American Colleges Bulletin, The Colleges Prepare for Peace, a symposium, XXX (1944), 5-47. "The Colleges Prepare for Peace," by Lord Halifax (pp. 5-11); "The College Program and the Returning Service Man," by E. J. McGrath (pp. 21-31).

Board of Education, City of New York. *Report on Training in the Armed Services with Implications for Postwar Education in New York City.* New York: Board of Education, 1946.

Fine, Benjamin. "Education in Review," *The New York Times,* February 10, 1946.

Miller, Francis T., et al., *The Complete History of World War II,* Chicago: Reader's Service Bureau, 1945.

Nash, Ray. *Navy at Dartmouth.* Hanover, New Hampshire: Dartmouth Publication, 1946.

"The Navy V-12 Program" *Journal of the American Association of Collegiate Registrars,* XIX (January 1944), 161-169.

Russell, John Dale (ed.). *Higher Education under War Conditions, Proceedings of the Institute for Administrative Officers of Higher Institutions,* Vol. XV. Chicago: University of Chicago Press, 1943. 160 pp.

United States Congress. House. Committee on Education. *Effect of Certain War Activities upon Colleges and Universities.* Report from the Committee on Education, pursuant to H. Res. 63. 79th Congress, 1st session. Washington: Government Printing Office, 1945. 57 pp.

United States Navy Department. *Manual for the Operation of Navy V-12 Units.* Navy V-12 Bulletin No. 200. Washington: Government Printing Office, 1944. 130 pp.

United States Navy Department. Bureau of Naval Personnel. Training Division. *Navy V-12 Curricula Schedules and Course Descriptions.* Navy V-12 Bulletin No. 101.

Washington: Government Printing Office, 1943. 94 pp.

Bureau of Naval Personnel 15,012 "Conference on the Navy V-12 Program at Columbia University, May 14-15, 1943."

Bureau of Naval Personnel. "The College Training Program," *U.S. Naval Administration in World War II*. Training Activity, Vol. IV.

Finance and Materials Division and the College Training Section, Bureau of Naval Personnel; War Department, Office of Chief, Training Unit Contracts, 1946.

Supplementary References of a More Recent Date:

Brinkley, David. *Washington Goes To War*. New York: Alfred A. Knopf, 1988.

Calvocoressi, Peter, and Wint, Guy. *Total War: The Story of World War II*. New York: Pantheon Books, 1972.

Hough, Richard. *The Greatest Crusade: Roosevelt, Churchill and the Naval Wars*. New York: William Morrow and Company, 1986.

Matloff, Maurice, Ed. *American Military History*, Army Historical Series. Washington, DC: Office of the Chief of Military History, U.S. Army, 1969.

Morris, Richard B., Ed. *Encyclopedia of American History*. New York: Harper and Brothers, 1953.

Northrup, Herbert R., et al. *Black and Other Minority Participation in the All-Volunteer Navy and Marine Corps*, Vol. VIII. Philadelphia: The Wharton School, University of Pennsylvania, 1979.

Schneider, James G. *The Navy V-12 Program: Leadership for a Lifetime*. Boston: Houghton Mifflin Co., 1987.

Sherwood, Robert E. *Roosevelt and Hopkins: An Intimate History*. New York: Harper & Brothers, 1950.

Stillwell, Paul. *The Golden Thirteen: Recollections of the First Black Naval Officers*. Annapolis: Naval Institute Press, 1993.

The Secret History of World War II: The Ultra-Secret Wartime Cables and Letters of Roosevelt, Stalin, and Churchill. New York: Richardson & Steirman, 1986.

Thompson, Robert S. *A Time for War: Franklin Delano Roosevelt and the Path to Pearl Harbor*. New York: Prentice Hall Press, 1991.

Weigley, Russell F. *History of the United States Army*, Enlarged Edition, Bloomington: Indiana University Press, 1984.

World Almanac and Book of Facts, 1993. New York: Scripps Howard Company, 1993.

Appendix A

Navy V-12 Production Records

Production Record as of September 10, 1944

1. Navy V-12 Production Record - World War II*

The magnitude of Navy training during the war was released by the Secretary of the Navy before the termination of hostilities on September 10, 1944:

During the past fiscal year ending 30 June 1944, the Navy trained 1,303,554 personnel, manning 4,063 new vessels - or 11 ships each day - plus more than 20,000 landing craft and keeping pace with the Naval Air Arm, which doubled the number of planes on hand.

The magnitude of the Navy's training task stems from the necessity of manning the world's greatest Naval force, predominantly with men who have had no previous sea-going experience. Of a total of 2,987,311 personnel in the Navy on 30 June, less than 12 percent were in the service prior to Pearl Harbor and 2,478,002, or approximately 83 percent, are members of the Naval Reserve.

The collapse of Germany will result in no curtailment of the Navy's training program. The continued successful prosecution of the war against Japan will require, according to present estimates, that the Navy continue to expand until it reaches a strength of 3,389,000 by 30 June 1945.

The complexity of the Navy's training activities is reflected in the fact that new personnel must be trained to proficiency in more than 450 enlisted specialties and petty officer ratings, which are indispensable to man, fight and maintain the highly complicated mechanism of a modern Navy.

The Navy now has a total of 947 schools with a daily average attendance of 303,000 personnel.

Up to the end of the 1943-1944 fiscal year, of this number 136 were basic and advanced air training schools, with an average attendance of 35,000 and a monthly output of 1,700. It is estimated that the Navy spends close to $30,000 on the training of each Naval aviator who is in training for 18 to 24 months.

The Navy's schools for training officers and officer candidates fall into two groups:

a. Six Naval Reserve Midshipmen's Schools have sent a total of 41,689 deck and engineering officers to duty assignment throughout the Naval establishment. These schools, established since 1940 for the training of officer candidates from civilian life and from the earliest ranks, are the Navy's principal source of young, seagoing officers and 95 percent of their graduates are serving at sea.

b. With the knowledge that Selective Service would in time sharply diminish or eliminate the supply of young men between the ages of 18 and 21 years upon which the Navy would have to depend for additional officer candidates, the Navy, on 1 July 1943, instituted the Navy College Program (V-12) for the preliminary training of young officer candidates. At this time, the Navy College Program (V-12) is operating 264 units at 202 colleges and universities and has a current attendance of 65,000 officer candidates. Since the establishment of the V-12 program, it has delivered more than 23,000 qualified officer candidates to the Reserve Midshipmen's Schools, Supply Corps Schools and Marine Officer Candidate's Schools. In addition to this number, 2,600 officers were commissioned directly from Naval Reserve Officer Training Corps, now a part of the V-12 Program, and the medical and dental schools have supplied the Navy with 1,400 doctors and dentists.

Of the Navy's training schools, 310 are devoted to the instruction of enlisted personnel.

Stoddard Hall V-12, Miami University, Oxford, OH, 1943. (Courtesy of Erling Larson)

Endnote:
*Bureau of Naval Personnel, Navy Department, Training Bulletin, NAVPERS 14923, Oct. 15, 1944, p.4. Excerpts.

2. Enrollment Statistics, 1943-1945

Naval College Training Program Enrollment Statistics
July 1, 1943 through December 31, 1945*

Cirriculum	Input	Output
Basic V-12	70,973	35,576
V-12 (a)	24,213	7,659
NAPP (V-5)	8,570	677
V-12 Engineering	27,333	9,634
Premedical and Predental	10,931	2,804
Pretheological	273	110
NROTC	9,200	7,415
Marine Corps	14,821	11,538
Theological	233	58
Dental	2,846	1,415
Medical	8,954	3,588
Total	178,347	80,474

Aggregate losses from attrition during this period were 44,004 trainees.

3. Number of Trainees Processed by V-12 Institutions
July 1, 1943-June 30, 1945

	Total Second Year	Two Year Total
Input:	20,369	123,905
Attrition	13,370	32,204
Output:		
To Midshipmen Schools	17,483	37,149
To Marine Officer Candidate Schools	1,521	6,585
To Medical, Dental & Theo. Schools	1,565	4,587
To Aviation Training	5,560	10,241
NROTC Commissions	2,413	5,007
Total Output	28,542	63,569

Appendix B

Outline of Curricula of the Navy V-12 Program
Regular Students

Type of Candidate Regular	Term Basic 1 2 3 4	Advanced 5 6 7 8	Further Training	Classification General & Special Service
Deck	X X X X		RMS -2-4 Mo.	Ens. D-V(G),D-V(S), C-V(S)
Pre-Med & Pre-Dent	X X X X	X	Med. or Dent. 21/4 yr	Lt. (jg) MC-V(S), DC-V(S)
NROTC Engineering	X X X X	X X X	Operational Training	Ens. E-V(S),CEC-V(S), O-V(S)
General	X X X X	X X X	Operational Training	Ens. D-V(G), D-V(S)
Business Admin.	X X X X	X X X	Operational Training	Ens. D-V(G), D-V(S)
Engineering, General	X X X X	X X	RMS - 2-4 Mo.	Ens. E-V(S), CEC-V(S), E-V(G)
Civil Engineering	X X X X	X X X X	RMS - 2 or 4 Mo.	Ens. CEC-V(S)
Construction Corps	X X X X	X X X X	RMS - 2 or 4 Mo.	Ens. CC-V(S)
Mechanical-Steam	X X X X	X X X X	RMS - 2 or 4 Mo.	Ens. E-V(S), A-V(S), O-V(S)
Mechanical-Int.Comb.Eng.	X X X X	X X X X	RMS - 2 or 4 Mo.	Ens. E-V(S), A-V(S), O-V(S)
Electrical Power	X X X X	X X X X	RMS - 2 or 4 Mo.	Ens. E-V(S), A-V(S), O-V(S)
Electrical-Com. & Preradad	X X X X	X X X X	RMS - 2 or 4 Mo.	Ens. E-V(S), A-V(S), O-V(S)
Aeronautical-Engines	X X X X	X X X X	RMS - 2 or 4 Mo.	Ens. E-V(S), A-V(S), O-V(S)
Aeronautical-Structures	X X X X	X X X X	RMS - 2 or 4 Mo.	Ens. E-V(S), A-V(S), O-V(S)
Physics Major	X X X X	X X X X	RMS - 2 or 4 Mo.	Ens. E-V(S), A-V(S), O-V(S)
Aviation, General (V-5)	X X Check Flight	Exam	Flight Training	Ens. A-V(N)-Oper. FlightTr.
Aerology	X X X X	X X X X	RMS - 4 Months	Ens. A-V(S)
Supply Corps	X X X X	X X	Supply Sc.-4 Mo.	Ens. SC-V(G)
Pre-Chaplain Corps	X X X X	X X X X	Thel. Seminary to degree, Chaplain School	Lt. (jg) ChC-V(S)
Marine Corps-Line	X X MC-Line 2	terms	Officers Cand. School	2nd Lieutenant
Marine Corps-Specl.	X X X X	X X X X	Officers Cand. School	2nd Lieutenant

V-12 and Marine V-12 trainees with advanced standing continued the same major as before entering. However, each candidate was required to meet the minimum requirements for each type candidacy. He remained under instruction the allowed number of terms and was processed the same as the fully prescribed regular candidates after V-12. Key to abbreviations: RMS - Reserve Midshipmen's School; D-V(G) and (S) - Deck and engineering officers, general and special; D-V(S) and (G) - Deck officer, general and special; E-V(G) and (S) - Engineer officers, general and special; A-V(N) - Naval aviation officers; A-V(S) - Aviation flight officers; MC-V(S) - Medical officers; SC-V(G) - Supply officers; DC-V(S) - Dental officers; ChC-V(S) - Chaplains; CC-V(S) - Engineer Officers (construction duties); CEC-V(S) - Civil engineers; O-V(S) - Ordnance Officers; and C-V(S) - Communication officers.

Appendix C

List of Naval District Offices

Naval District	Offices
First	North Station Office Bldg. 150 Causeway Street Boston 14, Mass.
Third	90 Church Street New York 7, N. Y.
Fourth	Building 4, Navy Yard Philadelphia, 12, Pa.
Fifth	Naval Operating Base Norfolk, Va.
Sixth	Charleston Navy Yard Navy Yard, South Carolina
Seventh	Headquarters, 7th Naval District Station Miami, Florida
Eighth	New Federal Building New Orleans, La.
Ninth	Naval Training Station Great Lakes, Illinois
Eleventh	Naval Operating Base San Diego 30, Calif.
Twelfth	Federal Office Building Civic Center San Francisco 2, Calif.
Thirteenth	Exchange Building 117 Marion Street Seattle, Washington
Potomac River Naval Command	Washington, D. C.

Appendix D

Navy V-12 and NROTC Units

V-12 Units

(B. Basic Curriculum; E-Engineering Curriculum; P-Pre-Medical and Pre-Dental Curriculum; M-Marine Corps Unit)

Alabama
  Howard College (B)
Arizona
  Arizona State Teachers College (B,M)
Arkansas
  Arkansas A&M (B, P, M)
California
  California Institute of Technology (E)
  College of the Pacific (B, P, M)
  Occidental College (B, P, M)
  University of California, Berkeley (B, E, P, M)
  University of California, Los Angeles (B, P, M)
  University of Redlands (B, M)
  University of Southern California (B, E, P)
Colorado
  Colorado College (B, P, M)
  University of Colorado (E, P, M)

Connecticut
  Trinity College (B, P)
  Wesleyan University (B, P)
  Yale University (B, E, P, M)
Florida
  University of Miami (B, P)
Georgia
  Emory University (B, P)
  Georgia School of Technology (E)
  Mercer University (B, P)
Idaho
  University of Idaho, Southern Branch (B, P)
Illinois
  Illinois Institute of Technology (E)
  Illinois State Normal (B)
  Northwestern University (E, P, M)
  University of Illinois (E, P)
Indiana
  DePauw University (B, P)
  Indiana State Teachers' College (B)
  Purdue University (B, E, P, M)
  University of Notre Dame (B, E, P, M)
  Wabash College (B)
Iowa
  Iowa State A&M College (E)
  St. Ambrose College (B, P)
  University of Dubuque (B, P)
Kansas
  Kansas State Teachers' College (B)
  University of Kansas (E, P)
  University Municipal, Washburn (B, P)
Kentucky
  Berea College (B, P)
  University of Louisville (B, E, P)
Louisiana
  Louisiana Polytechnic Institute (B, P, M)
  Southwest Louisiana Institute (B, P, M)
  Tulane University (B, E, P)
Maine
  Bates College (B, P)
Maryland
  Mt. St. Mary's College (B)
Massachusetts
  College of the Holy Cross (B, P)
  Harvard University (B, P)
  Massachusetts Institute of Technology (B, P)
  Tufts College (B, E, P)
  Williams College (B, F)
  Worcester Polytechnic Institute (E)
Michigan
  Alma College (B, P)
  Central Michigan College of Education (B)
  University of Michigan (B, E, M)
  Western Michigan College (B, M)
Minnesota
  Gustavus Adolphus College (B, P, M)
  St. Mary's College (B, P)
  University of Minnesota (E, P)
  College of St. Thomas (B)
Mississippi
  Milsaps College (B, P, M)
  Mississippi College (B, P)

Missouri
  Central College (B, P)
  Central Missouri State Teachers' College (B)
  Missouri Valley College (B)
  Northwest Missouri State Teachers' College (B)
  Park College (B)
  Southeast Missouri State Teachers' College (B, P)
  Westminister College (B, P)
Montana
  Carroll College (B, P)
  Montana School of Mines (B, E)
Nebraska
  Doane College (B)
  Peru State Teachers' College (B)
New Hampshire
  Dartmouth College (B, P, E, M)
New Jersey
  Drew University (B)
  Princeton University (B, P, M)
  Stevens Institute of Technology (B, E)
New Mexico
  University of New Mexico (B, E)
New York
  Colgate University (B, P, M)
  Columbia University (E, P)
  Cornell University (B, E, P, M)
  Hobart & William Smith College (B, P)
  Rensselaer Polytechnic Institute (E)
  St. Lawrence University (B, P)
  Union College (B, P, E)
  University of Rochester (B, E, P, M)
  Webb Institute of Naval Architecture (E)
North Carolina
  Duke University (B, E, P, M)
  University of North Carolina (S, P, M)
North Dakota
  State Teachers' College, Dickson (B)
  State Teachers' College, Minot (B)
  State Teachers' College, Valley City (B)
Ohio
  Baldwin-Wallace College (B, P)
  Bowling Green State University (B, M)
  Case School of Applied Science (P)
  John Carroll University (B, P)
  Denison University (B, P, M)
  Miami University (B, P, M)
  Oberlin College (B, P, M)
  Ohio Wesleyan University (B, P)
Oklahoma
  University of Oklahoma (E, P)
Oregon
  Willamette University (B, P)
Pennsylvania
  Bloomsburg State Teachers' College (B)
  Bucknell University (B, E, P, M)
  Franklin & Marshall College (B, P)
  Muhlenberg College (B, P, M)
  Pennsylvania State College (B, M)
  Swarthmore College (B, E, P)
  University of Pennsylvania (B)
  Ursinus College (B, P)
  Villanova College (B, E, P, M)

Rhode Island
  Brown University (B, E, P)
South Carolina
  Newberry College (B)
  University of South Carolina (B)
Tennessee
  Carson-Newman College (B, P)
  Milligan College (B)
  University of the South (B)
Texas
  North Texas Agriculture College (B, M)
  Rice Institute (E)
  Southern Methodist University (E)
  Southwestern University (B, M)
  Texas Christian University (B, P)
  University of Texas (E, P)
Vermont
  Middlebury College (E, P)
Virginia
  Emory and Henry College (B, P)
  Hampden-Sydney College (B, P)
  University of Richmond (B, P)
  University of Virginia (B, E, P)
Washington
  Gonzago University (B, P)
  University of Washington (B, E, P, M)
  Whitman College (B, P)
West Virginia
  Bethany College (B, P)
Wisconsin
  Lawrence College (E)

Marquette University (E, P)
University of Wisconsin (E)

NROTC Units

Units Established 1945:
  Alabama Polytechnic Institute
  Case School
  Columbia
  Cornell
  Dartmouth
  Illinois Institute of Technology
  Iowa State
  Miami University
  Oregon State
  Penn State
  Princeton
  Purdue
  Rochester University
  Stanford
  University of Idaho
  University of Illinois
  University of Kansas
  University of Louisville
  University of Mississippi
  University of Missouri
  University of Nebraska
  University of Utah
  University of Wisconsin
  Vanderbilt
  Villanova

Previously Commissioned:
  Brown
  Duke
  Georgia School of Technology
  Harvard
  Holy Cross
  Marquette
  Northwestern University
  Notre Dame
  Rensselaer Polytechnic Institute
  Rice Institute
  Tufts
  Tulane
  University of California
  University of California, Los Angeles
  University of Colorado
  University of Michigan
  University of Minnesota
  University of New Mexico
  University of North Carolina
  University of Oklahoma
  University of Pennsylvania
  University of South Carolina
  University of Southern California
  University of Texas
  University of Virginia
  University of Washington
  Yale

Appendix E

Summary of Navy V-12 Enrollment by Trainee Types
(Supplied by the Office of Director of Training, BuPers)

	Total Undergrad	Total^ Marines	NROTC	Undergraduate Level					Graduate Level			
				Upper Level Engineer	Pre Med & Dent	Pre Supply	Pre Theo	Deck	Lower Level	Med.	Dent	Theolo.
1 July 1943	72,919	11,460	5,132	253M 4,368N	1,268	--	--	2,058M 11,104N	9,149M 39,587N	4,742	1,925	--**
1 November 1943	67,623	10,965	4,572	242M 4,106N	1,173	--	7	1,365M 10,216N	8,758M 36,581N	4,863	1,660	--**
1 March 1944	65,019	5,453		1,600M 12,633N	3,893	476	47	3,895M 15,466N	22,851	5,986	1,853	55**
1 July 1944	67,182	3,653	5,276	821M 13,235N	3,080	854	157	2,832M 10,511N	30,416	6,204	1,816	121**
1 November 1944	50,500	1,858	6,333	657M 12,505N	2,478	1,298	205	1,201M 13,611N	12,212	6,348	1,717	174**
1 March 1945	39,525	1,808	7,536	357M 10,653N	2,040	1,198	186	1,451M 6,486N	9,618	6,619	1,695	193**
1 July 1945	33,483	1,874	9,086	354M 6,798N	1,226	791	105	Deck (NavSoi)#V-5 1,520M 9,378N	4,225	5,694	1,450	205**
1 November 1945	38,197	1,286	18,395	298M 3,761N	1	None	None	1,014M 8,503N	6,251	5,198	1,143	181**
1 March 1946	22,963	685	10,658	98M 1,190N	None	None	None	587M 5,106	5,324	None	1	None**

Legend:
M - Marines; N-Navy
^ - Total Marines included in total undergraduates
# - For the July 1, 1945 term, Navy deck candidates became Naval Science trainee preparatory to being transferred to NROTC Nov. 1, 1945.
** - Figures taken from Tra Div reports

Reasons for Termination of the Army AST Program

The fundamental weakness of the ASTP, at least while operating on a scale of 150,000 trainees, was that while causing endless administrative problems, it served no need recognized as urgent by most elements in the Army. It was one of the easiest items in the troop basis to sacrifice, once the demand for more combat troops became inescapable. On November 5, 1943, G-3, War Department General Staff, proposed a reduction of the ASTP to 30,000 trainees, largely in medical and related subjects, and the return to troops of the remaining 120,000. AGF dispatched its concurrence to the War Department on the same day. The troop basis published on January 15, 1944, showing signs of intervention and compromise, called for a gradual reduction of the ASTP to 62,190 by the end of 1944.

The condition of the ground arms, especially of the infantry, was by this time causing alarm. Divisions had been stripped of their infantry privates to provide overseas replacements. Many of these same divisions were scheduled for early movement for impending invasion of Western Europe. They had to be refilled with men already basically trained. The War Department judged also that the quality of enlisted personnel in the infantry must be raised. General Marshall, on February 10, 1944, informed the Secretary of War that 134,000 men, already basically trained, were required for the coming operations in France, and that the outstanding difficulty currently noted in our divisions is the number of non-commissioned officers who are below satisfactory standards of intelligence and qualities of leadership. He recommended withdrawal of all but 30,000 trainees from ASTP, offering the Secretary of War the alternative of cutting ASTP or of disbanding 10 divisions, three tank battalions, and 26 antiaircraft battalions.

ASTP was immediately liquidated. A large number of its trainees, almost overnight, became privates in the infantry. They had to start as privates because most units, with their former privates withdrawn as overseas replacements, had at least a full complement and sometimes a surplus of non-commissioned officers. It was expected and desired that the ASTP trainees would show their superiority over the older non-commissioned officers, win the ratings, and become leaders of small units.

For its trainees, the ASTP was a succession of disillusionments. None of their expectations was realized. Many of them, had they not been sent to college, would undoubtedly have gone to OCS, to the advantage, both of themselves and of the Army Ground Forces, though it is true that recruiting for ASTP came at a time when OCS quotas were declining.

p. 48, Qualified Problems of Enlisted Personnel in the Army Ground Forces (Study in History of Army Ground Forces), Number 5.

Following Norfolk, Virginia, 50th Navy V-12 Anniversary celebration, Lt. General Edward J. Bronars, USMC, and Commanding Officer Meldrim F. Burrill, LCDR, USNR, compare stories about the 1943 Navy V-12 program at Illinois State University, Normal, Illinois. General Bronars was in the First V-12 program at Illinois State University.

Wabash V-12, RearAdmiral Bill Thompson, USN, (Retired) former President of the United States Navy Memorial Foundation, chats with the two oldest V-12 Commanding Officers attending the 50th Navy V-12 Anniversary celebration at Norfolk, Virginia. Henry Herge, LCDR, USNR, and Meldrim F. Burrill, LCDR, USNRS (Retired). Both commanders are 88 years old and never missed an event!

UCLA V-12 on the roof of Bannister hall - House 9, Sepember, 1944. . L to R: John Giddens (San Diego, CA), from the fleet, was at Pearl Harbor; Stan Perkins (San Francisco, CA); Joe Godd (Sherman Oaks, CA). They sent me back to Southern California from Kansas; Tony Ellington (Azusa, CA). We three had been at Washburn. (Courtesy of Judge Joseph W. Goss)

Platoon RIII, Navy V-12 Unit -Caltech, Padadeua, CA, 1944. (Courtesy of Robert Y. Scapple)

First V-12 class at Princeton University, July, 1943. (Courtesy of Albert G. Ristan)

Note: All universities and colleges participating in the Navy V-12 Program were invited to send in materials for inclusion. Following are those who chose to participate.

Navy V-12
Special Stories

DePauw University

DePauw University received the first of the Navy V-12 participants in July 1943. The students, under the command of Lt. Commander William B. Dortch, lived in Rector and Lucy Rowland Halls. They attended regular college classes taught by the resident faculty.

The war had depleted the number of male students and the V-12 program allowed the college not only to continue financially but also allowed the college to continue its intercollegiate athletic programs.

"We have served the nation," President Wildman said about DePauw's experience with the Naval programs, "in helping to give training to its officer personnel and we have in turn been helped through some extremely difficult years by their presences on campus."

With the addition of the Navy V-12 program, new courses were added: meteorology, interpretation of maps, nautical astronomy and navigation, military German, Russian and "women in war". But because there were fewer students in the advanced courses, many were dropped: the honors program and the plan for senior comprehensive examinations was postponed.

In October 1945 the Navy V-12 men left DePauw University and the campus returned to its former routine.

Case Western Reserve University

Case Western Reserve (originally Case School of Applied Science) was the only college in Ohio to be chosen by the Navy as a site for their V-12 officer program.

In 1942 Case began discussions with the Navy in order to acquire a Navy V-12 program. The negotiations were difficult. The Navy had strict rules regarding the housing and messing of navel students. Case did not have dormitories or dining halls. In the end, Case solved the problem by buying the Garfield Exchange building. The building was remodeled and named "The Ship".

Originally Case had hoped for 1,000 men to be in the program. This was the original number set by the Navy but "The Ship" could only accommodate 250 men and that coupled with the elimination from the Navy's academic program all but elementary work in chemistry, metallurgy and physics allowed for the use of only a small number of faculty and staff.

Many of the V-12 students were former Case students who had been in the reserves and who had been ordered to active duty. Some were enlisted personnel. The V-12 students received $50.00 a month in addition to the full cost of tuition, housing, food and textbooks.

The students were expected to follow a strict code of conduct. This was a military code of conduct that expected its students to behave as gentlemen and officers.

In March 1946 the program ended. Seven hundred and fifty men had participated in the Navy V-12 program at Case Western Reserve University.

Bethany College

Prior to the men arriving at Bethany College, C.O. Lt. Sherman Henderson found the college in bedlam. "Three hundred double-decker bunks arrived only three days before the boys were due," he recalled. "All hands had to turn out to set them up. Most of the boys arrived wearing civilian clothes, and we had to outfit them all with uniforms. Trousers, middy jumpers, neckerchiefs, hats, shoes, rubbers, raincoats, bath towels, and underwear. What a job! But someone in Washington did a fantastic job of planning, because at the end we were left with only 20 uniforms that didn't fit."

On July 1, 1943 335 young men arrived at Bethany College to begin their first term. Some of the men were Naval ROTC, some were veterans of fleet service but most of the men were recent high school graduates.

Their course work consisted of hard work and discipline. They were up at 7:00am for calisthenics before breakfast. Classes were from 8:00am to 4:00pm with a break for lunch. The men had one hour for organized activity and then they had mandatory study time.

Except for naval subjects, the V-12 men took the same courses as the civilians but there was an intense need to succeed. Men who failed were often sent right into the fleet.

The first two semester courses consisted of: 16 hours Math, Physics, History, Engineering Drawing, and English. Those in the Pre-Med program or in advanced courses had an even tougher schedule.

While the Navy chose Bethany College for its excellent academic reputation, the school needed these Navy men. Enrolment was down. The V-12 program enabled the college to continue.

The program ended with the March-June term in 1945.

An editorial in the final issue of the *Service Men's News*, a newsletter published by Bethany for military service graduates stated,"The Navy Program brought 836 sailors to the campus. We truly hate to see them go—the Navy whites and the Navy blues have become so much a part of us. There has been no friction between Navy and civilians, Navy and townspeople, Navy and faculty; decidedly none between the Navy and the coeds. Romance has flourished and

Bethany is still a friendly town in which to live. But—the voice of the bugle must vanish and the rhythm of marching be stilled."

Emory University

On July 1, 1943 654 Navy V-12 trainees arrived on the campus of Emory University to begin their college course. Two hundred and fifty-four students began pre-med training and 400 students began a basic college course. Added to that the number of students in the School of Medicine who were "activated" and the total number of uniformed men on campus was 900.

The university provided all four of its dormitories to the Navy men. Civilian students lived in fraternity houses and in private homes. The V-12 men were fed exclusively at the university cafeteria while the civilians were fed at the Emory Grill.

While the Navy V-12 program was a serious endeavor it did have its lighthearted moments. When Captain J.V. Babcock, commander of the V-12 programs in Atlanta noticed stuffed sharks, embalmed cats and one live mule listed in instructional and upkeep costs for naval students he sent this official-sounding letter to Emory president Goodrich C. White:

"The commanding officer's responsibilities cause him grave concern for the availability and maintenance of the subject Stuffed Shark, Embalmed Cat and Live Mule for instruction of naval students as prospective naval officers.

It is requested that a monthly statement of the condition of the above instructional facilities be submitted, and that in particular a periodic physical examination be administered to the mule."

Emory President White sent this letter in response:

"Please let me assure you that I share with you the sense of grave responsibility for the maintenance of all instructional equipment used in the training of naval students...and this responsibility comprehends not one but numerous stuffed sharks, not one but numerous embalmed cats, and involves especially one, and only one, live mule. I am acutely aware of the fact that he must be kept 'live'. I am making extensive investigations as to the possibility of providing for regular and thorough physical examinations of the above-mentioned live mule....Is it likely that the navy would assign a doctor of veterinary medicine to us and provide necessary equipment not only for periodic examinations but for medical care and treatment in case of illness for the supervision of a proper physical training program?

"I venture to add that through oversight we failed to list 'Dooley' in the instructional equipment on which the navy should prop-

A group of Supply V-12 Students posing a few days before leaving the University of Washington for Harvard Graduate School of Business. Ira W. (Wes) Rimel is at the rear on the right. (Courtesy of I. W. Rimel)

U.S. Navy V-12 Officer Trainees at "chow". (Courtesy of Henry C. Herge)

Adlai Stevenson, Special Assistant to Secretary of the Navy and Lt. Meldrim F. Burrill, Commnding Officer of the Navy V-12 Program, Illinois State University, Normal, Illinois, 1943. (Courtesy of Meldrim F. Burrill)

Admiral A.S. Carpender, Commandant of the Ninth Naval District-Great Lakes and Lt. Meldrim F. Burrill, Commanding Officer of the Navy V-12 Program, Illinois State University, 1943. (Courtesy of Meldrim F. Burrill)

erly pay depreciation charges..." (Dooley was a skeleton in the anatomy department.)

When the program ended in 1945 over 2,000 Navy V-12 men had been trained at Emory University.

Harvard College

The Navy had announced on December 12, 1942 a plan for adapting college educational programs to its needs. The plan called for providing the needed educational fundamentals in a prescribed academic course load that combined college curriculums with course most needed by the Fleet.

On July 1, 1943 Harvard College welcomed 960 V-12 students to its campus. These students were not only faced with a military code of conduct that included 101 rules but with a course load that was strenuous at best.

The daily and holiday schedule consisted of the following:

Daily
0600	Reveille
0610-0615	Muster
0615-0635	Morning Exercise
0635-0710	Shower-clean quarters

0710	First breakfast formation-inspection
0710-0800	Inspection of quarters
0715-0730	First breakfast
0725	Second breakfast formation inspection
0730-0745	Second breakfast
0730-0800	Deposit laundry
0730-0900	Sick call
0800	Colors
	Sunday and holiday breakfast
0810-0900	First period class
0910-1000	Second period class
1010-1100	Third period class
1110-1200	Fourth period class
1210	First lunch formation
1215-1235	First lunch
1230	Second lunch formation
1235-1300	Second lunch
	Sunday lunch
1310	Third lunch formation
1315	Third lunch(men having 12 o'clock classes)
1330-1430	Sick call
1330-1430	Fifth period class
1430-1530	Sixth period class
1530-1630	Seventh period class
1630-1730	Eighth period class

1725	First supper formation
1730-1750	First supper
1745	Second supper formation
1750-1815	Second supper
1930-2250	Eliot Grill open
2300	Gates locked
2315	Taps

Wednesday—Liberty began at 1800 and ended at 2300.

Saturday—Liberty began on completion of last class or drill after 1200 and ended at 2100 Sunday.

When the Navy V-12 program ended in 1945, Harvard College had played a tremendous role in the training of our nation's Naval Officers.

Illinois State University

On July 1, 1943 291 Navy V-12 men arrived at Illinois State University. The university had seen the same lowering enrollment that many of the colleges and universities around the country had seen. These 291 men and the men who would follow helped to save this institution.

Meldrim F. Burrill was their commanding officer. Most of the men were housed

Navy V-12 Winning Football Team, 1943-44, Illinois State University, Normal, Illinois. (Courtesy of Meldrim F. Burrill)

Navy V-12 Basketball "Champions", 1943-44, Illinois State University, Normal, Illinois. (Courtesy of Meldrim F. Burrill)

Navy V-12 "Campion" Wrestlers, 1943-44, Illinois State University, Normal, University. (Courtesy of Meldrim F. Burrill)

in Fell Hall with the remaining being housed in Smith Hall. They were taught by the regular faculty and were allowed to participate in university extracurricular activities. Their course load included physics, English, advanced mathematics, chemistry, engineering, history, United States Naval history, economics and foreign languages. Those students who did well went on to midshipmen schools and were commissioned as ensigns and those who failed went straight to boot camp or fleet. There obviously was a strong desire to succeed.

The Navy V-12 program at Illinois State University was threatened one November day in 1943. Fell Hall was on fire! Everyone responded: the town's volunteer fire fighters; trucks from Bloomington and the Soldiers and Sailors Children School and the students. The men were able to retrieve most of their belongings and the damage was contained in the attic and roof.

After the damage was assessed, the townspeople and commanding officer went to work. After speedy telephone calls to the appropriate people, the repair materials were approved immediately and the State Architect was at the site by the middle of the afternoon. The men were temporarily housed in McCormick Gymnasium and in 24 hours the kitchen and dining halls were operating. Less than half a day of classes was lost.

The Navy V-12 program ended in June 1945. Six hundred and four men were trained at Illinois State University.

Lawrence College

In the Annual Report of the President to the Board of Trustees of Lawrence College dated June 23, 1945, President Nathan Pusey stated,"Political philosophers differ as to whether war represents a natural condition of man or only a temporary aberration from his normally pacific behavior. However this may be, colleges of liberal arts were clearly not intended originally to serve the interests of nations at war. And yet when wars have come, invariably they have been among the first of institutions to make their contribution. When a call was made at Lawrence for volunteers early in the Civil War, for example, we are told that the whole male enrollment of the class of 1864 and a considerable proportion of the faculty enlisted straight away. So it has always been, in four major wars."

"Already 1,825 sons and daughters of Lawrence have entered the services this time, and have acquitted themselves honorably on all fronts and in many different ranks and capacities."

Seven hundred and five men were trained at Lawrence College during the years of the Navy V-12 program.

In those two years the men took the same courses and faced the same code of con-

duct as other Navy 12 units but something occurred at Lawrence College that did not occur at any other Navy V-12 campus.

On a June weekend in 1944 the C.O. invited the fathers of the men to "live with them for 24 hours". Fifty-three fathers, from nine states including Oregon and New York, came to visit their sons.

They ate with their sons, bunked with them, went to calisthenics at 6:15 am with them and were the guests of honor at the final review.

When the program ended, it was agreed that all who participated had helped not only their country but their fellow man.

University of Louisville

A 1943 *Courier-Journal* article states that the University of Louisville faced "virtual extinction." The young men who normally would be attending the college were off to war and since the university depended on tuition for 60% of its funding, low enrollment really cut into the operating budget. But the Navy saved the day: specifically the Navy V-12 program.

The University of Louisville was the first institution to begin the Navy V-12 program. University of Louisville President Raymond A. Kent had been a member of the Naval Advisory Education Council that began the Navy higher education programs.

On July 1, 1943 475 young men arrived on the campus of the University of Louisville. Samuel V. Bell,a member of the first group of Navy V-12 officer candidates, remembers that day,"We arrived—from all over the country—in our civvies and no change of clothing because uniforms were supposed to be ready for us. But the uniforms didn't come for two weeks and we were restricted to campus until then."

In order to accommodate these 475 men, four dormitories had to be built. In a fund drive the college raised over $725,000 from businesses,the city of Louisville, and Jefferson County to build the dormitories. Without the dormitories there would have been no V-12 program and most likely no University of Louisville.

The Navy V-12 program and the Uni-

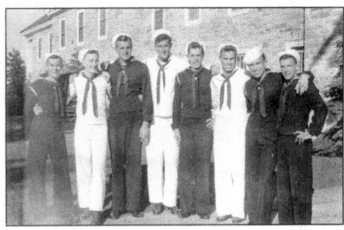

University of Louisville, V-12, "Roommates," 1943. L to R: Unknown, Unknown, Cabel Francis, Frank R. Fultz. (Courtesy of Frank R. Fultz)

St. Lawrence University, V-12, 1945, "Uniform Confusion". (Courtesy of Philip B. Jones)

versity of Louisville ended on June 20, 1946 but its influence on the men and this country exists today.

St. Lawrence University

In 1943, St. Lawrence University welcomed the Navy V-12 program to its campus. Even though the arrival of the Navy men threw the campus into turmoil, those who remember those days will tell you that they were a glorious time.

The university adapted to meet the needs of the Navy V-12 men. The class offerings expanded to accommodate the Navy's requirements but generally the Navy men went to class with the civilians. Saturday was an anticipated day. With the exception of muster at noon and inspection at 1pm, the men were free until midnight. Many participated in the intramural sports and some men published a gossip letter titled *The Billit Slip*. The band and choir were popular with the men and those involved with drama managed to produce skits popular with the student body.

In a November 2, 1949 article in *The Hill News* Vida Ragine comments on the time. "To average St. Lawrence women of the classes '45 to '48, the last war struck Canton when most civilian men left campus in 1942 and ended when they returned in the fall of 1946. Living conditions and social life were more than disheveled, ask anyone who was here. Old-timers and they are constantly getting scarcer—will tell you, as they look back on those glorious days, when women lived in Men's Res, when Pi Phi's lived in the SAE house, when the Alpha house was just another off-campus residence for civilian men, that those were the days, the Navy was here!"

Louisiana Tech University

With the motto "would be so nice to come home to" Louisiana Tech welcomed

Louisiana Tech University. Pictured are: Boyle, Aiken, Saloom, Vurnete, Fanquy, Moss, Lyons, Bagley, Chief Pearson, Ensign Belichek, Amos, Frnka, Bickhan, Bonnen, Hamilton, Guill., Bone, Allen, Ashley, Burns, Blackshear, Abernathy, Baker, Shabey, Brewer, Kohlers, Avery, Hendrix, Newton, Gaddy, Shettey, Calvin, Beach, Blatt, Dodson, Kirkpatrick, Child, Barr, Bond, Crawford, Bodron, Broussard, Andrews, Dalcombre, Barbay, Bailey, Blackstock, Blatt, Hattimore, Goodson, Brayboeuy, Perrymon, Hennington, Redmon, Branch, Baird, Douglas, Laborde, Beneke, Berry, Thompson, Boothman, McGraw, Mulligan, Barouse, Taylor, Brien, Ford, Biddison, Isles, Hamilton, Dupree, Cunningham, Hennington, Oneal, Alexander, Alleman, Black, Barefoot, Doss, Johnson, Speck, Evans, White, Barkar, Babin, Blume, Boudreaux, Holme, Benjamin, Owens, Myers, Merrideth, Kuhn. (Courtesy of D. R. Hamilton, Jr.)

Fellow V-12ers in front of Foster Hall, Southwestern Louisiana Institute. Front Row L to R: Don Gee, Russell Guill, Frank M. from Missouri. Top Row L to R: Clyde Hardy, Doyle Hamilton, L. Gillard. (Courtesy of Doyle R. Hamilton, Jr.)

men from the Navy V-12 program to its campus.

Lt. Commander Earl J. Aylstock wrote the following for a pamphlet entitled *Lest You Forget* given to the Navy V-12 men and

civilian students who were at Louisiana Tech during the Navy V-12 era.

"I am gratified at the privilege of commanding this fine group of officers and men. The relatively few separations from the V-

12 program for various reasons is indicative of the seriousness with which all trainees have strived to meet the objectives of the V-12 program.

The welfare and progress of all personnel within the limit of my responsibility, working in conjunction with the college authorities, will always be uppermost in my mind.

I wish to express my pleasure at having had the opportunity of serving my country in a locality where generous and kindly people are numerous. Wherever I may be I shall always cherish the memories of pleasant duty at Louisiana Polytechnic Institute."

Massachusetts Institute of Technology

In July, 1943, 910 men arrived at MIT to participate in the Navy V-12 program. Two hundred and thirty-eight of the 910 men were first year students who followed a course load prescribed by the Navy. The other 672 men were college transfers who were allowed to continue in the same fields they had followed previously.

The men were housed separately in the Graduate House dormitory. They attended classes with the civilian students, participated in extracurricular activities with the civilian students but they had separate physical training courses and student officers organization.

Contained in the 1943-44 Report of the President was this assessment of the program: "Our Navy V-12 program has continued undiminished and our quota for the term beginning in November had been reduced only about eight per cent. The Institute is proud to participate in this service program; it is soundly conceived and effectively managed by the Navy; the students in our unit are excellently qualified and their morale is fine."

Over 1,000 Navy V-12 men were trained at MIT during the two years the V-12 program was in operation.

Milsaps College

On May 13, 1943 Rear Admiral Jacobs, Chief of Naval Personnel, spoke at a conference at Columbia University. He stated his philosophy of the Navy V-12 program: "This a *college* program. Its primary purpose is to give prospective naval officers the benefits of a college education in those areas most needed by the Navy. We desire, insofar as possible, to preserve the normal pattern of college life. We hope that the colleges will give regular academic credit for all or most of the Navy courses, and we desire that college faculties enforce all necessary regulations to keep academic standards high."

And so on July 1, 1943 681 Naval and 192 Marine officer candidates arrived on the campus of Milsaps College.

According to Robert W. Donaldson, a member of the Milsaps College Navy V-12 program, "Included in the navy personnel were many (number not available) who entered the program through the Naval Aviation Cadet (V-5) recruiting program. We were assigned to the V-12 program due to there being a surplus of aviation cadets at that time. We were officially classified as V-12A. Many of us later elected not to enter the Naval Aviation Cadet Program (V-5) and were then integrated into the V-12 program."

The Navy V-12 program at Milsaps College lasted from July, 1943 to October 31, 1945.

Mississippi College

Three hundred and thirty-five Navy V-12 men reported to Mississippi College on July 1, 1943 on one of the hottest summer days on record.

The men went to classes with the civilian students. They were elected to student government positions and they participated in the intramural athletic program but because the college was a Southern Baptist school no dances were allowed on campus or in town. The men were undaunted, though. They went to Jackson or Vicksburg to dances there and on one August Tuesday night their C.O. even gave them liberty to attend a USO dance in Jackson. Once a month the V-12 men held variety shows. The shows featured their band, the 4.0 Skybirds.

Even though their social life was stifled, the Navy V-12 officer candidates made it through their program and Mississippi College can boast that it trained over 1,700 Navy V-12 men during the two years of the program.

Muhlenberg College

Muhlenberg College holds the distinction of being the only college of the 131 colleges and universities in the Navy V-12 program to begin classes at 0800 on Monday July 5, 1943. This occurred because of the tremendous cooperation between the community and the college and because people did what needed to be done.

One example of this cooperation occurred on June 30, 1943, the evening before 453 Navy V-12 officer trainees were to arrive. Commander Henry Polk Lowenstein Jr., John Wagner, college business manager and Levering Tyson, Muhlenberg president were inspecting the plant. They discovered crates of eggs that had been delivered after hours sitting on the dock. With no one else to help, they loaded the eggs into the mess hall refrigerator.

Other examples of this cooperation occurred when the alumni, the Women's Auxiliary, physicians and dentists from the community, and wives of faculty and staff pitched in to provided services that could not be provided for because of the wartime shortage of labor.

The Navy V-12 program changed Muhlenberg College during the time that it was in operation. Courses were adapted to fit Navy requirements, the Commons was enlarged into a mess hall, the Lambda Chi Alpha fraternity house was leased and used as an infirmary but as recorded in *Muhlenberg: The World War II Era*, "But a change that lives in memory more vividly than the others was the sounds of a military post that filled the air that summer of 1943." The marching of feet, the cadence count by Navy Chiefs, bugle calls all day, Reveille early in the morning, "colors" twice a day, mess call three times a day, mail call, fire call, "Tattoo" to end the day and "Taps" were all sounds that became familiar.

Muhlenberg was one of 54 colleges asked to extend the V-12 program into 1946. One of the last acts by C.O. Lt. Theodore Abel was to give Rusty Gizmo an honorable discharge. Rusty Gizmo was a Labrador retriever who had marched at the head of Saturday Reviews for two years as a mascot.

University of North Carolina

The University of North Carolina began its Navy V-12 program on July 1, 1943. Included in the 1,330 men to be trained were college students, reservists, sailors, high

La Tech, V-12, 1943-45. L to R: Robert W. Borders, Nathan Durhan.

school graduates, Marines and 230 NROTC students.

The college supplied the classrooms, laboratories, living quarters, dining halls, medical facilities and faculty. The men were housed in Swain Hall, five dormitories and fifteen fraternities. The Navy renovated all of the buildings.

The Navy V-12 men were required to take 17-20 hours of university classes including physics, trigonometry, calculus, English, engineering and Navy courses in navigation, ordnance, seamanship and communications.

According to a report written for the university "The University of North Carolina produced as many Naval Officers as any single institution in the country except for the Naval Academy."

In June of 1946 the Navy V-12 program ended and the trainees were placed into the NROTC program.

Oberlin College

On July 1, 1943 the first of the officer trainees in the Navy V-12 program arrived at Oberlin college. A little over a year before First Lady Eleanor Roosevelt had delivered a speech to 1,600 at Finney Chapel. She had admonished the audience that "We are still fighting the war to end all wars," and she warned the students against mental laziness: "...you can think lazily with high sounding words. Be a force. Make yourself count in the remaking of the world, both in the war and the reconstruction period." Oberlin College would strive to be that force that the First Lady had spoken of.

Seven hundred men arrived that first summer in 1943. Of that 700, 500 were transfers from 76 colleges including 100 from Oberlin. One hundred and fifty were recent graduates and 50 were from active service.

Everywhere there was a shortage of labor because of the war and circumstances were no different at Oberlin. Everyone pitched in to help. The college adapted its curriculum to meet Navy requirements, and the women formed the Campus Crew and the Library Stack-Dusters. The women were each paid 35 cents an hour to sweep floors, dust shelves and to clean the commons areas.

On February 24, 1946 the flag was lowered for the last time on the Navy V-12 program at Oberlin College. Two thousand men had been trained at Oberlin during the two and a half years it was in operation.

Ohio Wesleyan University

Ohio Wesleyan University welcomed 400 Navy V-12 trainees to its campus on July 7, 1943. The men lived in nine bar-

racks on the campus. These apprentice seamen were held to a strict code of conduct at all times.

The course of study was the same at each of the 131 colleges and universities that had a Navy V-12 program. The freshmen at-

University of North Carolina, Chapel Hill, Passing in Review. (Courtesy of Michael A. Kaufman)

Marquette University, Stratford Arms Hotel. Standing L to R: John Rivera, Bill Reed, Bin Randolph. Kneeling L to R: Al Rigling, Leon Schweda, Bob Thoresen. (Courtesy of Leon J. Schweda)

Miami University, July 1943, Oxford, Ohio. (Courtesy of E. J. Friederich)

18th Class Midshipman School, Columbia University, Marching T. Class, 1944. (Courtesy of Nelson Kasten)

V-12, UCLA. L to R: Stan Perkins (San Francisco), Supply, Dick Leonards (San Joaquin Valley), Pre Med, Bob Morton (Farmington, New Mexico, Supply. Picture taken in Sherman Oaks, CA, February 1945. Spring leave to San Francisco. Also aboard were Bill Schaupp (San Francisco), Pre Med and Joe Goss (Sherman Oaks), Supply. Two of us rode in turtle back with the exhaust. Dick Leonard's car had colorful history. Ownership of cars was prohibited on campus for V-12's. A room mate, Vince Holian (San Francisco) was restricted and got himself checked into sick bay so he could go to the dance Saturday night. He borrowed Dick's car and in the rain, ran into a pump at a gas station. He got caught at the dance and was kicked out.

tended four 16 week semesters taking courses prescribed by the Navy. The courses were part of the regular university curriculum and were also attended by civilian men and women. Upperclassmen were allowed to continue their previous field of study.

The seamen had their own inter-barracks sport competitions and they were allowed to participate in the intercollegiate athletic program as well.

With the closing of the Naval Flight Preparatory school in 1944 the V-12 men moved into Stuyvesant Hall from the various fraternity houses where they had been living.

According to a 1945 *The Ohio Wesleyan Log,* Ohio Wesleyan was "proud of the part it has had in cooperation with the Navy in this training program. The trainees assigned to the campus here have been of high quality and have made valuable contributions to university life. Scholastic attainments have been among the best achieved by any unit, anywhere."

The program ended in 1946.

University of Oklahoma

"It was the strangest mix of being a student and being in the Navy at the same time," retired Oklahoma Supreme Court Judge Don Barnes recalled.

"The NOTC and V-12 were the biggest group of overachievers I've ever seen, he said. "It wasn't an ordinary student body."

This group of "overchievers" attended classes with the civilian students and participated in the university's athletic program. The University of Oklahoma's football team was made up of almost all Navy V-12 and NOTC students.

"Most of the OU team members, coached by Dewey "Snorter" Lester, Dale Arbuckle and Bruce Drake, were Navy men with athletic prowess who had been sent to OU for military training, primarily in the Navy's V-12 program," said OU President emeritus George Lynn Cross in his book, *The Uni-*

UCLA, V-12. Pictured here are Tom (Duke) Ellington (Azusa, CA) and Jean Stryker (San Diego), USC-UCLA game, October 1944. (Courtesy of Joseph W. Goss)

Duke University, V-12 Unit, 1945 or 1946. L to R: Sol. Gruber, Dick Bisbe, Jack Geier, George Sinichko, Leon Winitsky. (Courtesy of Dick Bisbe)

versity of Oklahoma and World War II: A Personal Account, 1941-1946.

"The back field consisted entirely of Navy trainees," Cross said, "and the rest were either trainees or 17-year-olds too young for the draft."

While the Navy trainees greatly helped the university they also posed problems. Housing was so crowded that Navy students were placed in accommodations in rooming houses and in fraternity and sorority houses. The university asked the government for help to house the 900 trainees.

Even though times were difficult, the spirit of cooperation was everywhere. When the program ended in 1945 Roy Gittinger, the late dean of admissions and records and professor of History said, "When this period of storm and danger is over the university will again be ready to face the future with confidence and with ideals unchanged."

Penn State

In July, 1943 600 Navy V-12 officer candidates arrived on the campus of Penn State to begin a great adventure.

They were quartered in the 12 fraternity houses. They were to be issued white summer uniforms but there was a shortage of white uniforms so many of the trainees had to wear the hot blue uniforms for two weeks.

The men followed the curriculum that was standard in the 131 Navy V-12 colleges and universities. According to Robert D. Barron, an alumni of the Penn State Navy V-12 program, "It was equivalent to the basic engineering course at Penn State, except Physics was required the first year rather than Chemistry. In addition to the 17 hours of required work, which included

Dartmouth College, V-12, 1944-45, winter skiing. L to R: G. MacGillivray, 1st Marine Div.; "Fox" Fuller, 2nd Raider BN; "W.J." Owen (Steimke), 3rd Marine Div.; Al Southly. (Courtesy of G. C. MacGillivray)

Naval Organization, a strong program of physical education was included. A full 5 1/2 day per week schedule was maintained with additional time on Saturdays for Unit review and training for reviews."

The basic program consisted of four semesters followed by four months of Midshipmen's school. After Midshipmen's school the men would receive their commission as an ensign.

Classes were held in Carnegie Hall, Chemistry and Physics, Sparks, Agriculture Eng. and the old Chemistry Building. Men were trained in all the marching skills, manual of arms, Naval history and Naval organization. The Navy V-12 men dominated the Penn State athletic program.

In June, 1946 the Navy V-12 program ended.

Rice University

Rice University and Naval ROTC commander Captain T.A. Thompson welcomed 199 Navy V-12 officer trainees to the campus in July, 1943,

According to a newspaper story dated 7/1/43, "Because of its pre-eminence in engineering and technical fields, Rice was selected by the navy to train, exclusively, future engineering officers in the new V-12 setup."

The men were required to take a course load prescribed by the Navy. The men followed a strict code of military conduct and were on a routine that varied little. Their day began with calisthenics followed by breakfast and classes.

In the last paragraph of a letter received from the Navy Department to the last V-12 commencement, Rice was awarded a certificate of appreciation: "On this, the occasion of the last V-12 commencement at your institution, it gives me real pleasure to present the Rice Institute with this certificate as a tangible sign of the Navy's appreciation for all that the institute has done in the V-12 Program. Now that your share in the work is nearly completed, I should like to add just one thing more, the Navy's traditional phrase of approval, a hearty "Well done!"

Purdue University

Purdue welcomed 800 Navy V-12 officer trainees to its campus in July, 1943. These men like the men at the other 130 colleges and universities were selected after taking a qualifying test in high school. Other candidates were college level students and active duty students who met both the mental and physical qualifications.

The routine of the Navy V-12 students varied little day to day or college to university. The men were up at 0545 for calisthenics followed by cleaning the barracks and dressing. Breakfast was 0645 and classes followed.

The Navy V-12 students added greatly to the social life at Purdue. Prior to the Navy arriving the campus was almost deserted and certainly the number of males had dwindled. With the Navy on campus dances were held and the Sweet Shop and Union cafeteria were busy with romance.

The Navy was active in all of the extracurricular activities. They were on the football, basketball, baseball, swimming, track and wrestling teams. They participated in marching band and they even had a dance band that played at the USO in Lafayette.

In June, 1946 the program ended. In a letter to the trainees, Captain J.R. Hamley stated:"With the termination of the V-12 program at Purdue University on 23, June 1946, we close the door on our accelerated wartime Navy College Training Program. You have seen many changes in that program since it was started at Purdue on 1 July 1943. The grind has been tough, but I feel confident that here you have acquired sufficient academic and professional groundwork to permit a continuing development and acquisition of knowledge throughout your future years.

Regardless of the course which you have plotted for yourself, be it Active Duty, Inactive Duty, NROTC training, or civilian life, if you will continue to grow in mental stature and continue to apply those characteristics of sincerity, loyalty, devotion to duty, and a high sense of personal honor which we have tried to instill in you here, you cannot help but succeed.

I want to extend to you every good wish for the fulfillment of your hopes and expectations."

Navy V-12 Contribution to the Nation
by Capt. Bob Jones, USN (Ret)

Navy V-12 had long lasting results as it produced leaders for the top echelons of business and the professions. Lawyers, educators and engineers comprised the largest group. But the fields of medicine, dentistry, business, industry, advertising, journalism, sports, show business, and government service are well represented. More than 40 future admirals and 18 future Marine generals took part.

But perhaps the best known "alumnus" is Ensign Frank Thurlow Pulver of Mister Roberts. *Portrayed by V-12 Jackie Cooper on stage and V-12 Jack Lemmon, who won an Oscar, in the movies.*

Some Distinguished Navy V-12 Alumni

George Allen-Football coach
Warren Christopher-Secretary of State
Howard Baker-Senator, White House Chief of Staff
Angelo Bertelli-Heisman Trophy winner
Johnny Carson-TV personality
Louis J. Cioffi-TV newsman
Jackie Cooper-Actor, producer, director
Alvin Dark-Baseball player, manager
Jeremiah A. Denton, Jr.-Senator, Navy Admiral
Daniel J. Evans-Senator, Governor
Peter Hackes-TV newsman
Harry R. Halderman-White House Chief of Staff
Elroy Hirsch-Football great
Robert F. Kennedy-Attorney General, Senator
Bowie Kuhn-Commissioner of Baseball
Melvin Laird-Secretary of Defense
Jack Lemmon, III-Actor
Charles McC. Mathias, Jr.-Senator
James A. McClure-Senator
William Middendorf, II-Ambassador, Navy Secretary
Daniel Patrick Moynihan-Senator
Robert C. Pierpoint-TV newsman
Albert L. Rosen-Baseball player
Carl T. Rowan-Columnist, TV personality, Ambassador
Pierre Salinger-Newsman, Presidential Press Secretary
William Webster-Former Director CIO and FBI
Thomas G. Wicker-Columnist
Roger Williams-Musician, entertainer

Holy Cross V-12'ers and ROTCEES on leave at Joe Kings, NYC. (Courtesy of J. H. Curtin)

Navy V-12, University of Miami, Coral Gables, FL. (Courtesy of W. E. Browne, Jr.)

St. Ambrose College, Davenport, Iowa. Gerald Halterman is the last one on the right in second row holding glove. (Courtesy of Gerald Halterman)

Thomas (Tommy) Thompson took this picture of students in front of Martin Hall, 1943. (Courtesy of D.R. Hamilton, Jr.)

V-12 "Roll Call", St. Lawrence University, Winter 1944. (Courtesy of Philip B. Jones)

Judice Hall football team. Champions of S.L.D. intramural program. (Courtesy of D. R. Hamilton, Jr.)

V-12 program graduation at Middlebury College (Courtesy of K.H. Irwin)

Middlebury College, September, 1943. Standing L to R: Gillard, Howard, Guilbeau. Kneeling L to R: Grimes, Greer, Guill. (Courtesy of D.R. Hamilton, Jr.

The formal reception of Lt. Cmdr. Fox F. Taggart at Middlebury College. Fox is accompanied by Cmdr. William Baldwin. (Courtesy of D.R. Hamilton, Jr.)

Middlebury College. L to R: Doyle Hamilton and Jack Irving, August, 1943. (Courtesy of D.R. Hamilton, Jr.)

Middlebury College. (Courtesy of D.R. Hamilton, Jr.)

Middlebury College. (Courtesy of D.R. Hamilton, Jr.)

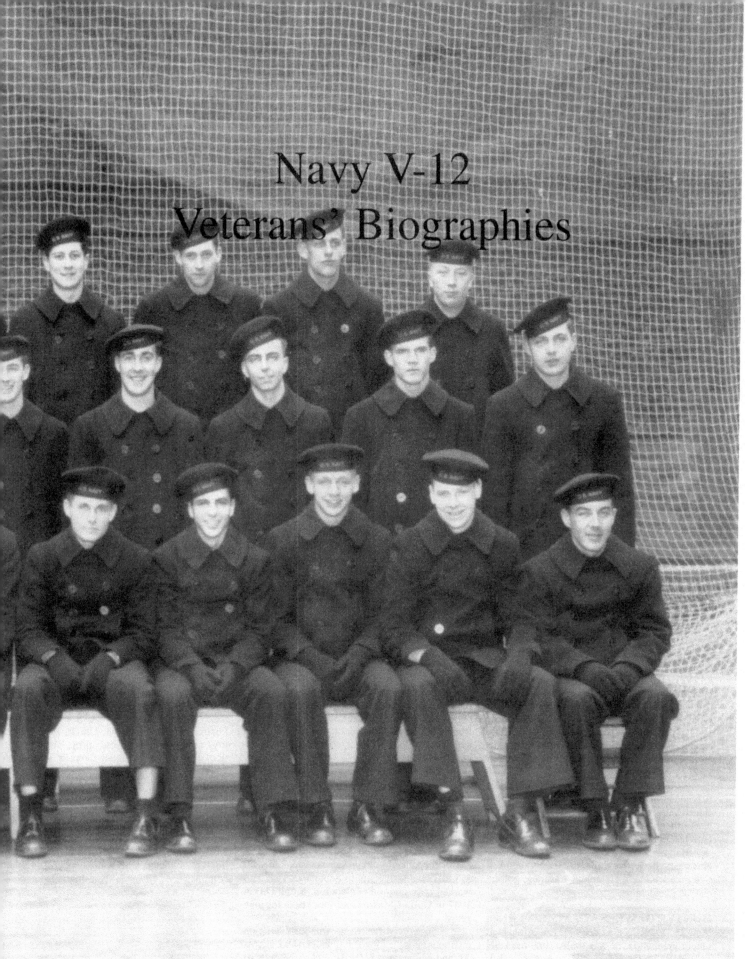

Navy V-12
Veterans' Biographies

H. NORMAN ABRAMSON (AVN CAD) was born March 4, 1926, in San Antonio, TX. He enlisted in the Navy V-5 Program in March 1943 and was assigned to the V-12 Program at Southwestern Louisiana Institute. He completed preflight at University of Georgia, primary flight training at Norman, OK, and was discharged as an aviation cadet in October 1945. He continued his aeronautical interests by working for Navy BuAer at Pt. Mugu, CA; earned his BS and MS degrees at Stanford University and later worked at Chance Vought Aircraft. After four years as associate professor of aeronautical engineering at Texas A&M, he received a Ph.D degree from UT-Austin and joined the staff of Southwest Research Institute (SWRI), San Antonio, where he remained for 35 years, rising to the position of executive vice president.

At the Institute he became known internationally for his expertise in the dynamic behavior of liquid propellants used in rockets and spacecraft. A specialist in the field of theoretical and applied mechanics, particularly in aeronautics, astronautics and structural dynamics, Dr. Abramson has published more than 75 research papers, as well as a well-known textbook, and serves on numerous state, national and international professional, scientific, and government advisory committees. He is a member of the National Academy of Engineering and holds top membership positions in a number of professional societies; he is listed in *Who's Who in America* and other similar publications and has received several national awards and medals for his professional work.

Since retiring from SWRI in 1991, he continues in an advisory capacity to a wide variety of professional and governmental organizations.

Abramson is married and has two sons.

LOWELL E. ACKMANN grew up in Huntley, IL (50 miles northwest of Chicago). He had two years of college before entering the V-12 Program at the University of Illinois on July 1, 1943. He completed the Program on Oct. 1, 1944, entered Columbia Midshipman School (on USS *Prairie State*) and was commissioned March 1945.

Ackermann was next sent to MIT Radar School until December 1945; he then decommissioned ships until his discharge in July 1946. He worked with Sargent and Lundy in Chicago, IL, until his retirement in 1987.

Happily married to Dorothy Collier for 45 plus years; they have three married children and five grandchildren. He enjoys traveling, golfing and his grandchildren.

HUNTER DEWEES ADAMS (LT) was born May 23, 1925, in Philadelphia, PA. After high school graduation, he was drafted into the USN and reported to Great Lakes, IL on Aug. 11, 1943. He was assigned to Navy Pier Diesel Engine School, Chicago, IL, in October 1943; reported to NAAF, Mayport, FL in December 1943; entered V-12 Program on July 1, 1944, and reported to Duke University, Durham, NC.

Adams was discharged as F1/c at Bainbridge, MD on June 25, 1946; he re-entered Duke University NROTC Program, graduated BSME and commissioned ensign on June 6, 1947.

Affiliated with Surface Division 4-5, Camden, NJ and trained new recruits. The Division won the James V. Forrestal Trophy for the best division nationally for 1949. He was later assigned to: Div. 4-36, Philadelphia, PA; Div. 5-2, Baltimore, MD; and Div. 6-20, Charlotte, NC.

LT Adams completed nearly 21 years reserve service and retired May 23, 1985. His awards include the American Theater Medal, WWII Victory Medal, Naval Reserve Medal and Armed Forces Reserve Medal.

Adams created Petrolube Corp. in 1973 and continues as corporate president. He married Elizabeth Foulon in 1948; they have twin children, Eric and Cindy, born in 1954. Currently essentially retired.

JAMES MILLER ADAMS (LCDR) was born Sept. 9, 1924, in Cleveland, OH. He entered Case, September 1942, enlisted, June 1943, V-12, November 1943. Received BS(ChE) in April 1945 and commissioned ensign, Columbia Midshipmen's School, August 1945.

Reported to USS *Casablanca* (CVE-55) as assistant navigator. He made three voyages to Maui, Noumea and Yokosuka; transferred to USS *Whythe* (APB-41) as gunnery officer in Tsingtao; made one voyage to Inchon, then stateside for mothballing at Galveston.

He was released from active duty in August 1946 at Great Lakes; transferred to Retired Reserve as LCDR in July 1971; attended Virginia Polytechnic Institute; and received BS (biology) in 1948, MS (Ch.E) in 1949, and Ph.D (biochemistry) in 1954.

Following various research positions, most of his career was spent with U.S. Customs Service, regional director, Baltimore Lab, 1969-1982; chief, Research Branch, New York Lab, 1982-86; and is author or co-author of 40 or more scientific papers or publications.

Currently resides in Joppa, MD, with his wife Lorraine Waryold. They have seven children and, so far, seven grandchildren.

THOMAS B. ALBRIGHT was commissioned and received BS in EE at Oklahoma University in January 1945. He entered Tokyo Bay as part of naval occupation forces in August 1945. Received MS at Texas University in September 1947 and worked in frontier oil exploration (Mobil Oil) in 1947 in Gulf of Mexico and four states for nine years.

1956-worked for North American Aviation on frontier missile jobs of Jupiter, Navaho, GAM-77, and Minuteman I, II, III; and space craft of Apollo, Shuttle and XB-70 aircraft. He singularly saved the Air Force $21 million by eliminating equipment; got laid off by successor Rockwell Int.

Designed optimum failure detection method with Milco Int.

Post 65-age, sans pensions: authored book *21st Century Blueprint* (1994); developed ENVIROECONOMICS method for peaceful world trade; copyrighted future instruction using Think Links and Links of Learning for semi-automated classroom; and completing next book *21st Century Turfs, Powers and Solutions*.

Albright is widowed and has one son and two grandchildren.

ELVIN F. ALEXANDER (ENS) was born Nov. 10, 1922, and enlisted in the USN Jan. 12, 1942. Stations and schools include: Algiers, LA (1942-43) Naval Station; Southwestern Louisiana Institute, Lafayette, LA (1943-44); Louisiana Tech, Ruston, IA (1944-45); Pre-Midshipman School, Princeton University, Princeton, NJ (1945); Northwestern University, Abbott Hall, Chicago, IL (1945).

(July 30, 1945) was commissioned ensign, USNR and married on the same day; (August 1945) U.S. Naval Sub Chaser Training Center, Miami, FL; (December 1945) was released to inactive duty at Great Lakes, IL.

Married 48 years to Marian C.; they have two children, Dr. Gary C. (five years Navy CTR2/c) and Mrs. Kathy Chmiel; two grandsons (one deceased); two granddaughters; and one great-granddaughter.

After 41 years he retired from Consolidated Ink Co., St Paul; attended St. Cloud State University, MN, BES degree in 1994; and currently works part-time for U.S. Postal System Postmaster Relief, Silver Creek, MN. He enjoys hunting, fishing, bicycling, and has participated in three bike-a-thons a year since 1988.

CHARLES D. ALLEN JR. (CAPT) was born Dec. 23, 1925, Brooklyn, NY. He entered the Navy V-12 Program at Princeton University on July 1, 1943; transferred to NROTC at Yale University in February 1944; graduated and commissioned ensign USNR in October 1945; and commissioned ensign USN in July 1946.

Allen served in destroyers, destroyer escorts, cruisers, LST and ice breaker over next 25 years. He commanded USS *Goldsborough* (DDG-20), USS *Norfolk* (EDL-1) and USS *Belknap* (CG-26) and taught NROTC at Brown University. Other shore billets were mostly in guided missile R&D including program manager for point defense missiles and NATO SeaSparrow, plus tours in Arms Control Agency and Joint Staff. He is a graduate of Armed Forces Staff College and National War College.

Captain Allen retired in 1971. His awards and medals include the Bronze Star, Joint Meritorious Service, Navy Meritorious Service and Vietnamese Navy Cross.

Has consulted for the Navy since his retirement on long range planning; president of Charles Allen Associates, Inc.; lived in northern Virginia since 1971 and now resides in Alexandria, VA with his wife, Carolyn V. Titus. They have four children and two grandchildren.

BEN ALMANY JR.

BEN ALMANY JR. (LTJG) was born Nov. 3, 1923, in St. Clair Shores, MI. He joined the USMC in 1943. Duty stations and schools include: V-12 Program, Western Michigan University; Columbia Midshipman, Miami, FL; Philadelphia Fire Fight School, Algiers, New Orleans, LA; and Columbia University.

Participated in the Pacific Theater, primarily in New Guinea and Philippines. He decommissioned USS *Clairbourne* at Yokosuka Naval Base and turned it over to the Japanese immediately after the war ended. He trained a Japanese crew of 65 (at sea in Tokyo Bay) for one week prior to turnover. They were told the event was necessary for Japan to start its re-birth. Discharged in 1946.

He met his wife, Joyce, while attending Western Michigan Univ. They have a son who is a cardiologist. Almany is retired from Ford Motors; his last assignment was compliance engineer manager.

DANIEL G. ANDERSON

DANIEL G. ANDERSON (LCDR) was born Oct. 5, 1921. He attended University of Washington (1939-40), drafted in 1943 (enlisted USN), sent to boot camp, then to Metalsmith School, Norfolk, then to V-12 at University of Washington (mechanical engineer), March 1, 1944. He was commissioned February 1946, trained at Davisville, RI, assigned to the USS *Montpelier* (CL-57) and released in July 1946.

He returned to the University of Washington to receive a degree in mechanical engineering. After receiving degree in 1947, he was employed by the USGS Water Resources Division (WRD) in Tacoma, WA, as a hydraulic engineer and hydrologist. In 1950 he joined a USNR unit in Tacoma. He was transferred in 1956 to the WRD Research Section in Washington, DC, where he joined a USNR unit.

Anderson completed 22 years of satisfactory service and retired as lieutenant commander in 1966. Most of his career involved research in hydrology. Those studies included low and flood-flow frequency analysis; snow and ice studies including Chamberlin Glacier, Alaska and the Ellesmere Ice Shelf in Canada; effects of urbanization on floods; delineation of flood hazard areas; bridge site analyses; and applications of remotely sensed data obtained from aircraft and spacecraft.

He retired in 1980 and resides in Vienna with his wife, Bernice. They have four children and five grandchildren. He currently farms five acres of vegetables, selling the produce at five local farmer's markets.

FRANK ALBERT ANDERSON

FRANK ALBERT ANDERSON was born Oct. 14, 1925, in Moline, IL. He attended schools in Illinois, Minnesota, Oregon, California and Wisconsin; entered Navy V-12 in June 1943 at Marquette; graduated BSEE in March 1946; received officer training at Newport and on board USS *Denver* (CL-58); and was discharged in July 1946.

He worked for Remler Co., the city of San Francisco and Dalmo Victor Co. while serving as Ready Reserve instructor. Active duty in November 1951 for Korean Conflict. He was electronics officer in USS *Bradford* (DD-545) and discharged in May 1953.

Anderson worked for Color Television Inc., Eitel McCullough, Varian Associates and Lockheed Missiles and Space Co. while serving in Standby Reserve. He held various positions dealing with electronic equipment: design, testing, reliability, quality control, technical writing, manufacturing, production engineering and production control. He completed 20 years Navy Reserve and 26 years Lockheed, retiring in 1986.

He and wife of 43 years, Barbara, reside in Portola Valley, CA. He is busy with real estate, hobbies, sports, square dancing and personal growth groups. He and Barbara have three married children and six grandchildren.

LESLIE J. ANDERSON

LESLIE J. ANDERSON (AS, MIDSHIPMAN, S1/c, QM) was born Aug. 4, 1923, in Portland, ME. He enlisted in the USNR in December 1942 with active duty in July 1943; attended Columbia University Midshipman School, June 1944; Great Lakes Quartermaster School, 1945; assigned to "The Cat" USS *Colahan* (DD-658), Halsey's Pacific Task Force.

He participated in action in Okinawa and Japan. His memorable experience was reunion with V-12 roommates on their ships. Anderson was discharged Jan. 10, 1946 (AS, Midshipman, S1/c, QM). His awards and medals include the American Campaign, Asiatic-Pacific with two Battle Stars, Occupation Service (USN) and the WWII Freedom.

Earned BA from Bates College, Lewiston, ME (1948) and taught high school business and social studies in Massachusetts, Connecticut, Guam (in Pacific), Aruba. MEd, Fitchburg (MA) State College (1951); director, economic education, Montclair (NJ) State College (1962); curriculum consultant (family finance) Institute Life Insurance (1964); Ph.D, University of Oklahoma (1970); director, Adult Consumer Economic Center; professor, Economics, Rose State College, OK.

Retired in 1986 and presently adjunct professor, Economics, Park College, Tinker AFB. Anderson is married and has three daughters and two granddaughters.

CHARLES L. ANDREWS III

CHARLES L. ANDREWS III (LT) was born in Fort Leavenworth, KS base hospital. He enlisted Dec. 24, 1943; attended Navy V-12 Program, Bates College, Lewiston, ME (1944); Navy V-12 USNROTC Engineering, Tufts College, Medford, MA (1944-46); commissioned ensign USNR, Tufts College with BNS degree in 1946.

He reported on board the USS *Richard B. Anderson* (DD-786) in San Diego, CA in 1946. As part of the Pacific Fleet, they spent four months on battle maneuvers attacking the Hawaiian Islands. The best part was a week of R&R on the island of Hawaii.

Received honorable discharge in June 1947 and used GI Bill to attend Tufts College, graduating in 1948 with BSME degree. He joined an Active Reserve Unit in Providence, RI and studied the new missile gunnery technology and taught small arms at the indoor range.

During his Active Reserve, he made a two week tour to Puerto Rico on board the USS *Johnnie Hutchins* (DE-360). In November 1950, he received orders to report to USS *New* (DDE-818). By the end of tour while attached to the 6th Fleet, the ship had visited Tripoli, Mers El Kebir, Naples, Palermo, Venice, Soudas Bay (Crete), Monaco and Cannes. He was stationed ashore in charge of the Fleet Shore Patrol in Izmir, Turkey; did some shore bombardment training outside Trieste, Yugoslavia; and last stop was Tarragona, Spain (they were first U.S. ship since Spanish Civil War to enter the port).

Returned to Norfolk and received orders to proceed to Astoria, OR to USS LSSL-58 as XO; on to San Diego, Pearl Harbor, Midway Island and Yokosuka, Japan. He flew back to the States in July 1952 and was honorably discharged for second time. In 1964 he was promoted to lieutenant, USNR, and retired with 16.5 years in 1967.

He decided on a career in the plastics industry and after 25 years of plastics engineering work, he started his first company in 1968, The Carolina Color Co. in Salisbury, NC; in 1972 he started Megacolor Inc. in Statesville, NC and sold it in 1984. He started Andrews Products Inc. in Statesville, NC, in 1987 and is still running the company with his wife and partner, Cindy.

JOHN S. ANDREWS (LTJG) was born Jan. 19, 1924, in Corry, PA. He entered the Navy V-12 Program on July 1, 1943 at Bucknell University after two years of college at Penn State. He was commissioned at Columbia USNRMS in October 1944, then attended Officers Diesel Engineering School at Cornell University. He served on an LCI(L) in the Pacific Theater from April 1945 until separated from active duty as lieutenant junior grade in July 1946.

Returned to Penn State for a BS in chemical engineering in 1947. In 1951 he received a BS in business management from Rutgers University. He joined Johnson and Johnson in 1948 and worked for 37 years in various positions in Research and Development until retiring in 1985.

His wife, Margaret, passed away in 1986; son, Richard, passed away in 1989; son John Jr. and his wife, Adrienne, have two children, Nathan and Ashley; and his daughter, Carolyn, lives in Vermont. Andrews resides in Lavallette, NJ and enjoys boating, fishing and traveling with Joan McGrath, his companion and long-time friend.

ROBERT L. APPEL JR. (CAPT) enlisted in the USN in May 1943. He entered the V-12 Program on July 1, 1943, at Wesleyan University in Middletown, CT and was commissioned an ensign in May 1945 at Northwestern Midshipman School. He served on board the USS *Sicily* (CVE-118) and was released to inactive duty on July 3, 1946. He was recalled to active duty in November 1950 and served in the USS *Palau* (CVE-122). Released to inactive duty in October 1952, he participated in the Ready Reserve while not on active duty until 1970.

Appel earned the following degrees: BA, Michigan State; MA, University of Michigan; and Ed.D, University of Colorado. His positions include: teacher, Muskegon Public Schools; associate professor, North Park College; vice president, Rock Valley College; president, Elgin Community College; academic vice president, Montgomery College; and professor, Montgomery College. He retired June 1, 1970.

ROBERT B. ARDIS (CAPT) was born in Bay City, MI in 1925. He entered the Navy V-12 Program at the University of Michigan in July 1944 and was commissioned in February 1946. He received BSEE and MSEE degrees from University of Michigan in 1946 and 1947 and JD degree from New York University in 1951.

Ardis joined AT&T Bell Laboratories in 1947 as a member of its patent attorney staff. He retired in 1988 as general attorney, Intellectual Property Matters. He is currently doing patent work for several major corporations with a former colleague from Bell Labs and uses a computer to edit and prepare editorial commentary for the newsletter of the local chapter of The Retired Officers Association.

70

He lives with his wife, Catherine, in Mountainside, NJ.

ROBERT E. ARNOLD (LCDR) was born Dec. 22, 1925 in Wadsworth, OH. He joined the service on May 18, 1943. His military location, stations and schools include: COMPHIBGRU-4; USS LST-220; USS Cavalier (APA-37); Fleet Marine Force; Columbia University, Midshipman School; and V-12 Program Denison University, Granville, OH.

As a Navy officer he was sent to Marine Corps School, Quantico, VA, then assigned to Fleet Marine Force, Pacific as a liaison officer. He was with Staff, Commander Amphibious Group 4, 6th Fleet in Mediterrean and spent three years as gunnery officer in the Korean War. He retired from the USN in 1970.

Arnold is married and has four children. He retired from Northern Telecom Digital Switching Division.

RUSSELL H. BABCOCK (LTJG) enlisted in the USN in 1943 and entered the V-12 Program at Tufts College as a student in Civil Engineering. He received his BS degree in June 1945 and was commissioned from Midshipmans School, Columbia University. He served as instructor in trigonometry at Navy Pacific University Pearl Harbor, T.H. He was assigned to the Service Division 101 and served at Shanghai, Hong Kong and Tsingtao.

Following his discharge from active duty he transferred to the Civil Engineering Corps. He achieved the rank of lieutenant junior grade, CEC. Babcock entered GSE Harvard and received an SM in sanitary engineering in 1947. He worked for the Foxboro Co. as division manager from 1950 to 1970; C.E. Maguire as chief engineer from 1970 to 1976; and as a consulting engineer in private practice to the present time.

Served many years as a captain in USPHSR; is author of over 50 articles and one book; P.E. in numerous states; and is a past president of the New England Water Works Association.

Married to Ellen Clewley and they have five children and four grandchildren. They presently live on Cape Cod and enjoy the scenery, weather and boating.

ALDEN O. BAILEY (COL) was born on Aug. 17, 1924 in Pasadena, TX. He entered Southeastern Louisiana Institute, V-12 Program on July 1, 1943 and was commissioned a second lieutenant USMCR in June 1945. He served with the 3rd Marines on Guam and 1st Bn., 7th Marines in North China.

COL Bailey, USMCR, was released in September 1946 and recalled in January 1951. He served in Korea with the 1st Amtrac Bn. and was released in June 1952.

Earned a BS degree from Sam Houston State Teachers College in 1947 and a MEd. from the University of Houston in 1955. Bailey taught school, then owned and operated a hotel and restaurant.

A past president of the Pasadena Chamber of Commerce, past president of the Rotary Club, past vice-president of the School Board, past district governor of Rotary International, and past member of the Area Council, Boy Scouts of America.

He resides in Pasadena, TX with his wife, Dorothy Jean Roberts. They have three children and five grandchildren and are members of the First Methodist Church. He retired in 1975 as a colonel, USMCR.

PAUL O. BAILEY (RM3/c) enlisted in the USNR on Dec. 29, 1941. He attended boot training at Great Lakes, IL; RM School at Indianapolis; Compool School at Norton Heights, CT; and then assigned to the Armed Guard Center in Brooklyn, NY. He first served with Port Directors and operated radio station NAD6 at Woods Hole, MA.

Bailey was assigned to the USS *Beaconhill*, which carried 10,000 tons of airplane fuel to Murmansk, Archangle and Molotovsk, USSR. He was in convoy PQ23. Arriving on his 20th birthday in 1943, he served eight months above the Arctic Circle. There, he endured 16 combat missions, crossing the Arctic Circle 16 times, while refueling vessels in the area. They came within four feet of an enemy mine, narrowly missing it, and fought off numerous submarine and air attacks in and out of port.

Memorable experience was when Mrs. Roosevelt came to the Armed Guard Center to welcome them back home. He completed the V-12 Program at Williams College, then went to Tufts to Midshipman School. While at Tufts, the war ended. On Nov. 13, 1945, RM3/c Bailey received his discharge. He received a letter of commendation from the Secretary of the Navy, Frank Knox.

Worked for the Ed Grace Co. for 30 years in the mechanical contracting and engineering business in Lafayette, IN. He retired from the Ed Grace Co. to operate his own electrical service business. Bailey is married and has two children.

On Oct. 8, 1992, he was presented the Soviet Jubliee Medal by Ambassador Vladimar Lukin, along with a certificate signed by President Boris Yelstin, commemorating 40 years victory in The Great Patriotic War.

HOWARD H. BAKER JR. (LTJG) was born on Nov. 15, 1925 in Huntsville, TN. He enlisted in the USN on July 2, 1943; attended the University of the South, Sewanee Tulane University and Midshipman School, Fort Skylar, NY.

LTJG Baker was discharged on June 25, 1946. He was a former United States Senator, majority leader, minority leader, and President Reagan's Chief of Staff. He is now a practicing attorney.

NICHOLAS J. BARBAROTTO (CAPT) was born in San Francisco, CA on Feb. 25, 1923. He attended the University of San Francisco and was called to active duty, July 1, 1943, with the Marines while enrolled in the Navy V-12 Program. He was assigned to the College (now University) of Pacific in Stockton, CA, which he attended for two semesters. He later completed boot camp at Parris Island; assigned to the Special Officer Candidates' School at Camp Lejeune in May 1944; graduated on Sept. 30, 1944; and was commissioned a 2nd lieutenant in the USMCR.

Joined the 4th Mar. Div. in November 1944 at Maui, where the division returned following the Saipan-Tinian Operation, and was assigned to B Co., 1st Bn., 24th Regt. He saw action as a platoon leader on Iwo Jima where he was wounded twice.

Returned to the States and on July 22, 1945, while on a 30-day recuperative leave, he married his hometown fiancee, the former Tamara Rusanoff, who had followed him into the USMC and was a corporal stationed at Marine Headquarters in Washington, DC.

In November 1945, he returned to inactive duty in the Reserves and was awarded the Purple Heart with Gold Star. He enrolled in the University of San Francisco School of Law the following year and was graduated and admitted to the California State Bar in June 1949. He then entered private law practice in San Francisco.

Recalled to active duty in March 1951, during the Korean War and attended Naval Justice School in Newport, RI, after which he served as assistant post legal officer and trial and defense counsel on general courts-martial at Camp Pendleton. Capt. Barbarotto was released to inactive duty and received an honorable discharge in 1952.

Married for 50 years and resides in South San Francisco. He has three children: Wayne, Duane and Vicki Lynn; and four grandchildren: Brigette and Brett Caprio, Jason and Joshua Barbarotto.

Although semi-retired, he continues to actively engage in the general private practice of law in San Francisco and enjoys good health while maintaining a close attachment with the members of his family.

FRANK A. BARNES (LTJG) was born on May 27, 1922, in Meeker, OH. He enlisted in June 1942 in the V-2 Program and attended Cal Tech in Pasadena, CA. He was stationed aboard USS *Prairie State* from July 1944 to June 1945 as instructor in electrical engineering and aboard USS *Ticonderoga* (CV-14) from July 1945 to July 1946 as electrical officer.

A memorable experience was helping to put the USS *Ticonderoga* into "moth balls" from January to July 1946. LTJG Barnes was discharged in July 1946.

Attended MIT Graduate School, September 1946-June 1951; employed by MIT, Cambridge, MA, in guided missile engineering, September 1946-June 1955; RCA Aerospace Systems Division Burlington, MA, in lunar module engineering, June 1955-December 1973; and by C.S. Draper

Lab, Cambridge, MA, in satellite engineering, February 1974-December 1987.

Retired in January 1988 and currently resides in Weston, MA with his wife, Beatrice (Ahlgren). They have a daughter in Worchester, MA and a daughter and three grandchildren in Chelmsford, MA.

Z.B. BARNES (LT) was born on Aug. 8, 1925, in Lakeland, FL. He entered the V-12 Program on July 1, 1943, as a pre-med student; served at the naval hospital, Key West, FL, November 1944-September 1945; Midshipman V-12 unit, Emory University School of Medicine, Atlanta, GA, September 1945-December 1945. He received his MD degree from Emory University in 1950; internship was at University Hospital, Minneapolis, MN, 1950-55; and residency at Kennedy VA Hospital, Memphis, TN, 1951-52.

Barnes served as a medical officer during the Korean War, LT, MC USNR, USNH, Jacksonville, FL, medical officer 15th Naval District Headquarters Panama Canal Zone and medical officer for the naval station, Rodman, Canal Zone, 1952-54. His surgical residency was from 1954-59 at Kennedy VA Hospital, Memphis, TN. He was certified by the American Board of Surgery, Fellow American College of Surgeons and had a private practice of surgery in Montgomery, AL until 1984 when he retired.

He and his wife, Frances Shaw Barnes, M.D. still live in Montgomery and have two daughters, Beryl and Roxanne. He enjoys hunting, etc.

ROBERT D. BARRON (AERM3/c) was born on Dec. 4, 1925, in Friedens, PA. he enlisted in the USN on May 10, 1943; attended V-12 Program at Penn State; and stationed in Washington, DC and Pearl Harbor.

A memorable experience was working in the Office of Naval Intelligence in the weather section. AERM3/c Barron was discharged on April 12, 1946.

Barron graduated from Penn State in 1949 with a BS degree in metallurgy. He is retired from ITT Corporation. He and his wife, Margaret, have four children: Kevin, Karen, Keith and Katie.

ROBERT E. BATEMAN (LCDR) enlisted ONI, 13th ND in August 1941. He was selected for the Naval Academy in June 1942 but failed the eye exam. He was appointed to V-12 at Wabash College, transferred to Purdue in April 1943 and graduated with BSAE in February 1946. He served at Moffet Field and transferred to Ready Reserve 13th ND in April 1946 and Retired Reserve in August 1963.

Bateman joined the Boeing Co. in April 1946 and served in a wide variety of technical and management positions, retiring as a corporate VP in 1988. Prior to that he was VP, Boeing Marine Systems; producer of 43 military and commercial hydrofoils utilized worldwide. He owes V-12 and Purdue a great deal and has devoted time and resources in supporting the "system." LCDR Bateman has served on the boards of Navy League, Naval War College, Navy Memorial, Naval Aviation Museum, Purdue President's Council, the Seattle Museum of Flight and the Seattle World Affairs Council.

Prized among the honors he has received are the SecNav Meritorious and Distinguished Public Service awards and an Honorary Doctor of engineering degree from Purdue. He and his Purdue wife, Sarah, live in Seattle and have three children and four grandchildren.

JAMES M. BAUMGARDNER (CDR) was born on Jan. 26, 1927, in Emmitsburg, MD. He enlisted in the Navy V-5 Program in January 1944 and was a V-5 in the V-12 Program until June 1946. He completed college in the Navy NROTC Program in June 1947 at Villanova College. He served aboard the *USS Albany* (CA-123); *Joseph P. Kennedy Jr.* (DD-850); NADO Charleston; MCB-1; Bureau of Supplies and Accounts; *USS Randolph* (CVA-15); Supply Center, Norfolk; George Washington University, where he received his MBA; Ordnance Supply Office, Mechanicsburg; *USS Mars* (AFS-1); Naval Air Systems Command.

Baumgardner saw action in Vietnam and the Suez Canal. CDR Baumgardner retired on June 30, 1967. CDR Baumgardner, USN, RET, awards include the WWII Victory Medal, American Theater Medal, Navy Occupation Medal and the Vietnam Service Medal.

His civilian career included: computer systems supervisor, Sinclair Oil Co.; corporate computer manager, B.P. Oil Corp.; Eastern Region Manager Computer Systems Amoco Corp., systems development supervisor, Amoco Corp., Chicago. He retired from Amoco in November 1986.

Currently living in Yorktown, VA and serving as volunteer public relations coordinator for the Watermen's Museum in Yorktown. He and his wife, Betty Jean (Baird) have three children: Jo Ann Avellone, school teacher in Lake Bluff, IL; Lisa Lee Freeman, school teacher in Richmond Hills, GA; and Charles, LCDR SC USNR, Hampton, VA; and nine grandchildren.

ROY E. BAXTER (1st LT) was born on Oct. 15, 1921, in Paris, TX. He enlisted in the USMC in February 1942 and attended Louisiana Tech, Ruston, LA. His duty stations include Parris Island, SC; Camp Lejeune, NC; Quantico, VA; South Pacific; San Diego, CA; Marine OCS.

He was in the 1st Mar. Div., 5th Reg., 1st Bn., Co. A and saw action in Okinawa and the South Pacific. Memorable events were the Okinawa fleet landing; the kamikaze; and the fact that Ernie Pile slept in his platoon on the eve of his death.

Baxter was discharged on March 26, 1946, to inactive duty and resigned in 1951 due to a service connected injury. He was awarded the Purple Heart. He had attained the rank of captain in 1951, but the injury at Okinawa prevented his promotion, thus "1st lieutenant."

Married on Jan. 18, 1947, and they have one daughter and one son. Baxter worked 40 years, VP, Purchasing, U.S. Brass, Plano, TX, retiring in 1986. He is now golfing, ranching and since 1977 has been the leader of eight reunions of LA Tech V-12.

WILLIAM D. BAYLESS (TUB) (1st LT)

was born on July 22, 1925, in Denton, Denton County, TX. He enlisted in the USMC, 4th Mar. Div. in February 1943 and attended Louisiana Tech University. Other duty stations included: Parris Island; Camp Lejeune, NC; Oceanside, CA; and Quantico, VA.

A memorable experience was preparing for the Japan invasion in July and early August 1945. Bayless was discharged in December 1945 and stayed in the inactive Reserves; he was recalled to active duty in June 1951 for the Korean Conflict and went directly to Camp Pendleton, Oceanside, CA where he joined the 8th Replacement Draft. However, he was recalled back to Quantico for further officer training. 1st Lt. Bayless only received stateside awards.

He is the owner of Bayless Insurance, Inc. in Denison, TX. His son, David Jr., has joined him in the insurance business and his daughter, Brandy, is a flight attendant for American Airlines. He and his wife, Patsy, reside in Denison, TX.

MORRIS J. BECKMEYER (S2/c) was born

on May 7, 1925, in Snohomish, WA. He enlisted in the Navy in February 1944 as a V-5 NAVCAD. He attended Minot State Teachers College and the University of Washington as a V-12(a) trainee. He entered the flying program in July 1945; S2/c Beckmeyer resumed civilian life in July 1946 and graduated from the University of Washington with a major in accounting. He obtained the master of business administration degree and is a certified public accountant.

Beckmeyer's career was devoted to 34 years of Federal Service (including Navy). He served in various technical and administrative capacities with the Internal Revenue Service (IRS), the Small Business Administration, and the Agency for International Development (AID), within the continental United States and overseas. He served in the foreign service programs of IRS and AID for over 10 years. His career called him to duty in Tacoma and Seattle, WA; Portland, OR; San Francisco and Los Angeles, CA; El Salvador; Dominican Republic; South Africa and Washington, DC.

After retirement from IRS in 1980, he worked for Arlington County, VA for a number of years. Morris and his wife, Patty, live in Bellevue, WA. They have four sons and three granddaughters.

LEO W. BEDELL (LTJG) was born on Sept.

24, 1921, in Oklahoma City, OK. He enlisted in the USN Amphibious Force on Aug. 17, 1942. He attended the V-12 Program at John Carroll University, Cleveland, OH and Midshipman School at Northeastern University. He was assigned aboard LCI 449, Pacific. Battles participated in were Saipan (Mariana Islands), Iwo Jima and Okinawa.

Memorable experiences included: supporting demolition team on Feb. 17, 1945; two days before D-day at Iwo Jima when their ship was hit by three mortar and artillery shills. They had about 70% casualties, 20 killed and 20 wounded (Bedell was the only officer who was not killed or wounded); the suicide planes at Okinawa; and being in a typhoon on the way to the Philippine Islands.

LTJG Bedell was discharged on Dec. 18, 1945. His awards included the Silver Star, Presidential Unit Citation and Unit Commendation Ribbon.

Bedell married on May 4, 1946, and has 10 children (seven boys and three girls). At the present time he is semi-retired from his job as insurance agent.

JOHN F. BEDINGER (ENS) was born on Jan.

26, 1925, in Worsham, VA. He entered the Navy V-12 Program on July 1, 1943, at the University of Richmond, VA. He transferred to Duke University V-12 on July 1, 1944, and received a BA degree in physics. He was commissioned ensign at Ft. Schyler, NY, attended Midshipman School in November 1945 and served in the Pacific Theater aboard USS ATA-193 and on shore bases in Hawaii. Bedinger was discharged to inactive service on Aug. 13, 1946.

He was instructor of math and physics at St. Helena Extension of College of William & Mary, 1946-1948; received an MA in physics from University of Virginia in June 1950; physicist, conducting research in upper-atmosphere at USAF Cambridge Research Center, 1951-1958; GCA Corp., Bedford, MA, 1959-1976; scientist at Environmental Research and Technology, Concord, MA, 1977-1982. He is the author of numerous publications in scientific journals and at international meetings and presently a registered representative for First Investors Corp., from 1982.

He currently resides in Framingham, MA with wife, Ruth, and keeps close contact with two sons, John and Paul, and grandchildren.

RICHARD G. BEHLER (LTJG) was born on

July 30, 1925, in Plymouth, MI. He enlisted July 1, 1943; attended Western Michigan University; Midshipman School, Columbia; served on USS LST-334 and USS LCI (R) 648. He participated in action in Okinawa.

Behler was released to inactive duty in the USN on July 28, 1946. LTJG Behler received all the regular service ribbons.

He is currently retired and living in a small community in northern Michigan. After release from the Navy, he earned a degree in economics from Michigan State University. All of his working career was spent in the Human Resources Department of Wayne County, MI. In 1978 he retired after 30 years of service of which the last nine were spent as the county personnel director. Since retirement he has worked as a consultant for the Michigan Municipal League.

He and his wife, Marjorie, have two sons, Christopher and Jonathan.

EARL M. BEHNING (LT) was born June 13,

1924, in Deer Creek, MN. He enlisted in the USN in April 1943; entered the Navy V-12 Program at Carson Newman College in 1943. He received a BS degree in 1946 and graduated from the University of Minnesota Dental School in 1949.

Commissioned a lieutenant in the USN Dental Corps in June 1949; served at the USNH, Bethesda, MD; then USNH, Oakland, CA. In 1951 he transferred to the Navy Dental Clinic, Pearl Harbor, HI; served during WWII, 1943-45; the Korean War, 1949-52. He achieved the rank of lieutenant.

Behning practiced dentistry in Austin, MN, 1952-89 and retired in Lake Hubert, MN. He married Suzanne and they have two daughters, Cindra Sue and Jill.

WALLACE B. BEHNKE JR. (LT) was born

Feb. 5, 1926, in Evanston, IL. He enlisted in the USN in 1944; entered V-12 Program at Northwestern University; transferred to NROTC and was commissioned ensign USNR in 1945. He served in YMS-294 in WWII; active Naval Reserve Division 9-21, 1947-50; Korean War (earned two Battle Stars) aboard the  USS *New Jersey* as E-Div. officer from 1950-52 with rank of lieutenant.

His principal career was with Commonwealth Edison Co. where he became vice chairman in 1980. He retired in 1989 and is currently a consulting engineer.

A director of Duff and Phelps Utilities Income, Inc., he served on the Boards of Directors of Commonwealth Edison Co., Calumet Industries, LaSalle Bank Lake View, Paxall Group, Tuthill Corp., Standard of America Life Ins. Co., Argonne National Laboratory, Northwestern Memorial Hospital, United Way of Chicago and Illinois Institute of Technology.

His professional honors include Fellow in Institute of Electrical and Electronics Engineers, National Academy of Engineering, Electrical Industry Man of the Year for 1984 and the James N. Landis Medal of the American Society of Mechanical Engineers.

SAMUEL V. BELL JR. (LCDR) entered the

V-12 Program and the University of Louisville in June 1943. He was commissioned ensign USNR at Fort Schyler, Bronx, NY, November 1945 and stayed on active duty until 1964. He attended advanced electronic and guided missile schools and served in destroyers, cruisers, aircraft carriers and amphibious vessels.

LCDR Bell, USN, RET, received four Battle Stars in Korea (1950-51) plus the Navy Commendation. He finished school with a degree in electrical engineering at the University of Louisville in 1958 while still on active duty. Commenced academic career teaching electrical engineering at the University of Louisville in 1964 and completed his Ph.D from the University of Kentucky in 1977.

He retired in January 1993 and was appointed Professor (Emeritus) of electrical engineering. Married Carolyn C. Woolery of Louisville in 1946 after two years of dating while in the V-12 Program. They have four children and two grandchildren

EDWIN M. BELLES JR. (LTJG) was born Jan. 19, 1924, at Devils Lake, ND. He entered the USNR V-1 Program on Dec. 2, 1942; attended the University of Washington and graduated 1n 1945, Metallurgy. He was stationed at Puget Sound Naval Shipyard, Bremerton, WA as metallurgical lab officer.

LTJG Belles was discharged Sept. 1, 1955, and retired from the Reserves in 1967. His awards and medals include the WWII Victory and American Campaign. AEC Hanford (1952-56), Lockheed Missiles and Space Co., 1960-86. He retired in Portland, OR, and is involved in education.

VERNON BRADLEY BENNETT (LT) was born July 28, 1924, Cincinnati, OH. He enlisted Aug. 22, 1942, V-1 Program at University of Notre Dame and graduated Jan. 18, 1945, from Northwestern. Other duty stations include the USS *Guadalupe* (LCT-897), Philippines, Okinawa and Korea.

His memorable experience was the two typhoons in September 1945 in Okinawa. Bennett was discharged Sept. 22, 1946. The Navy was a wise choice for him. To this day he loves the water, the discipline and camaraderie of it all. LT Bennett retired from the Reserves in 1967. He received the Philippine Liberation.

Married Lini, Hague, Netherlands in December 1950. They have four children. Currently working a lime grove with 200 trees on two acres in Fallbrook, CA.

WILLIAM P. BENNETT (LTJG) was born Feb. 28, 1924, in Monrovia, CA. He enlisted in December 1942; attended Westminister College, Fulton, MO and Columbia University, NY, NY.

His memorable experience was serving as XO on the USS LCS (L) (3) 90 in the Pacific and participating in action at Okinawa.

LTJG Bennett was discharged in July 1946. He is married and has four boys (one deceased) and one girl. He is a citrus (avocado) grower.

CHARLES J. BERKEL was born in Muskegon, MI on May 10, 1925. He entered the Illinois Institute of Technology, George William V-12 Unit, in July 1943. He transferred to the University of Illinois in 1944; graduated in January 1946 with a BS in civil engineering and was commissioned ensign. Assigned to USS *Mattaponi* A041 as a deck officer in Fusan, Korea and traveled throughout the Far East stopping at many major Asian ports. He was released to the Naval Reserve in September 1946.

After a few months working for the city engineer in Muskegon, he joined a national foundation company, Intrusion-Prepakt, Inc., in Chicago as their technical engineer on a major project. He advanced to regional manager in Kansas City in 1953. He started Berkel and Co. Contractors in 1959 and is now the chairman residing in Shawnee, KS with his wife Antoinette. They are the parents of a son and two daughters. They also have one granddaughter.

Berkel and Co. has grown to a national company with offices in Kansas City, Atlanta, Baltimore, Detroit and Houston, employing upwards to 200 persons. In 1966 he purchased a Beechcraft airplane and obtained his pilot license including an instrument and multi-engine rating. He traveled throughout the States, both business and pleasure, accumulating over 7,000 hours.

ALVARO D. BIAGI was born June 24, 1925, Luzerne, PA. He entered the University of South Carolina in the V-12 Program in 1944, received BS in electrical engineering and was commissioned ensign on June 19, 1946. Received his MS in management engineering at Long Island University in 1970.

He was associate lab. dir., ITT Federal Labs from 1948-62; VP, Eng. Republic Electronics, 1962-72; president, Republic Electronics, 1972-75; and president of A.D. Biagi Industries, Inc. from 1976-1992. He participated in the development of short-range navigation system known as TACAN (Tactical Air Navigation) and its civil counter-part VORTAC/DME.

Biagi belonged to the IEEE, AAAS and Air Force Association.

He was married to Raffaella R. and they had two children, David R. and Eric P., and two

grandchildren, Stacey and Eric Paul. Alvaro D. Biagi passed away Dec. 5, 1992.

ROBERT PAUL BIELKA (CDR) was born Feb. 19, 1925, Berkeley, CA. He enlisted in November 1943 in the V-12 Program at the University of Washington in Seattle, WA, November 1943-November 1944; Radio Material School (RMS) and electronic training from December 1944-December 1946; and various ACDUTRAs related to Intelligence activities. He was commissioned ENS 1105 USNR on Aug. 8, 1948.

Active duty on USS *Dade* (APA-99) from December 1945-February 1946 and USS *Bon Homme Richard* (CVA-31) as electronic repair officer from April 1951 to April 1953. Bielka retired as commander, 1635, USNR on July 1, 1975.

His civilian employment included: 1949-1950, Boeing ground to air pilotless aircraft project; 1950-1951, Southern Pacific Railroad; 1953-1956, Boeing Co. Engineer Bomber Weapons Staff Unit; 1956-1958, Crosley Div. AVCO Mfg. Co., reliability engineer; and from 1956-1986, was supervisor in 767 and 747 Reliability And Maintainability Engineering. He retired in August 1986. Currently he is owner of Bielka's Repair Shop which provides repairs and equipment for model railroad construction and layouts.

Married to Muriel Maxine and they have three children: Robert Bruce, Janice Maxine Dressler and Barbara Ann Wans.

CHARLES BILL JR. was born in Cleveland, OH on May 23, 1925. He entered the V-5 Program in January 1943; transferred to the V-12 Program in July 1943; sent to Asbury Park, NJ, pre-midshipman school in October 1944 and Harvard Supply Corps School in February 1945. He was commissioned and assigned to the Great Lakes Training Station, Treasure Island and Cavite, P.I. Naval Base. He worked with prisoner interrogation and graves registration.

Discharged May 1946, the work he will never forget was tracing missing naval personnel.

Bill received BA in economics, Baldwin-Wallace, June 1947; MA in finance, Bowling Green University in June 1949. He was an instructor in economics at Cuyahoga Community College in Cleveland for 12 years.

He lives with his wife, Mary, in Lakewood, OH. They have three children and three grandsons.

They are independent consultants in pensions and theater, are avid readers and he bike rides 50 miles every week.

RICHARD E. BISBE (DICK) (LTJG) Duke BS-ME, opted for a Navy V-12 commission after a futile three semester wait at Muhlenberg for a V-5 pre-flight opening. As a June 1946 graduate and ensign, he then rode the destroyer USS *Perry* (DD-844) for a year in the Med. as damage control officer/assistant engineer.

Raising a family and his DuPont career were interrupted by a January 1951 Navy recall as lieutenant junior grade. A two year tour on the USS Cony (DDE-508) included carrier plane guard and shore bombardment duties around the Korean peninsula. This tour started through the Panama Canal, returned via Suez, and in seven months logged 62,000 miles. He ranks qualifying as an officer of the deck, underway, as his most interesting and satisfying experience.

His 39-year DuPont working career was almost equally mobile, retiring in 1986. Procuring critically scarce petrochemicals during the Arab oil embargo was his job in Geneva, Switzerland from 1973-78. In a second career, 1986-93, he helped U.S. businesses gain international quality certification by providing management consulting services.

His immediate family consists of Swiss born wife, Silvia, two daughters and a son (who served in the USMC) and eight grandchildren. He is now retired in Pinehurst, NC with good health and mediocre golf scores, otherwise "steaming as before."

STEPHEN E. BLACKMON (ENS) was born Nov. 19, 1935, in Washington, GA. He volunteered for Navy V-5 in March 1944 and spent 16 months at Mercer University, Macon, GA. With the war over in Europe, he was sent to the NROTC Program at Duke University, Durham, NC. In June 1946, he graduated with a BS and was commissioned ensign.

How well he remembers the first session of orientation when they arrived at Duke. Capt. Kawalski told them, "This is the Annapolis of the South and it is tough here. Two-thirds of you won't make it; before a year is up, the guy on your left and the one on your right will be gone, or will it be you?" Somehow Blackmon made it.

In June 1947, he graduated from the University of Georgia with an AB in economics. He began work in the family department store and beef cattle operation, becoming a partner in 1960. He retired in 1989.

He married Eleanor Boyd of Durham, NC; they are the parents of four children: Myra, Steve, Phillip and Jeffrey, and grandparents of eight children.

RAYMOND E. BLAIR (XO) was born in Winston-Salem, NC. He attended Catawba College and signed up with the naval program in October 1942. Called to active duty in the V-12 Program, Newberry College in July 1943; June 1944, ordered to Asbury Park, NJ for pre-Midshipman School; September 1944, assigned to Midshipman School at Northwestern University, Tower Hall Div.; and was commissioned ensign, USNR in January 1945.

Ordered to naval training in Miami, FL; shipped out of San Francisco in March 1945 to report to Amphib. Div. in the Philippines. He was assigned to LCT Flot. #5 in the Manus Islands. The group was ordered to New Guinea to work with the Australian Army and Air Force. The Japanese surrendered in October 1945 and in November 1945, the LCT group was ordered back to the Manus Island for decommission of unit and all the boats were sunk in the bay.

January 1946, ordered to Amphib. Group #1, Shanghai, China and assigned to LST 910, becoming XO in February 1946. After four months with the Manchura Project, he received orders to return to the States and was discharged in July 1946.

September 1946, he entered the University of North Carolina and graduated with BS in commerce in May 1949. Began work with the NCR Corp. in sales and held 10 different sales management in the southeastern States from 1956-83 when he retired. He now enjoys golfing, fishing and his beach-front home in Kitty Hawk, NC.

Married in 1948 and has three children and six grandchildren, living in Virginia and Kentucky. Having the opportunity to attend the 50th anniversary of the V-12 Program and seeing again many of his V-12 friends will certainly be a wonderful experience that will be remembered and cherished forever.

WILLIAM C. BLANKS (LTJG) was born Oct. 14, 1923, in Greensburg, PA. He joined the USNR in 1942; attended Columbia University; and V-12 Program at Williams College. He served in the USS *Tucson* (LST-801).

Blanks was discharged April 10, 1952. He is currently working independently as a geologist and petroleum engineer to find oil and gas. He and his wife, Violette, have two daughters, two sons and three grandchildren.

CLAY S. BLECK (LTJG) was born March 26, 1923, in Spokane, WA. He enlisted in the USN in 1942; served in LSM-81 in the Pacific; and participated in action at Okinawa. Bleck attended New York Midshipman School-Columbia and V-12

Program at Colorado College. His memorable experience was the kamikaze sorties.

Released from active duty on July 24, 1946, as LTJG USNR (D). Graduated from Stanford University with BA in economics and political science in 1947. He entered the machinery business in Portland, OR, in 1948; purchased Dodge-Plymouth Dealership, Spokanne, WA in 1952; organized Dodge City, a retail auto dealership in 1962; and opened Precision Motors for sales and service in 1968. He founded Honda City auto dealership and assumed Chevrolet dealership in Spokane in 1970 and is still active in both.

Bleck is an active polo player (35 years), Spokane Polo Club. He married Dianne Welsh Bleck Ph.D Assoc. Professor, Eastern Washington University.

EDWIN E. BLY (LT) was born April 13, 1926, in San Francisco. He enlisted in V-12 Program on Dec. 8, 1943; attended University of California (Berkely) for eight months in 1944; BS, civil engineering from University of Illinois (1947); and MBA from Stanford (1949).

Discharged from service at Great Lakes in 1946 and recalled as officer in Korean War. He served in Coronado, Japan and Korea (ACB-One attached to 1st Marines and Task Force 94). Lt. Bly retired from USNR in 1965 after 22 years of service. Received the Expert Piston and Battle Star for Korea.

He had various engineering, marketing and marketing research positions from 1949-1965. From 1965 to present, he has owned and operated as general partner or as owner in real estate property in California and Texas.

Married in 1956, separated in 1983, he has five children and five grandchildren. He enjoys diving, dancing, skiing bridge, traveling, and made a parachute jump and several bungee jumps in the last few years.

NORWOOD BRUCE BONEY JR. (LCDR) was born Aug. 22, 1924, in Kenansville, NC. He entered Newberry College, V-12 Program on July 1, 1943, and commissioned ensign in January 1945 at Northwestern Midshipman School, Chicago, IL.

Served in the South Pacific as officer in charge USS LCT-159 and was mostly attached to Australian army in New Guinea. Later he was assigned to Guam and USS *Tombigbee* (AOG II), serving as communications officer. This ship provided water for small amphibs and to citizens of Nagasaki during occupation. He was released to inactive duty at Camp Shelton, VA in 1945 as apprentice seaman/LT MC.

Attended the University of North Carolina, Chapel Hill and received AB in economics in 1947 and LLB in 1950. He served as attorney for Mortgage Dept., Jefferson Standard, Greensboro, NC for five years; then as vice-president for Kansas City Title for several years; and most of career with Lawyers Title of North Carolina as president and then chairman. Following retirement in 1992, he still serves as legal consultant.

Presently resides in Charlotte, NC and Myrtle Beach, SC. He has a son and two grandchildren who live in Tucson, AZ and a daughter on Sanibel Island, FL. Boney retired from USNR as lieutenant commander after 27 years of service.

MARVIN C. BONN (ENS) was born Jan. 27, 1924, in Stark County, Canton, OH. He was accepted into the Navy V-12 Program in 1943 at IIT George Williams College and later transferred to Duke University. He was assigned to USS LST-220 and later to USS LCI-549 while participating in Joint Task Force One Operation Crossroads to test effects of atomic bomb on naval vessels at Bikini Atoll, Marshall Islands.

Transferred at Guam to USS LST-1135 and duty that led to sailing throughout the Philippines including Luzon, Cebu and Palawan. He achieved the position of chief engineering officer before leaving the ship to be separated from active duty in 1947 to inactive Naval Reserves as ensign, USNR, until receiving an honorable discharge in 1959.

Worked for Illinois Northern Utility Co., Dixon, IL, and after 39 years and two company mergers, he retired from the Nuclear Stations Division of Commonwealth Edison Co. in 1986.

He and his wife, Laura, live in Seminole, FL; they have three sons: Bob, Mike and Rick

MARVIN L. BOOTE was born Dec. 15, 1925, enlisted in the V-5 in October 1943 and called to V-12 in March 1944. He moved to Notre Dame Midshipman in July 1945 and was discharged in May of 1946.

His best Navy remembrance, among many, was meeting and marrying the finest lady in the world while at St. Ambrose, Notre Dame. They have four fine children: Debby is a CPA in Los Angeles; Bud is in business in Akron, OH; Brad is his business partner; and Lisa is a chiropractor in Hollister, CA. He will always remember his 27 months in the Navy as a great experience and a dividing point in his life. He still returns to St. Ambrose occasionally.

Graduated from the University of Minnesota in 1949 and was employed by H.J. Heinz in 1950; Phillips Petroleum Co. in 1955; started his own

business in 1969 and is currently the chairman of Martrex Inc. (chemical and petroleum trading company). He and his wife live in Excelsior, MN, a suburb of Minneapolis. They do some traveling, particularly to warmer climates during the winter months as Minneapolis is very cold.

ROBERT WILLIAM BORDERS, MD (LT) was born Nov. 17, 1925, in Stratton, CO. He joined the USN on June 23, 1943 and was discharged March 31, 1956, as apprentice seaman/LT MC. He graduated from University of Kansas Medical School in 1949; residency in anesthesiology at Duke University, 1952-54; and has practiced anesthesiology in North Louisiana since 1954. He is a member of American Board of Anesthesiology (1959) and Fellow of American College of Anesthesiology (1955).

Participated in the Korean War, 1951-52. He will always remember his year of sea duty aboard the USS *Jason* (ARH-1) as the only medical officer.

He and his wife Emma Sewell Borders have five children: Robert, district manager for a retail firm in New Orleans; William, a periodontist in Shreveport, LA; Blaine, a cardio-thoracic surgeon in Monroe, LA; Jonathan, age 16; and Anne (deceased). They also have nine grandchildren. They live at their country home, Shiloh Farms, in the hill country of northeastern Louisiana. They raise cattle, have horses, pets, and, of course, a tennis court. He expects to retire in three years and looks forward to touring castles.

Borders has been an avid tennis player and has won numerous local and worldwide amateur tennis tournaments, including being the Louisiana State Men's Singles Champion in 1962 and the Southern Father-Son champion in 1968 with his son, William.

ROBERT M. BOYNTON (S1/c) joined the service July 1, 1943. Military stations and schools include: University of Illinois; Great Lakes Naval Training Center; Yeoman demobilization interviewer training, San Diego; Camp Shoemaker Discharge Center, CA; Toledo Discharge Center, OH; and Grosse Isle, MI. No Midshipman School, no battles and he never even saw a ship.

S1/c Boynton was discharged May 3, 1945, from Great Lakes. Today he is retired and doing baseball research. He is emeritus professor of psychology at the University of California at San Diego, where he is active in the Emeriti Association and represents the psychology department in the academic senate.

Boynton married Alice Neiley on April 9, 1947; they have four children and five grandchildren.

JAMES L. BRADLEY (LT) was born Dec. 1, 1921, in Clinton, AR. He joined the service June 20, 1942, and was stationed at U.S. Amphib. Base, Ft. Pierce, FL; AKA-66, South Hampton (1944-46) 16th Class; Staff, Commander Transport Div. 14 (1950-52); Midshipman School, Columbia

University. He participated in the Iwo Jima Invasion and Okinawa Invasion in WWII.

Memorable experiences include Iwo Jima Invasion; training for D-day landing; bombardment of island; transporting casualties back to ships; kamikaze air attacks on ships and the high cost in casualties for victory.

Bradley separated from the USN after WWII on May 28, 1946, and attended the University of Arkansas and received BS degree in 1947 and MS degree in vocational agriculture in 1948. He taught vocational agriculture in Pulaski County, AR from 1947-1950. While living in Little Rock, AR, he was in the Naval Reserve and called to active duty Oct. 13, 1950, for the Korean War. Separated from the service on Sept. 4, 1952; his medals/awards include the Asiatic-Pacific with two stars, Victory Medal and Victory Ribbon for WWII, Korean Service Medal and United Nations Service Medal for Korea.

Married Linda Lou Carrick of Little Rock, AR in June 1949. They have two children, Joyce Bradley Babin (an attorney in Little Rock) and Dale Carrick who still lives at home. Was employed as sales representative for Central Soya Co., Ft. Wayne, IN and worked in Mississippi, Kentucky, Alabama and Arkansas from October 1952 until March 1985. After retiring from Central Soya Co., he was a sales rep. for Mid-South Industrial Sales, Fort Smith, AR until May 1991. Now lives in Fort Smith, AR and is enjoying retirement.

RICHARD F. BRANCO (LTJG) was born June 11, 1926, in Holstein, IA. He enlisted in the USN in June 1944; attended NROTC, University of Notre Dame. LTJG Branco separated from the Navy in July 1946.

Branco practiced law for 30 years. He was appointed Iowa District Court Judge and served for seven years.

Married Gladys Cassells on Jan. 30, 1948, and they have two children. Richard Branco passed away in 1987.

ALFRED W. BRANDT JR. (ENS) was born July 23, 1922, in Gary, IN. He joined the USN in

September 1942. His military locations/stations/schools include Fort Pierce, FL; Yokosuka, Japan; Midshipman School, Notre Dame University, Class of Oct. 26, 1944; V-12 College, Indiana State University.

Brandt trained for Operation Olympic (Invasion of Japan), but thanks to the big bomb, there was no invasion. He visited Hiroshima and Nagasaki. Contracted TB and spent three and a half years taking the "cure." Ensign (DL) Brandt was discharged July 28, 1946.

He married Marilyn Nye; they have sons, David, Alfred and Garry, and daughter Barbara. He is president of Florida Maids on Call, Inc.

JEAN ROLAND BRAUWEILER (ENS) entered the V-12 Program at University of Notre Dame on July 1, 1943. He was commissioned ensign in February 1945 from USNROTC V-12 at Notre Dame. He served as officer in charge USS LCT-996 based at Saipan, Mariannas Islands. Primary activity was to off load bombs for B-29s from munitions ships anchored off Saipan and Tinian. He was released to inactive duty at Chicago in June 1946.

Brauweiler received BSME from Notre Dame in 1948 and MBA from University of Chicago in 1950. He was field engineer for Western Precipitation, 1950-54; engineer economist for Standard Oil Co. in Indiana, 1954-63; head, Operations Research for Kraft Foods, 1963-74; managing consultant for Northwest Industries, 1974-85.

Following retirement in 1985, he became independent consultant for Management Information Systems. He currently resides in Arlington Heights, IL with his wife, Joan. They are the parents of four daughters and five sons and grandparents of 19 children.

JAMES E. BREEN was born in Port Henry, NY on Nov. 18, 1922. He entered the USN on April 19, 1943; went to boot camp at Sampson, NY; Hospital Corps School, Portmouth, VA; and Navy Hospital Jacksonville, FL. He was accepted for V-12 Program, Mercer University and Louisiana Polytech, Princeton University Pre-midshipman School, Midshipman School Northwestern University and Advanced Officer's Training School, Miami, FL.

Served on the Kuala Gulf as deck and gunnery officer. He transferred to LST-1015, Shanghai, China and served as deck and communications officer. The 1015 was engaged in transporting Nationalist Chinese troops from Hong Kong and Shanghai to Chinwangtao in northern China.

Released from active duty on June 26, 1946,

and received BA (education), Louisiana Tech; MA (Ed. Admin.), NYS Teacher's College. He taught social studies in Crown Point, NY for eight years; and in Germany, Italy and France for six years with U.S. Army's Dependents School System. He returned to the States and taught in Farmingdale School System for 18 years.

Breem and his wife, Jean, have been retired since 1952 and reside in Port Henry, NY. They have three children: one daughter is a teacher, another is a nurse and their son is a lawyer. They have four grandchildren.

HERBERT M. BRIDGE (RADM) native of Seattle, WA, joined the Navy after graduating high school in June 1942 and entered V-12 from the fleet in October 1943. He received his commission at Columbia Midshipman School and was fighter director and finally navigator of the USS *Breton* (CVE-23) and served in the Pacific and in Asia.

Returned to school in 1946, received his BA degree and joined his family's jewelry business. Staying in the Reserve, he was recalled to Korea and commissioned TACRON 5, serving in forward air control and support of amphibious operations.

Remained in the Active Reserve and was promoted to rear admiral in 1976. He had numerous commands including the Naval Reserve Readiness Command and 45 ships of Military Sealift, Far East, headquarters in Yokohama, extending to the Arabian Gulf and the Indian Ocean. His decorations include the Legion of Merit with Gold Star in lieu of 2nd award.

Co-chairman of his family owned business, Ben Bridge Jeweler, now grown to 46 stores in six states, he has been past head of Seattle's Chamber (its Downtown Association) and is on numerous boards including the Naval Academy Foundation and the Washington Mutual Bank.

He and his wife, Shirley, married in January 1948; they have two sons, one a captain in the USNR and the other an ordained rabbi. They also have four grandchildren, all living in Seattle.

GARDNER R. BROWN (CDR) was born Nov. 3, 1927 in Holden/Sterling, MA. He enlisted in the USN in 1944 and entered the V-12 Program at Dartmouth College. He served with the 4th Mar. Div. from November 1944 through December 1945.

Served in combat marine service in the Pacific Theater, February 1945; entered the Submarine Force in 1946 serving on *Cubera, Amberjack* and *Harder*. Selected in 1953 as initial crew member for Nuclear Power Program, he entered Union College, Schenectady, NY, and received BS degree in 1955. He served at both STR and SIR Protypes and on *Nautilus* and *Seawolf*.

Then served in the Division of Naval Reac-

tors from 1957 to 1971 at the Schenectady, NY and Idaho Operations Offices and at Elective Boat Division, Groton, CT, performing prototype and initial reactor refuelings, chemical processing, dismantling, disposal and overhauls of nuclear reactor plants. Also worked new construction first on a class submarine reactor plant, test programs and sea trials. He left the service in January 1970. His awards/medals include the Asiatic-Pacific, Korea, Asiatic Occupation, Purple Heart, three Good Conduct and Expert Rifleman.

After naval retirement, he worked for 18 years in the commercial power generation industry for Northeast Utilities and Potomac Electric Power Co. He currently owns and operates his own engineering business, primarily with independent power producers in the geothermal and biomass power generation.

He and his wife, Sondra Gillice Brown, have three sons.

ROBERT D. BROWN (LTJG) was born Aug. 20, 1924, in Albert Lea, MN. He joined the USN in 1942 and completed the Navy V-12 Program at the University of Dubuque and Midshipman School at Cornell University.

Served aboard the PC-1588 at Iwo Jima. His most memorable experiences were typhoons, a determined depth charge attack on a sonar contact off Iwo Jima, and a middle of the night request to accept the surrender of a Japanese destroyer at sea between Iwo Jima and Saipan. The sonar contact proved to be a pinnacle on the bottom, and the destroyer was sent on to Saipan to turn itself in, since the war had ended.

Service aboard PGM-31 was in the spring of 1946 while the ship was en route to Bikini Atoll for atomic tests. He achieved the rank LTJG, USNR.

Brown and his wife, Patricia, have three children: Susan, Robert and Corey. He was awarded the Ph.D degree by the University of Minnesota in 1955 and retired as professor of industry and technology, Northern Illinois University in 1985. His achievements include authorship of six books and two years' service as a University of Chicago/Ford Foundation consultant in East Pakistan.

HOWARD S. BROWNE (CAPT) was born July 1, 1925, attended V-12 University of Oklahoma, July 1943; USNH, Norman, OK, 1944; Med School, Northwestern, October 1944 (RAD USNR December 1945), graduated 1949.

ACDU USNH, Bremerton, 1949; MSTS NORPAC, 1950; USNH San Diego, 1950-51; 1st MARDIV (3/11 & A Med), Korea, 1951-52 (Commendation Medal with V, Presidential Unit Citation and ROK PUC); MCRD San Diego, 1952-55; NAVFAC, London, 1955-57; USNH, Oakland, 1957-61; USNH, Newport, 1962-67, private practice ORS, Newport, RI, 1967-79. He achieved the rank of CAPT MC USN (RET).

1979- dir. courses for Harvard residency training faculty; med expert, social security hearings; independent med examiner. He is married and has four children. His hobbies are military history and uniforms, heraldry, computers and court tennis. He is commander of the Ven Order of St. John in the British Realm.

BERNARD S. BROWNING (RADM) enlisted in 1943 and entered V-12 Program at the University of Richmond. He was commissioned Nov. 1, 1944 and served in the Atlantic Fleet on USS *Mount Vernon* (AP-7) and the Pacific Fleet on USS *Lavaca* (APA-22). Was selected for graduate school earning an MBA at Harvard University in 1947. Subsequent duty stations included Naval Air Station, San Diego, Navy Ships Store Office, Brooklyn, CINCNELM Staff, London and BUSANDA Washington, DC.

Resigned from regular Navy in September 1954 (retired from Reserve in 1975) and devoted next 35 years to small business and franchising. He founded General Business Services Inc. in 1962 and served as CEO through 1985. He was inducted into Franchising Hall of Fame in 1983. Browning received Meritorious Service Award for service as chairman, National Advisory Council, Small Business Administration.

He and Adeline Rogers were married in 1955; they have four children: Frances Elaine, Virginia Diane, John Scott and Lawrence Rogers.

EDWARD AUGUST BRUNS, P.E. was born May 6, 1926, Floral Park, NY. He enlisted Aug. 25, 1944 in Electronics Technicians Program and attended school in Michigan City, IN; Del Monte, CA; Corpus Christi, TX; V-12 interview Aug. 6, 1945 (emerged to read A-bomb headlines).

In first V-12 class at Vanderbilt University, Nashville, TN, September 1945-June 1946; transferred (GI Bill) to Rensselaer Polytechnic Institute, Troy, NY; commissioned ensign in June 1948 (Holloway Plan); and graduated June 1949 as electrical engineer. Received ham and commercial radio licenses at RPI.

He worked at RCA, Sperry, Grumman and has been self-employed since 1969; treasurer, Green Mountain Club (hiking); founder, Park Bridge Civic Assoc; second place, 1994 Triathlon. The V-12 helped tremendously in developing leadership and self-reliance and in giving a brief glimpse into southern living.

RICHARD S. BULL JR. (S2/c) was born Jan. 21, 1926, in Chicago, IL and enlisted in November 1943. His military stations, ships and schools include Illinois State Normal and Columbia University; Great Lakes, Treasure Island, USS *Dorchester* and others.

Bull received all the usual medals and was discharged in June 1946. He attended Yale and received BA in economics, JD and LLM at NYU School of Law. Married Lois on July 19, 1950; they have five children and nine grandchildren. He retired in 1991 as president of a company in Chicago, IL.

JOHN E. BURDOIN (LCDR) entered the V-12 Program in July 1943. He transferred to the Naval ROTC, University of Washington in February 1944 and was commissioned in June 1945. Burdoin served in destroyers in Atlantic Fleet until released from active service in September 1946.

Recalled to active service in August 1950, he served in destroyer escorts in Pacific Fleet and participated in the evacuation of United Nations Forces from Chinnampo, December 5, and Hungnam, North Korea, Dec. 24, 1950. He was released from active service in September 1952.

He graduated from University of Minnesota, BOFEE, August 1947 and worked in construction of commercial power plants and steel mills for 10 years; 1958 was employed by Lawrence Livermore National Laboratory until retiring in July 1976; August 1976 was employed by U.S. Nuclear Regulatory Commission in Washington, DC and Walnut Creek, CA until retiring in June 1990.

He and his wife, Dorothy, live in Dublin, CA with five children and seven grandchildren, all living in the Greater San Francisco Bay Area.

HERBERT R. BURGETT (LTJG) was born in Des Moines, IA on Aug. 30, 1923. He enrolled in the V-12 Program while attending Drake University in 1942 and was assigned to the first V-12 class at St. Ambrose in July 1943. The school was small, lacking some facilities specified by the Navy such as a swimming pool on campus. In keeping with the Navy's aim to use the current curricular organization of colleges, several classes were taught by priests.

Commissioned Ensign USNR (D)L with the 19th class at Columbia University Midshipman School in August 1944, and was assigned to AO-63 USS *Chipola* serving as navigator and 3rd Division officer. The *Chipola* participated in the Iwo Jima and Okinawa campaigns, and refueled the carrier attack forces during their raids on the Japanese home islands. After Japan surrendered, the *Chipola* served as station tanker in Tokyo Harbor and made a run to the Persian Gulf for petroleum products before LTJG Burgett was discharged in July 1946.

Completed his studies in education at Drake, and was employed as a sales representative in Des Moines until both his son, Steve, and daughter, Cindy, were through college and married. Since 1976 he and his wife, Marie, have lived in Sarasota, FL where he was employed in management by a major department store chain until retirement in 1988.

EDWARD W. BURKE JR. (ENS) entered the Navy V-12 Program at Newberry College on July 1, 1943. Attended Notre Dame Midshipman School, June 1944 and commissioned ensign in October 1944. Attended MTB (PT Boat) School, Melville, RI in November 1945, joined MTB RON 17, Philippines in January 1945 and decommissioned PT boats at war's end. Assigned to minesweeper AM-316 in Japan in January 1946 and discharged to Naval Reserve in May 1946.

Graduated from Presbyterian College in June 1947; received Ph.D in physics from University of Wisconsin in 1954. Was professor of physics at King College from 1949 to present. Fullbright Professor, University of Chile, Satiago, Chile, calendar year 1959. Received Pegram Award (outstanding physics professor, southeastern U.S.) in 1973. Astronomy research on variable stars.

Retired in 1991 and currently resides in Bristol, TN with his wife, Julia S. They are the parents of a son, Edward, employed by Lockheed, Colorado Springs, CO and daughter, Julia Burke Torbert, employed by Bell South in Atlanta, GA.

G.L.D. BURNETT (ENS) was born Feb. 13, 1925 in Lynchburg, VA. He enlisted in the Navy Jan. 4, 1943, and attended V-12 Program, University of Rochester. Other stations include Jacksonville, FL; Yellow Water, FL; Cecil Field, FL.

His memorable experiences include: flying, SBD and PBYs and sports and education at University of Rochester. Discharged Oct. 26, 1946, his awards include the Good Conduct, Victory Medal and American Theater.

Married Jean Gibbon and they have four children and nine grandchildren. Received DDS from Medical College of Virginia and still practicing dentistry in Rochester, NY.

JAMES VIRGIL BURNETT (CAPT) joined the Navy in 1943 and entered the V-12 Program at Southwestern Louisiana Institute at Lafayette. After completing his pre-dental requirements, he was sent to Naval Hospital in New Orleans and received his DDS degree in 1948 from Baylor Dental College in Dallas.

In 1950 he was recalled as a dental officer for two years during the Korean War and was stationed in San Diego, CA; Orange, TX; and White Sands Proving Ground, NM. Returned to private practice in Fort Worth, staying in the Naval Reserve and retired after 34 years as captain.

Took additional educational courses at TCU, Baylor, Texas University, Louisiana State University and University of Tennessee; was active in the dental profession and went through the chairs of the Fort Worth District Dental Society; and was honored in 1982 by being named Fort Worth Dentist of the Year and in 1983 named Texas Dentist of the Year. He has received numerous other civilian honors and awards.

Married Patti Joyce Dean and they have four children: Jim Jr.(wife Stephanie), DeAn, Patrick (wife Julie) and Michael; and six grandchildren: Jamie, Amelia Grace, Matthew, David, Christopher and Ann Marie. He and Pattie traveled a great deal and attended many NATO Reserve Officer Assoc. meetings in different NATO countries. Patti developed cancer and passed away in 1986. Since then, he has married a lovely lady, Janie Read Gunn. Janie has a son, Bill, married to Maryann and they have two children Noble and Molly.

MELDRIM F. BURRILL (LCDR) was born in Mattawamkeag, ME, but claims California as his home state. Commanding officer, Navy V-12 Program; Illinois State University, Normal, IL, July 1, 1943-45. "The Captain" opened the unit after coming from the USNR Midshipmen's School at Columbia University.

"ISNU did an outstanding job! The president, instructors and regents not only lived up to the Navy V-12 contracts, but went above and beyond in their individual efforts to contribute to the success of the program."

LCDR Burrill, USNR, lives in Irvine (Turtle Rock), CA. He participated in the Navy V-12 Colloguim in 1989 in Washington, DC and also in the 1993 Navy V-12 50th anniversary celebration in Norfolk, VA.

As a civilian, he devoted 42 years to the field of education in the Los Angeles Unified School District; served 32 years as director of California Federal Saving and Loan (5th largest in nation); and as director of Networks Electronic Corp. He has one daughter, Maralou Harrington, four grandchildren and seven great-grandchildren.

ALBERT JACKSON BUSH (CAPT) was born Jan. 22, 1923 in Irwin County, GA. Joined the Marines in December 1942. His military locations and stations include: Quantico, VA (1945); Paris Island; Camp Lejeune; Duke University, V-12; Camp Pendleton; Camp Beaumont, Ohau, 1st Marine Div.; USS *Oglethorpe* (1952-53) as

combat cargo officer; and Tientsin, China.

His memorable experiences include: when promoted to captain; repatriating Japanese soldiers back to Japan; playing for Fleet Marine Force All Stars (1945); coached and played

78

Tientsin area championship basketball team (1945-46).

Discharged first time in August 1946 and second time in April 1953. Bush is in the auto business; he is married and has three children and five grandchildren.

ALLEN W. BUSH was born in Kalamazoo, MI in 1923. He enlisted in the USMC in January 1942 and served on FunaFuti in the Pacific with the 5th Defense Bn. He was assigned to V-12 at Western Michigan in July 1944 and later to Princeton in November 1944.

Discharged in June 1946 and commissioned the next day. Recalled to active duty from February 1951 to May 1952.

Received BS from Western Michigan and MA from Michigan. He taught and coached at several high schools before joining the Michigan High School Athletic Association from which he retired as the executive director in 1978.

Bush and his wife, Lois, live at a lake near Kalkaska, MI. They have two sons, the older is an attorney and the younger works for the U.S. Government and is a lieutenant colonel in the USMCR.

MALCOLM S. BYERS, M.D. (LCDR) was born April 5, 1923 in Nickerson, NE. He enlisted in the service in March 1942; attended V-12 Program at University of North Carolina; Naval Convalescent Hospital, Glenwood Springs, CO; and College of Medicine, University of Nebraska.

Participated in the Korean Conflict as senior medical officer in the USNS David C. Shanks. He will always remember his shipmates on the old Shanks.

Discharged from active duty Nov. 2, 1952. Married June P. Peters, M.D. in 1949. After 42 years of family practice, he is now retired and enjoying life

JAMES J. CALLAHAN III (LT) graduated from Holy Cross, Harvard and Columbia Midshipman School. He was born in Worcester, MA; enlisted in 1943 and commissioned in 1945. He served in the Pacific Theater on Kitkun Bay and Bunker Hill aircraft carriers. He was released to inactive duty in 1946, completed his Harvard education and joined the Trane Co. as an air conditioning engineer.

Recalled to active duty in 1951, he was assigned to Tactical Air Control Sqdn. 6 during the Korean War and released to inactive duty in 1953. He continued part-time Reserve duty until released to Retired Reserve in 1962.

Callahan joined the Bell System in 1957

where he held engineering management and executive level positions with Western Electric, AT&T and Bell Communication Research. He retired in 1987 and currently is a consulting engineer and arbitrator for construction disputes. A licensed engineer in New Jersey and other states he received the Outstanding Leadership Award, American Society of Mechanical Engineers, Real Property Administrator, *Who's Who in Real Estate,* and is author of engineering texts and papers.

His community activities include serving on the local school board, United Fund and various volunteer organizations. He is a member of the Telephone Pioneers, has been an avid golfer for many years and enjoys traveling, painting, wood working and magic entertainment for children.

He lives in Chatham, NJ with his wife, Helen. They have three sons: James (golf professional), Kevin (engineer) and Brian (a chef).

ARTHUR F. CALNAN (FLT SURGEON) entered the Navy V-12 Program on March 1, 1944, at Tufts. Discharged Nov. 1, 1945, after studying pre-med, he went to Medical School in Army ROTC. After internship, he spent two years as a flight surgeon in the USAF during the Korean War. He has earned the distinction of serving in three branches of service.

Calnan specialized in ophthalmology and practiced in Boston and its South Shore. He had the honor of starting the Ophthalmology Dept. at the Lahey Clinic and being chairman for many years.

He is fading part-time into the twilight of his professional career, but has no thought of immediate retirement. He has a biography in *Who's Who in the East* and realizes he is in deep debt for what the three services have given him to achieve what he has.

EUGENE V. CALVELLI, M.D. (1/LT) was born Oct. 9, 1923, in Port Washington. He enlisted in V-1, Dec. 2, 1942; active duty in V-12 from July 1, 1943-Dec. 22, 1945, at Dartmouth College; USNH Great Lakes, IL; and New York University College of Medicine. He achieved the rank ensign HP.

Enlisted in the USAR on Dec. 19, 1950, and was recalled to active duty on July 12, 1951; Med. Fld. Svc. Sch., Ft. Sam Houston, TX (July-August 1951); USAH Ft. McClellan, AL (August 1951-August 1952); POW Hospital, Koje-do, Korea (September-October 1952); 3138 AU (Hosp. Trains), 8th Army, Korea (October 1952-July 1953), commanding officer, May and June.

He was discharged in July 1953 as first lieutenant. His medals include the WWII Victory,

American Theater, Korean Service with three BS and United Nations Service Medal.

Earned his AB at Dartmouth (1944); MD at NYU College of Medicine (1947); and served his residency in medicine at Bellevue Hospital, NY, NY (1947-51). He was in the private practice of internal medicine in Port Washington, NY (1953-87) before retiring to Orlando, FL in August 1987.

Married Harriet Heffernan in 1950 (she passed away in 1981); they had six children. Calvelli married Jacqueline Loughe Burke in 1982 and has two step-children.

WILLIAM C. CALVERT (COMM OFFICER) was born in LaVerne, CA in 1942. He enlisted in the Naval Reserve in 1942 and entered the V-12 Program at the University of Redlands in California in 1943.

Graduating from the Midshipman School at Columbia University, he attended the Officers Communication School at Harvard University and then served as a communications officer on the USS *Steamer Bay* (CVE-87), receiving the Asiatic-Pacific Campaign Ribbon with two Bronze Stars.

Earned HIS BA, MA and Ph.D degrees (Pomona College, Stanford, UCLA). Calvert has two daughters. He is a retired psychology professor and lives with his wife, Barbara, in New Jersey.

JAMES VERNON CAMP (PFC) was born Aug. 8, 1924 in Union, SC. He enlisted March 13, 1943, in the USMC after a football and basketball season as a freshman at college. He attended boot camp at Parris Island; Sea School at Portsmouth, VA; boarded the USS *Cabot* to Hawaii in the fall of 1943; transferred back to the States in January 1944, and

went to V-12 at University of Carolina, where he was the starting HB on their football team.

In the Spring of 1945, he was sent to Camp Lejeune to await transfer to OCS at Quantico. Discharged in September 1945, he went back to UNC Chapel Hill and was starting HB in 1945, 46 and 47. In 1946 they played Georgia in the Sugar Bowl. He graduated in 1947 and received his Masters in 1948.

Played football as a grad student and one year of professional football with the Brooklyn Dodgers. He was a football coach for 21 years: 4 years at UNC (Cotton Bowl in 1950); one year at Mississippi State; 7 years at Minnesota (Nat. Champs in 1950 and Rose Bowl in 1961); 6 years as head coach at George Washington; and 3 years at UCLA.

Joined the Chamber of Commerce, Durham, NC (1970) as industrial recruiter and retired in 1990. He and his wife, Carol, have four children.

WILLIAM W. CANNON (LT) was born in Salt Lake City, UT on April 10, 1925. He entered the Illinois Institute of Technology V-12 unit on July 1, 1943. He was hit by a taxicab while walking

to the physics lab in August 1943 and sat out most of two semesters, recovering from serious injuries, in Wesley Memorial Hospital (sick bay, Navy Pier, Chicago). He received his BS degree in civil engineering and was commissioned ensign in February 1946.

Because of no openings in the Civil Engineer Corps at time of his graduation, he had a choice of Supply Corps or Line and he chose Supply Corps. He entered the last class of Naval Supply Corps School and he is a graduate of the School of Business, Harvard University. Completed school and he was separated to inactive duty in June 1946.

Cannon was in the Naval Reserve for 13 years and retired as lieutenant. He received his MS degree from University of Utah in 1950 and stayed there as instructor in engineering mechanics for three years. He was president of Salt Lake Stamp Co. (1952-85); division manager, Zions Securities Corp. in Utah and Hawaii (1985-94); served on Salt Lake School Dist. and state of Utah Boards of Education; LDS Mission President in Hawaii/Micronesia and retired in 1994.

He and his wife, Margery, have six sons, two daughters, and 15 grandchildren. They reside in Salt Lake City.

HERBERT M. CANTER (CAPT) was born July 13, 1925, Syracuse, NY. He entered the V-12 Program at Cornell University on July 1, 1943, after having attended Syracuse University for one year. He was commissioned at Columbia University Midshipmen's School on Nov. 2, 1945, and served on the USS *E.C. Stanton* (AP-69) and in BuDocks, Arlington, VA, until released from active duty in June 1946.

Returned to Cornell University and he received his BS in EE in June of 1947. Operated a family-owned theater equipment business, theaters and drive-in-theater, until October 1950 when he was recalled to active duty for the Korean Conflict. He served on USS *Piedmont* (AD-17) and attended the School of Justice at Newport, RI. Canter was released from active duty in August 1952, attended Syracuse University College of Law and received his JD Magna Cum Laude in June 1955.

Served in the Dept. of Justice, Washington DC, under Attorney General's Honor Recruitment Program, 1955-56; returned to Syracuse for private practice with various firms, 1956-87; was specialist in Transportation Regulatory Law, representing trucking and bus companies in practice before the Interstate Commerce Commission in Washington and the New York Dept. of Transportation in Albany, NY.

Served in Reserves as CO, NR MOB TEAM 3-5(S), Group Cdr. NR Group Cmd. 3-25 and as mobilization augmentee in the Office of Deputy Assistant Secretary of Defense for Reserve Affairs in the Pentagon. He retired from the Naval Reserve as captain in July 1985. He served as trustee and vice-president, Temple Society of Concord, Syracuse, NY, 1938-88 and currently maintains law offices in Fayetteville, NY.

Resides in Fayetteville with his wife, Terry. He has three sons and two daughters, all of whom are grown, and one granddaughter.

WILLIAM I. CARGO (LTC) entered the V-12 Program directly from high school. He enlisted in the Navy in April 1944, graduated high school in June and reported to Purdue University on July 1. On Nov. 1, 1944, he transferred to Wabash until Nov. 1, 1945, when the V-12 unit was decommissioned and they were transferred to the University of Illinois.

Discharged at Great Lakes on June 26, 1946, he enrolled at the Ohio State University and also in NROTC-Marine option. In June 1948, he graduated with a BS in civil engineering and a minor in naval science. He was the first Marine option graduate of the Ohio State NROTC. He was commissioned a second lieutenant in the USMCR in October, after a 6 week course at Quantico, during the summer of 1948. He started work for the Ohio Dept. of Highways in June and received military leave to go to Quantico.

Started with a consulting engineering firm in 1950 and was called to active duty in October 1950 in the 1st Special Basic Class at Quantico; then assigned to the 2nd Engr. Bn., 2nd Mar. Div. at Camp Lejuene and filled billet of battalion engineer officer.

Returned to his previous employment in April 1952 and started a new firm. He was president of that firm from 1960-90 when he retired. Still remains active as the Ross County bridge engineer and also remained active in the USMCR. He was CO of the 75th Rifle Co. in Portsmouth, OH from 1959-61; became a member of a VTU in Columbus until transferred to the Retired Reserve in July 1975.

Married to Jessie Ruth and they have two sons, Stephen and William, and two daughters, Sharon E. Cavey and Theresa E. Cargo.

MARK D. CARLS was born in Nuremberg, PA on Dec. 10, 1925. He entered Penn State College V-12 Program on Nov. 1, 1943; was selected as a candidate for electrician engineer degree after two semesters; transferred to Columbia University on

July 1, 1944 (didn't complete course); and served aboard the USS *Horner* in the Asiatic-Pacific Theater.

After leaving the Navy, he graduated from Penn State College with BS degrees in both science and pre-med. He was employed by Wyeth Inc. for one year in pharmaceutical sales and by E.R. Squibb for 40 years in sales and clinical research. He served as Baltimore Division sales manager and as clinical research manager, and served 10 years as treasurer of Lutherville Recreation Council. He retired from Bristol Myers-Squibb in January 1991.

Carls sailed aboard the USS *Wasp* as a tiger plankowner in December 1991; qualified as Merchant Marine officer in June 1994 as master of inland waters, USCG, and is currently planning historical cruises on the Chesapeake Bay.

GILBERT M. CARPENTER (LTJG) was born Nov. 13, 1925, in Birmingham, AL, and was called to active duty July 1, 1943. His duties, stations and schools included: V-12 Howard (Samford) Tulane University; Bayonee, NJ; Naval Supply Corps School and was Navy Supply Corps Paymaster.

Discharged in 1945, he graduated from the University of Alabama with BSC in 1946, and worked in the profession of investments and real estate. Married over 46 years, he has three children and seven grandchildren.

STANLEY H. CARPTENTER (COL) was born Jan. 21, 1926, in Hattiesburg, MS and enlisted in the USNR V-6 in January 1944. After boot camp he transferred to V-12 Unit at Tulane; was NROTC battalion commander; member of Kappa Sigma, Phi Beta Kappa, Omicron Delta Kappa; awarded a BS in math and made 2nd lieutenant, USMC in June 1946.

Basic School (3rd BSC), followed by 8th Marines tour and sea duty as CO, MD, USS *Breckinridge* (AP-176), then 1st Mar. Div. as amtrac platoon leader. He was wave commander of 1st Wave ashore Beach Blue II, Inchon (Sept. 15, 1950) and 1st Wave of Han River crossing (Sept. 20, 1950).

Designated naval aviator on Oct. 31, 1952, and returned to Korea with VMA-251 in June 1953. He joined VMO-6 in January 1954, then USNPGS in July 1954 and awarded a BS in engineering in 1956 and aeronautical engineer (AeE) from CalTec in 1957. At Edenton, NC he was XO, maintenance officer of H&MS-14 before closing the base and going to NOTS, China Lake as MC Liaison Officer (January 1956-December 1961).

He made Far East tour as XO/CO of MABS-12 (Iwakuni, Philippines, Thailand) followed by assignment as CO, VMA-331 (May 1963-April 1965). Squadron's A4Es flew over 11,000 accident-free hours, including a 10-month tour aboard USS *Forrestal* (CVA-59). He won CNO Safety Award (FY '65); was Marine aide to assistant secretary (R&D) June 1965-June 1968, then to Vietnam as OIC, Chu Lai AB and CO, MWSG-17.

80

Assigned to staff, Director of Defense Research and Engineering, September 1969-June 1971, then as Chief Air Ops. Div. and Asst. Chief of Staff (Firepower and Manpower) of Development Center, Quantico.

Retired as colonel in August 1974 and worked as aero engineer/safety consultant, picking up MS (Safety Mgt.) from USC in 1982. He was named executive director, Marine Corps Avn. Assn., January 1988 and served until March 1993.

EARL H. CARROLL (LTJG) enlisted June 30, 1943, in the USNR Supply Corps; ASTC, UCLA, and Harvard. Duty as supply and disbursing officer on (APA-192).

Carroll was discharged on Aug. 23, 1946. His medals and awards include the American Area and Victory Medal. He is a U.S. District Judge.

JOHN FREDERICK CATLETT (SN) was born March 24, 1926, in Romney, WV. He enlisted Feb. 12, 1944 in the USNR. Stations, ships, and schools include: Bethany College, Franklin and Marshall Colleges, Princeton University; and USS *Saugus*.

Discharged on Sept. 14, 1946, his awards and medals include the American Theater and Victory Medal. He graduated from West Virginia University in 1950.

Married, Catlett has three children, eight grandchildren and four great-grandchildren. He is retired from General Telephone Co. of Florida.

ARMANTE J. CERRO (LTJG) was born Oct. 26, 1924, in Boonville, NY. He was assigned to Union College, Schenectady, NY, V-12 unit in November 1943. Commissioned ensign in March 1945 at Fort Schuyler Midshipman School, NYC. His first assignment was at USNTC Miami, FL for antisubmarine training, amd he completed the course in May 1945.

Overseas duty was spent aboard YMS-463, minesweeping in the Pacific, mostly in the Marshalls and Marianas.

Postwar, his ship was the first ship assigned to Bikini, minesweeping the atoll and making preparations for "Operation Crossroads," the Bikini bomb test. He served as executive officer (LTJG) until May 1946; then returned stateside for release from active duty.

Graduated from Albany College of Pharmacy in 1948; served as President of Hess Pharmacy Inc. until February 1987 when the business was sold to Fay's Inc.; and served as trustee of the Pharmacy College for nine years.

Currently residing in Boonville, NY with his wife, Lucille; they are the parents of three married daughters: Janice is a teacher, Kathleen is a therapist and Patricia is an administrator with an environmental association. They also have one granddaughter.

JOHN LIVINGSTON CHAFIN was born Jan. 31, 1924, in Chatham, VA. He is a graduate of Hargrave Military Academy and attended the

University of Cincinnati and the University of Louisville V-12 Program.

Completing the V-12 Program, he received a commission at Camp Endicott, RI as a heavy equipment officer in the 75th, 93rd and 143rd Naval Construction Bns. and served in the Philippine Islands.

After his discharge in June 1946, he returned to the University of Louisville and received a BCE degree. He worked with Louisville & Nashville Railroad, Maintenance of Way Dept. (1947-63) and with Rohm & Haas Chemical Co. as project engineer (1963-1975).

John Livingston Chafin passed away and is survived by his wife, Martha Ellen Paulsen Chafin.

GEORGE L. CHALMERS (LT) enlisted in the Naval Reserve in December 1942; was ordered to the V-12 Program, Rensselaer Polytechnic Institute as a sophomore in July 1943 and graduated with a BS in aeronautical engineering in 1945. He attended Midshipman School at Notre Dame and was commissioned in November 1945.

His tour of duty was in Maui, Hawaii until inactive duty in July 1946. He maintained his status via correspondence courses until he was 3[?] years old, then accepted permanent commission a lieutenant, USNR, Inactive.

From 1946-67, he was employed at a wholesale hardware business; 1961-67, manager of same; 1967-85, was self-employed as a door and hardware specialist, sold the business in 1985 and retired in 1986. He is past chairman of the Bangor Republican City Committee; past director of YMCA; Masonic bodies; Paul Harris Fellow Rotary Club; and in 1992, he created a paper and forwarded it to U.S. Senator Chohen, that promoted a theory as to how magnetic hydromatics forces could be used to develop a space ship.

He and wife, Rita, have a son, daughter and two grandchildren.

CARL CHANCE (CDR) was born in Earl County, GA on Jan. 27, 1926. He entered the Navy V-12 Program at Milligan College, July 1, 1943; attended the University of the South, Feb. 1, 1944; USNRPMS Asbury Park; and was commissioned ensign at Columbia University in April 1945.

Transferred to the Regular Navy in 1946. His duty stations included: SCTC Miami; YP-62; USS YP-619; atomic bomb tests, Bikini Atoll, 1946 aboard USS *Clamp* (ARS-33); USS *Orteck* (DD 886); Navy Supply Corps School, Bayonne; USS *Fresno* (CL-121); HU-2, ZP-2, Lakehurst; fleet activities, Sasebo, Japan (Korean War); Navy Dept (BUSANDA); AEC Bettis Plant, Pittsburgh (Rickover Naval Nuclear Power Program); USS *Enterprise* (CVAN-65) and the naval shipyard in

Philadelphia. He retired Oct. 1, 1965 as commander, SC, USN.

Graduated from the University of Pennsylvania with ABA in 1966; BA in economics in 1969; from Glassboro State College with MA in 1980; and advanced graduate studies at Virginia Tech. He was assistant business manager, Community College of Philadelphia; business manager, dean of business; and professor at Gloucester County College, Sewell, NJ. He was promoted to full professor in September 1983 and retired as professor emeritus on July 1, 1989.

Since retirement, he handcrafts strip canoes and wherries for friends and is a tax counselor and instructor in IRS/AARP tax aide and TCE Program. He resides in Turnersville, NJ with his wife, Dorothy L. They are parents of a daughter, Melissa (flight attendant with U.S. Air), and son, Raymond (flight engineer and FAA certified A&P mechanic).

ROBERT FOSTER CHAPMAN

was born April 24, 1926. He entered the V-12 Program on July 1, 1943, at Emory and Henry College; transferred to NROTC at the University of South Carolina in March 1944, where he graduated with a BS degree and was commissioned in 1945. Assigned to Apra Harbor, Guam, he commanded YW-92 and took it from Guam to Operation Crossroads at Bikini Atoll in 1946.

His memorable experience was commanding the YW-92 and being referred to as the "Old Man" when he was just 19 years old and not yet shaving.

Chapman returned to the University of South Carolina Law School for a LLB in 1949 and an honorary LLD in 1986. He practiced law in Spartanburg, SC (1949-51); was recalled to active duty (1951-53); and served on staff of COMNAVMARIANAS. He returned to his law practice (1953-1971); was chairman of South Carolina Republican Party (1961-1963); appointed U.S. District Judge by President Nixon in June 1971; appointed U.S. Circuit Judge for the 4th Circuit Court of Appeals by President Reagan in September 1981; and was given the Patriots Award by the Congressional Medal of Honor Society in 1985.

Chapman is married to Mary Winston Gwathmey and they have three sons.

MELVIN L. CHEESMAN

(CAPT) was born in Lakin, KS on Sept. 14, 1924. He enlisted in the USMC in 1943 and was admitted to the Dartmouth V-12 in 1945 from the Cherry Point, NC Marine Air Station where he served as a technical sergeant. He served in almost all the USMC stations located in the States.

Discharged in 1946, he graduated from Dartmouth in 1948 and received his commission. In 1950 he was recalled to the Korean Conflict where he served with the 1st Mar. Div. as a 1st lieutenant and platoon leader. He was awarded the Combat Bronze Star Medal.

Returned to civilian life in 1952 and was discharged as a captain in the USMCR in 1958. He was an educator all his life and returned to college to earn two master degrees and a Ph.D. He is a single man and at the time of this publication and is living in retirement in Olympia, WA.

EDWARD CHEPPA

(LTJG) was born July 4, 1925, in Freeland, PA. He entered the V-12 Program directly from Freeland High School. On Nov. 1, 1943, he started his first two terms at the University of Louisville, transferred to the University of Virginia for the remaining six terms, from where he graduated June 4, 1946, with a bachelor of electrical engineering and at the same time received his commission as ensign, USNR.

Cheppa stayed on active sea duty for a year after graduation. His first brief assignment was on the USS *Mt. Olympus*. In September 1946 he was transferred to the USS *Bowditch* at Bikini. In January 1947 he was assigned to the USS *Crescent City* (APA-21) that had a Caribbean run, operating from Bayonne, NJ. He was released to inactive duty in June 1947 and continued in the Standby Inactive Reserve for over 20 years.

After returning to civilian life in 1947, Cheppa worked as a control systems design engineer (primarily for electrical power plants) for Westinghouse Electric Corp., Pittsburgh, PA, from where he retired in 1990 after 43 years. During his later years with Westinghouse, he worked on assignments for Naval Sea Systems. He has an MS degree from the University of Pittsburgh, is a licensed professional engineer, a senior life member of the IEEE, and a member of the American Society of Naval Engineers.

He and his wife, Anne, live in Pittsburgh, PA. They have three children (all University of Pittsburgh graduates) and three grandchildren. They enjoy traveling and he is again taking up golf. In November 1993 he and Ann attended the Navy V-12 50th Anniversary Reunion at Norfolk, VA.

ALEXANDER BUCHAN CHURCHILL (ALEX)

(ENS) was born in Berlin, NH in June 1925. He entered the Dartmouth College V-12 Program in July 1943; transferred to Bucknell University V-12 in March 1944; was commissioned ensign and received a BS degree in electrical engineering in February 1946.

Released to inactive duty in June 1946 in Boston, he was recalled in December 1951 and assigned to the USS *Laffey* (DD-724). He served as electronics officer and later as engineering officer. The *Laffey* was part of Carrier Task Force 77, east coast of Korea, and participated in the blockade of Wonsan Harbor, during which *Laffey* was in the "longest ship-to-shore gunnery engagement of the war." He returned to the USA (Norfolk via Suez Canal) to inactive duty in Philadelphia in September 1953. He resigned from the USNR in June 1956.

Earned his MS in electrical engineering from Carnegie Mellon University in 1948; and his MS in computer science from Washington University, St. Louis in 1978. His career was primarily in aerospace engineering at Westinghouse, ITT Labs, General Dynamics and McDonnell Douglas. He finished his career as a program manager for NASA Program, Astronomy Dept., University of Arizona.

Churchill retired in 1990 and lives in Tucson, AZ. He is married and has two children and five grandchildren.

WAYLAND D. CLANCY (OKIE)

(LCDR) was born in Carter, OK. He entered the V-12 Program in 1943 and attended the University of Illinois and DePauw University. He was commissioned ensign at Ft. Schuyler, New York Midshipman School in 1945. He served in the USS *South Dakota* (BB-57) in the Pacific, was released in 1946 and recalled in 1951. He served aboard the USS *Corregidor* (CVE-58), operating under MSTS in the Atlantic and Pacific until released in 1952.

Graduated from Oklahoma University in 1955 with a BS in electronics engineering. He was employed with the General Electric Co. in Syracuse, NY; Oklahoma City, OK; and Johnson Space Center in Houston, TX until 1971. He was employed with the Federal Aviation Administration in Oklahoma City, OK and Albuquerque, NM until retirement in 1986. He served in the active Naval Reserve until retirement as lieutenant commander in 1985.

With his wife, Doris, they own and operate their own business, Visi-Comm Systems, in Albuquerque, engaged in sales and service of visual and tactile communications and other assist devices for the deaf and hard of hearing.

WILLIAM J. CLARK

entered the V-12 Program on July 1, 1944, and was discharged on July 20, 1946. He earned his BA at Baldwin-Wallace College; MA at Western Reserve University; and Ph.D. at Ohio State University. Clark's 38 years as

an educator included roles in teaching, coaching and administration.

From 1948 to 1963 he served Ohio high schools in Lorain, Painesville, Cleveland, and Bay Village, teaching science and mathematics and coaching basketball, football and baseball. Clark was awarded a National Science Foundation Fellowship Award and was elected president of the Cuyahoga County Basketball Coaches Association.

His college tenure included four years at Mansfield (PA) State College and 10 years at Baldwin-Wallace. Clark's basketball teams earned two trips to the NAIA National Tournament and were OAC finalists twice. He served as administrative assistant to the president, public relations director, HPE division chairperson, assistant athletic director, and associate professor. He was elected president of the Ohio Athletic Conference Basketball Coaches, chosen on the UPI Small College Basketball Rating Board, and was consultant to the Ohio Dept. of Education.

In 1977 Clark became principal of Ashtabula (OH) Harbor High School. The school earned a state-awarded Certificate of Educational Excellence, and served on the State Athletic Committee of the OASSA as president of the Northeastern Athletic Conference and the Ashtabula County Touchdown Club, and as vice-president of the Ashtabula Area Middle Management Assn. He retired from education in 1986.

Clark resides in Pagosa Springs, CO, with his wife, Glenda, also a retired educator. They have a son, Jim, in Walton Hills, OH, and a daughter, Chery, in Key West, FL. Dr. Clark spent five years as a manufacturers representative for Colton Technology in the nuclear industry and has continued active as chairman of the Pagosa Fire Protection District Board and vice-president of the Archuleta Water Company Board.

EDWARD N. CLARKE enlisted in the USN and entered the V-12 Program at Brown in 1943. He served as a public works officer at the 7th Fleet Hospital on Samar in the Philippines, and as a deck officer on the attack transport, USS *Arthur Middleton* (APA-25). After naval service, he attended graduate school at both Harvard and Brown, receiving his Ph.D. in physics from Brown in 1951.

Clarke is one of the founders of the United States semiconductor industry and is co-founder of National Semiconductor Corp. He manufactured semiconductor components for the nation's military and space programs, including NATO's F-104 fighter plane, the Minuteman ICBM and the Apollo Moon Landing Program.

Entered his third career at age 62 as professor of engineering and science at Worcester Polytechnic Institute in Massachusetts, leading students in projects concerning the applications of solar energy, including the design, construction, and racing of solar-powered racing cars. He is a fitness buff, largely through distance swimming and downhill skiing.

KENNETH V. CLEWETT was born in Pomona, CA on June 3, 1923. He entered Redlands University V-12 Program on July 1, 1943, and was commissioned ensign in June 1944 at Columbia Midshipman School, NYC. He attended amphibious training at Coronado, CA in 1944 before serving in the Pacific Theater as deck officer on USS *Kenton* and LSM-492 in Philippine and Okinawa engagements.

Released to inactive duty in 1946, he earned his AB from Stanford University in 1947, and subsequently was in California State service for over 30 years, retiring as executive director of Patton Hospital in 1978. He was vice-president (administration) at Southern California College, 1978-84; vice-president (administration) and has been director of Public Relations at the University of the Nations in Hawaii since 1985.

Resides in Kailua-Kona, HI with his wife, Margery. They are parents of three sons, one daughter and grandparents of seven girls and three boys. Their sons live in Austria, Spain and California and their daughter lives in California.

JOHN KEELY CLIFFORD (LTJG) enlisted in the USN on Dec. 17, 1942. He entered the V-12 Program at Yale University and served as deck gunnery officer aboard the USS *Chestatee* (AOG-49) in the Asiatic-Pacific Theater and Philippine Liberation.

He completed his degree from Yale University, Class of 1948, and graduated from Georgetown Law School in 1951. He then practiced law in Washington, DC until becoming involved in real estate development in Annapolis, MD. He was Annapolis' first urban renewal director and also was a driving force in many marine-related businesses in Annapolis.

Clifford recently completed a journey to his hometown of Rochester, NY via the Erie Barge Canal aboard his 65' Feadship motor vessel *Fifty/Fifty*.

He and his wife, Tara, live on the South River just outside of Annapolis. His daughter, Darby, lives in Warsaw, Poland with her husband, Jarl, and their children, Philip and Maria. His son, John, lives in Annapolis with his wife, Eileen, and son, Parker. His youngest daughter, Keely, is an Army Reserve officer and lives in Annapolis.

JAY PHAON CLYMER JR. (LTSG) was born in Fleetwood, PA on July 31, 1923. He entered Tufts College V-12 unit on July 1, 1943 and graduated in October 1944. He attended Miami Steam Engineering School and was awarded a BS in mechanical engineering from Penn State

College. He attended Cornell Midshipman School and was commissioned ensign in February 1945. He was assigned to PC-485 performing air-sea rescue duty from Pearl Harbor.

His memorable experience was being the first naval vessel to enter the harbor, Island of Lanai, Dec. 7, 1945, as guests of the Dole Pineapple Corp. and citizens of Lanai. Also memorable was the 45 foot tidal waves in the Fall of 1945 that hit the Hawaiian Islands; Naval Air located victims and Clymer and crew of PC-485 rescued them from the sea.

In July 1946, the PC-485 was decommissioned in Portland, OR and he was released from active duty. He joined the organized Naval Reserve, Lancaster, PA and Tucson, AZ. He was employed by Bell of Pennsylvania (now Bell Atlantic) for 38 years and became a division manager of General Services, Pennsylvania and Delaware. He retired in February 1985.

He resides with his wife, Jeanette, in Wayne, PA; they have one son, two daughters and seven grandchildren. Clymer enjoys traveling, fishing, boating and gardening for recreation. His father was LTJG Jay P. Clymer, aboard the battleship *Oklahoma* during WWI.

RICHARD W. COAKLEY was born in Havre de Grace, MD on Feb. 28, 1926. He enlisted in the Navy in 1943 and entered the V-12 program at Franklin and Marshall College. He was commissioned and ensign, USNR, at Northwestern University (Tower Hall) in 1945. He attended Advanced Line School in Miami and served as CIC officer, USS *Vella Gulf* (CVE-111), in the Asiatic-Pacific Theater. He also served as Welfare and Recreation officer, Quarters K, Arlington, VA. He was released to the Inactive Reserve in September 1946, entered the University of Maryland and received a BS in chemical engineering in 1950.

Employed by El du Pont de Nemours & Co. as supervisor at Dana, IN; Aiken, SC; and Niagara Falls, NY (1950-1956) in the production of nuclear materials and chemical intermediates. He joined Olin-Mathieson Chemical Corp. as senior engineer for production of high-energy fuels. He was employed by Dow Chemical Co. as senior technical engineer in production and research of synthetic textile fibers (1960-1969). Coakley was employed as chemical engineer at the Naval Weapons Station, Yorktown, VA in 1977, for research and development of explosives and weapons. He retired, as senior process development engineer, from the Naval Surface Warfare Center, Indian Head, MD in 1993.

He was elected to the Board of Supervisors, James City County, VA in 1967 and served as vice-chairman, 1969-70, and as chairman in 1971.

Coakley and his wife, Martha, live in Williamsburg, VA where their twin daughters, Virginia and Janice, were born in 1960. Martha retired as RN from the Eastern State Hospital in 1985; Virginia graduated from Christopher Newport University in 1989; and Janice received her BS in Zoology from Colorado State University in 1991.

VICTOR LEE COFFEY JR. (LTJG) was born May 2, 1922, in Salina, KS. He entered the V-1 Program at Regis College on April 17, 1942; entered Colorado University V-12 Program in July 1943; and was commissioned ensign in January 1945 at Northwestern Midshipman School, Chicago, IL. He attended advanced line officer training (USNTC) in Miami, FL.

Coffey served in the South Pacific on the USS SC-1325 as navigator, ASW officer and gunnery officer in Ulithi, Guam and Saipan, until she was decommissioned at Treasure Island, CA. He was then reassigned to the USS PGM-20 as navigator, 1st lieutenant and ship officer in the Philippines with a minesweep unit and weather service.

His memorable experience was the picking up of dead Japanese soldiers off the island of Yap for intelligence information and delivering it to Ulithi.

Coffey was released to inactive duty as lieutenant, junior grade, on June 29, 1946. He attended the University of Colorado and received a BS degree in business in June 1948. He then attended Denver University and received a MBA degree in management on Dec. 9, 1950.

He was an industrial engineer at Gates Rubber Co. for three years, then appraiser and mortgage loan officer with Empire Savings and, later, appraiser and loan solicitor for Kassler Mortgage Co. He developed two housing subdivisions and office buildings with Coffey Investments.

He and his wife, Bette, and son, Victor III, reside in Denver, CO, and their daughter, Betti Anne, lives in Dallas, TX.

DONALD DENNIS COLE

was born in June 1925 in Phillipsburg, PA and moved to Elmira, NY when one year old. He enlisted in the Navy V-5 Pilot Training Program in April 1943. He was called to active duty, Nov. 1, 1943, at the University of North Carolina, Chapel Hill, as a V-12A Cadet. After two terms the Navy closed down the V-5 Program and all V-12As automatically transferred into V-12.

Cole continued at UNC until assigned in March 1945 to Columbia University Midshipman School; he graduated in July 1945 and was commissioned ensign. He was assigned to the light cruiser, USS *Astoria,* and served in several capacities, concluding his career as assistant navigator.

His memorable experience was when the Navy Dept. sent the *Astoria* to Portland, OR in June 1946 to participate in its 75th Anniversary Celebration, then to Astoria, OR; Tacoma, WA for celebrations; and lastly, to Vancouver, B.C. for its Diamond Jubilee Celebration.

Discharged from active duty in August 1946; he received a BS degree from Columbia University in June 1948; a Doctor of Law degree in June 1951 from Cornell Law School, and he has practiced law in Elmira from 1951 to the present.

Cole and the former Donna Albertson were married in January 1956. They have three daughters, one son and two grandchildren.

PRENTISS H. COLE

(LTJG) was born May 25, 1924, in Fargo, ND. He enlisted in April 1943 and entered the V-12 Program at Minot State Teachers College in North Dakota and Midshipman School at Northwestern University.

His military locations and stations included Asbury Park, NJ; Chicago, IL; Pacific-Philippine Islands; and he saw action in the Leyte Gulf and Borneo.

A memorable experience was scrounging ammunition and provisions from the Army on Halama Nara Islands for the Borneo invasion. Also memorable was being commanding officer of LCI-748 at the age of 21.

LTJG Cole was discharged in June 1946. He received the Pacific Theater with star.

Cole and his wife, Joyce, have two daughters (both married to ex-Navy men). He retired as chairman of Cole Papers, Inc.

DAVID YOUNG COLLINS

(1LT) was born in Beloit, WI on June 10, 1927. He entered the Navy V-12 Program (also in the band) at the University of Notre Dame, 1944-46; Great Lakes, 1946; and decommissioned LSTs 726 and 979 in San Francisco Bay.

Collins was discharged in 1946, attended the University of Wisconsin and obtained a BS in civil engineering. From 1948-49 he was an instructor in surveying; obtained an SJD Law degree from the University of Wisconsin in 1952; was 1st lieutenant JAGC, U.S. Army; and was on the staff of Judge Advocate General's School, Charlottesville, VA, 1953-54.

Served on the State Bar Board of Governors, 1980-84; was president, life director and trombone player, Beloit Janesville Symphony Orchestra; officer and director, Stateline United Way. Cole has been in the general practice of law in Beloit, WI from 1954 to the present.

Resides in Beloit, WI with his wife, Annabelle. Their daughter, Sara, is a Hawaii state archeologist in Honolulu, and their son, Philip, is a practicing lawyer in Dallas, TX.

RICHARD H. COMER

(LTJG) was born in Halifax County, VA on Sept. 26, 1923. He enlisted in December 1942 and entered V-12 duty at Berea College, KY in July 1943. He was commissioned, NTS, Camp MacDonough, Plattsburg, NY in June 1944.

He reported to Commander, Eighth Amphibious in Naples, Italy for amphibious landing (LCT) training in July 1944. He saw his first combat action in the early morning of Aug. 15, 1944, when the battleship, *Texas,* fired the first salvo from its

14-inch guns directly over his small LCT near St. Tropez, France.

LTJG Comer was honorably discharged in August 1946. He was awarded the WWII Victory Ribbon and the European-African with star.

Returned to the University of Iowa and Pennsylvania State for graduate studies in physics under the GI Bill. He joined the Army's Ballistics Research Lab (BRL) and spent some 30 years doing R&D on interior ballistics of guns. He performed and supervised some of the "seminal" research on the use of liquid propellants in gun (LPG) systems. He was elected a "fellow" on the BRL scientific staff and also taught physics courses for the University of Delaware.

Received the Army's decoration for Meritorious Civilian Service for "exceptional leadership, numerous direct technical contributions and outstanding achievements in the area of interior ballistics." After retiring he has kept busy consulting in physics and LPG systems; studying and lecturing on genetics and the world-wide "Human GENOME Project; golfing; and traveling Europe and China.

He and his wife, Doris, still reside in their original "hand-built" house near Havre de Grace, MD. Both of their daughters are married and have earned advanced degrees.

BOB CONKWRIGHT

(LTJG) was born Feb. 27, 1925, in New York, NY. In June 1943 he enlisted and entered the V-12 Program at Bates College and Harvard (in the same class as Jack Lemon and Bob Kennedy) and was commissioned at Cornell.

He was engineering officer on LCIs 704 and 766 and went to Okinawa, Saipan, Tinian and Tiensen, China to transport supplies to the Marines and carry Japanese prisoners to Taku. They had to leave Tiensen because of the communist offensive and sailed to San Francisco and then decommissioned at Charleston.

Conkwright resigned his commission (LTJG) in September 1946 and left the USNR in April 1948. He received awards/medals for the American area and the Asiatic-Pacific area.

After the war he spent time on carriers and cruisers, helping to improve operations with new technology, and he was active with AWACS. He received a BS in math and physics from Bowdoin College, then became a manager at Martin Marietta and Westinghouse. He owns a small export/import corporation in Baltimore, MD and works with the Japanese.

Conkwright married in 1948 and has a daughter, son, two granddaughters and one grandson.

DR. W. HOWARD COOK

(ENS) was born June 3, 1925, in Earlham, IA. He enlisted July 1, 1943, and entered the V-12 Program. He earned a BS, MS and Ph.D. in engineering from the Illinois Institute of Technology, Chicago. Cook was stationed on the USS *Oxford* and the USS *Kan Ka Kec.*

He was discharged in July 1946. His awards/

medals include the WWII Victory Ribbon, Asiatic-Pacific and the American Theater Ribbon.

Dr. Cook had 30 years' experience in research development, operations, planning and management of energy projects. He was employed by Stanford Research Institute, General Electric Co., Argonne National Laboratory, Linde Air Products Co. and Allis Chalmers Co.

Dr. Cook passed away on March 5, 1991; he is survived by his wife, Esther, and four sons: Peter, Howard, Stuart and Philip.

ROBERT L. COOPER (CAPT) was born June 9, 1925, in Hauford, CA. He entered the V-12A Program at Montana School of Mines on July 1, 1943. He transferred to V-5 in February 1944, completing tarmac in Seattle; flight preparatory at California Polytechnic and St. Mary's preflight before leaving the program. He was discharged in May 1946 as an aviation mechanic.

After receiving a Bachelor's degree in aeronautics from San Jose State College, he re-entered the USN as a naval cadet and earned his pilot's wings and commission in April 1952. Following a duty tour with a fighter squadron, he returned to the civilian arena and the aerospace industry as a rocket propulsion engineer for Douglas Aircraft and TRW Systems on the Thor and Minuteman Ballistic Missile programs.

Having remained active in the Naval Air Reserve, he retired in 1985 with the rank of captain. In 1988 he retired from Huiet-Wesson Foods where he had been the corporate security officer. He currently resides on the southern Oregon coast where he and his wife own and operate a Bed and Breakfast.

WALTER G. COOPER (LTC) joined Platoon Leaders Class at the University of Mississippi and was commissioned 2nd lieutenant USMCR upon graduation. He was called to active duty in 1940. After a brief tour of duty with the USN, he was assigned to the Marine Barracks, Washington, DC.

Cooper was ordered to Duke University to Navy V-12 in July 1943 as officer-in-charge of 600 Marine students. In 1944 he was assigned to the USS *Chilton* in the Pacific Theater. He proceeded to Leyte with re-enforcement via Pearl Harbor, Eniwetok and Ulithi.

He took part in the initial preparations for the invasion of Okinawa, landed in Ie Shima and was hit by kamikaze in April 1945. He returned to San Francisco for repairs and was back just before the Japanese surrendered. Cooper transported troops to Korea, China and Manilla. He carried National-

84

ist troops to Manchuria and transported troops to the West Coast from Nagaya, Japan and Saipan.

Cooper returned to Duke University after the war as personnel director. He married Louise and they had two daughters, Dr. Betty Epanchin and Mrs. Barbara Lowery, and three grandchildren.

WILLIAM GORDON CORYEA was born in Seymour, IN on Jan. 15, 1925. He entered Wabash College V-12 Program on July 1, 1943, and was commissioned ensign on May 24, 1945, at Northwestern University Midshipman School, Chicago, IL.

Coryea completed Destroyer School at Norfolk, VA and Tactical Radar School at Hollywood, FL. He served in the occupational forces in Japan as CIC officer aboard Radar Picket Destroyer, *Neuman K. Perry.* He was discharged from the USN on June 1, 1946.

Graduated from Indiana University School of Law in 1951 and serves as attorney for, and director of, First Federal Savings Bank of Marion, IN. He and his wife, Karol Ann, reside in Marion, IN and have three children and four grandchildren.

THOMAS R. COSTELLO (LTJG) was born Jan. 11, 1924, in Atchison, KS and raised in St. Joseph, MO. He enlisted in the Navy in 1942 and entered the V-12 Program, Park College, Parkville, MO. He attended Midshipman School, Abbott Hall, Chicago and was assigned as 1st lieutenant in LST-1038.

LTJG Costello did service in the Philippines, Guam and Tokyo after the bomb hit. He was discharged in 1946.

After the war he moved to Oklahoma City and sold life insurance and commercial real estate until he retired. He and his wife, Norma, have eight children, 22 grandchildren and one great-grandchild. They are all unusually smart and exceptionally good-looking people. His present hobby is the Toastmaster's Club.

JAMES H. COWEN (AVCAD) was born Aug. 12, 1926, in Greensburg, PA. He enlisted in the USNR on Aug. 15, 1943; entered the V-12 Program at Milligan College, Chapel Hill; preflight at Glenview NAS; primary at Corpus Christi; and advanced at Kingsville. He was discharged May 6, 1946.

He earned his BME from the General Motors Institute, Flint, MI, in 1950 and MBA from Wayne State University in Detroit, MI in 1972. He worked for General Motors as senior engineer superintendent and field service manager. Also, he worked for Ford Motor and Rockwell International. He

was self-employed as president and CEO of Jim Cowen and Associates, manufacturers representative to the auto industry, until he retired.

Cowen married Barbara Krem in 1948 and they spend six months in Bradenton, FL and six months in Clarkston, MI. They have three children and eight grandchildren.

CLIFFORD J. CRAFT III (1LT) was born Jan. 6, 1925, in Philadelphia, PA. He ran away from home when he was 17 and enlisted in the USMC in August 1942. After boot camp at Parris Island, Craft attended Aviation Machinist's Mate School at Jacksonville NAS, where he graduated with the highest score in the school's history and was promoted to corporal.

Because of the encouragement he received from his commanding officer, Craft was persuaded to enter the V-12 Program in May 1943. Initially, Craft attended Duke University but eventually was transferred to the University of Michigan where he graduated with a BSE(EE) degree. Craft served in the USMCR until January 1951 when he was honorable discharged to engage in research for the Army Signal Corps.

Craft earned his MSE in electrical engineering in 1949 from the University of Michigan; MBA in industrial engineering in 1954 from Wharton School, University of Pennsylvania; MS in 1976; and his Ph.D. in 1984 in accounting from the University of Southern California.

President, 1968-1970, and founder of Management Software Development Corp., Los Angeles, CA. He developed a financial management simulation model of a dealership for Ford Motor Co.; performed a review of data processing requirement for LTV, Dallas, TX; and is consultant to National Clearing Corp.

Vice-president, 1970-1973, of Science Management Corp., Washington, DC and was responsible for management development programs, cost control systems, systems analysis and operations research. He developed a management training program for the National Association of Security Dealers.

Senior operations research specialist, 1974-1990, with Northrop Aircraft Div., Hawthorne, CA. He published papers on *Learning Theory Analysis Directed at Enhancing Instructional Systems Development* and *Improving Pilot Situation Awareness Skills Training.* He wrote a report on cost and training effectiveness analysis entitled, *Training Device/System Design Process,* and an Air Force Commendation Letter, regarding the report, was sent to Northrop's CEO.

Craft was emeritus professor of accounting, 1973-present, California State University, Los Angeles; teaches advanced accounting courses: intermediate accounting and managerial accounting; was director of the Bureau of Business Research; and was founder of the school's quarterly magazine, *Business Forum.*

He is an independent management consultant of Craft & Associates, Malibu, CA; director, Atlantic Pacific Pizza Bakery, Costa Mesa, CA; and reasurer of Retired Public Employees Assoc. of California.

Craft has published numerous articles and five books including: *Responsibility Accounting, Flowcharting and Programming Business Problems, Top Management Decision Simulation, Management Games* and *An Examination of the Decisive Decision Style in Tasks Using Accounting Information.* He is a member of numerous professional organizations and has many honors; Sigma Xi, Omega Rho, etc.

In June 1948, he married Carolyn Denny of Flushing, NY and together they raised five daughters: Carol, Jeannie, Gerianne, Dorette and Teresa; one son, John; and 11 grandchildren. Unfortunately, the Crafts just lost their home and all their worldly possessions in the firestorm which raged through Malibu, CA in November 1993, so at a time in life when most people are looking toward a peaceful retirement, the Crafts are busily engaged in building their first new home.

DALTON LEE CRISWELL (SGT) was born July 12, 1921, in Throckmorton, TX. He enlisted in June of 1943, attended the V-12 Program at La Ploy Tech, and served with the 4th Mar. Div., 24th Regt.

He was stationed at Quantico, San Diego Marines, Hawaii and was in the 2nd Wave at Iwo Jima.

His memorable experience was when Co. A, Mortar Plt. set a record for the number of flares fired in one night.

Criswell was discharged in the Spring of 1945. On April 3, 1993, he and his wife, Kathleen (Harrington), celebrated 50 years of marriage. They have two daughters and one grandson. He is retired from coaching and teaching, but still ranches in Andrews, TX.

JOHN W. CRONIN (RADM) enlisted in the V-12 Program in July 1944. He was commissioned from the NROTC Program at Georgia Tech in 1946 and served in the USS *Holder* (DD-819) until released from active duty in 1947.

Recalled to active duty in 1951 and was designated a naval flight officer and flew with VS-801, aboard various CVEs and CVLs in the Atlantic Fleet. He was released from active duty in 1953 and served mostly in intelligence billets in the Naval Reserve.

Selected for Flag Rank in 1979 and he served as director, Naval Reserve Intelligence Program. His personal decorations include the Legion of Merit and the Navy Commendation Medal. The Secretary of Defense cited him for distinguished service as a member of the Reserve Forces Policy Board.

During 45 years with the Connecticut Mutual Life Insurance Co., he served as general agent of one of the country's largest agencies in Philadelphia. Upon retirement he was asked to fill a chair in management education at the American College in Bryn Mawr, PA. He and his wife, Anne, live in Radnor, PA.

FRANK A. CROSSLEY was born in Chicago, IL on Feb. 19, 1925. He enlisted in January 1944 and was commissioned ensign DL USNR on Nov. 2, 1945 from the USNRMS, Columbia University. He served in the USS *Storm King* (AP-171), a troop transport. Non-U.S. ports made were: Honolulu, HI; Guam; Manila, Philippines; and Panama Canal Zone.

Received a BS in chemical engineering in 1945; released from active duty on Aug. 1, 1946; received his MS in 1947 and Ph.D. in 1950 from Illinois Institute of Technology.

Crossley married Elaine J. Sherman on Nov. 23, 1950, and they have a daughter, Desne Adrienne. He taught at IIT, 1948-1949, and Tennessee Agricultural and Industrial State University, Nashville, 1950-1952. He was senior scientist of IIT Research Institute, 1952-1966 and has held various staff and managerial positions at Lockheed Missile & Space Co., Sunnyvale, CA, 1966-1986. He held staff and director positions at GenCorp Aerojet, Sacramento, CA, 1986-1991.

Crossley contributed 60 articles to technical journals and symposia and was awarded eight patents. He is the inventor of the Transage titanium alloys, and is listed in various who's who, including *Who's Who in America.*

RICHARD A. CROSSLEY (LTJG) was born May 3, 1923, in Providence, RI. He joined the USNR on Feb. 8, 1943, and entered the V-12 Program at Brown University, Providence, RI and Trinity College, Hartford, CT. He graduated from Midshipman School, Abbott Hall (Northwest University) with the 15th Class.

LTJG Crossley was discharged on June 26, 1946. He received the European and Victory Medals.

He is married and has five children and 10 grandchildren. Crossley is retired.

JOE B. CROWNOVER (MAJ) joined the USMCR in 1942 while a junior at Texas A&M College. He entered the V-12 Program in July 1943; was sent to Arkansas A&M College; OCS at Quantico, VA and graduated as a 2nd lieutenant, USMCR in April 1944. He received a regular commission in 1946.

Crownover served in combat at Iwo Jima and Korea. He had duty at Stockton, CA; Crane, IN; Washington, DC; Quantico, VA; Camp Pendleton, CA; Okinawa; San Diego, CA and retired as a major in 1964.

Graduated with a BS degree from the University of Maryland in 1957 and with an MS from SSIU, San Diego, in 1971. From 1964-71 he was

a programmer, systems analyst, operations manager and manager of training. 1972-93 he managed his consulting business and taught at SDSU and Grossmont College.

He lives in San Diego, CA with his wife of 47 years. Their four daughters also live in San Diego.

PHILIP E. CULBERTSON (ENS) was born Aug. 19, 1925, in Pullman, WA. He enlisted in the USNR in August 1943; entered the V-12 Program at Georgia Instite of Technology; was stationed at Mojave NAS, Pt. Mugu, CA and made ensign in June 1946.

Following his discharge in June 1947, he attended the University of Michigan where he earned his Master's degree in aeronautical engineering.

After 13 years with General Dynamics Corp. in San Diego, CA, he joined NASA. Duties over the next 23 years included: deputy associate administrator for Manned Space Flight; associate administrator for Space Station; associate administrator for Planning and Policy and general manager of NASA. Culbertson married the former Shirley Caskey in August 1950. They have two children, Camden and Philip Jr.

He retired from NASA in 1988 and has been a lecturer and consultant since. He resides in McLean, VA.

CLEVE A. CULLERS (LCDR) was a petroleum landman in Abilene, TX. He earned his BA from Abilene Christian University; entered the V-12 Program at TCU; and attended Midshipman School at Tower Hall in Chicago in 1944. Cullers has been active in the "oil patch" since finishing college after WWII.

He was born in Oklahoma, raised on a farm in Jack County, TX and has lived in Abilene, where he served as CO of Naval Reserve, City Council, church deacon, director of Oil & Gas Assn., Mental Health, United Way, TROA and Kiwanis lieutenant governor with 38 years' perfect attendance.

Culler is the father of four children and calls his wife "Jiggs." They have traveled extensively the last 18 years, including one trip around the world.

DONALD E. CUMMINGS (LT) was born in Aliquippa, PA on Jan. 11, 1925. He enlisted in the service on May 23, 1943, served in the V-12, V-5 and NROTC at Drew, Colgate, University of Georgia, College of the Holy Cross and Yale, where he graduated with a BSME in 1946.

Cummings is a survivor of the *Hobson* and *Wasp* collision on April 26, 1952. He was discharged in January 1946 and recalled from April 1952 to March 1954 for the Korean Conflict.

He is a management consultant. Married since Jan. 31, 1953, he has seven children. His son, Mike, is a naval aviator (LT, USN).

ROBERT L. CUMMINS (ENS) was born in Enid, OK on March 17, 1926. He enlisted in May 1943 and entered the V-12 Program on July 1, 1943, at Milligan College. He transferred to Duke eight months later and graduated from Ft. Schuyler Midshipman School on March 6, 1945.

After completing two and a half months of advanced training in Miami, he was assigned to the US PCE-898, a weather ship operating out of Guam, M.I. His assigned transportation was the *Indianapolis;* they left San Francisco on July 16, 1945, with the Hiroshima atomic bomb aboard. They delivered it to Tinian, M.I. on July 25, 1945, and he left the ship the next day in Guam to await the arrival of the PCE-898.

On or about July 30, while en route from Guam to Leyte, and a few minutes after midnight, the *Indianapolis* was hit by two Japanese torpedoes and sank in 12 minutes, resulting in the largest loss of life from a ship sinking in the history of the Navy.

Served onboard the PCE-898 and reached the position of XO. During this period of 10 months, the ship rode out three typhoons, the worst being in October 1945. It also took part in the Crossroads Operation, the first peacetime test of atomic weapons.

After being discharged in July 1946, he became a partner in the family construction business and is presently chairman and CEO of the Cummins Construction Co., a highway-heavy contractor specializing in the production and installation of "hot mix asphalt" in Oklahoma and Arkansas.

He is married with two children and five grandchildren. He and his wife have cruised the East Coast, Bahamas, Gulf of Mexico and inland waters for the past 25 years.

WILLIAM J. CUNNINGHAM (FSO) was sworn into the USN on Dec. 8, 1943, at Los Angeles, CA. He entered the Navy V-12 Program at Washburn Municipal University, Topeka, KS on March 1, 1944. He was assigned to NROTC Unit at the University of New Mexico, Albuquerque, in October 1944. He was commissioned ENS, USNR on June 24, 1946; reported to USS *Incessant* AM-248 at Terminal Island in July 1946 as first lieutenant and supply officer.

The *Incessant* was decommissioned at Bremerton, WA in November 1944 and

Cunningham was sent to Harbor Defense School at Tiburon, CA. He was assigned as XO to USS *Cohoes* AN-78 in February 1947 and was released to inactive duty in June 1947. He completed his BA and MA degrees in political science at the University of New Mexico.

Entered U.S. Foreign Service in 1949 and served in Prague, Paris, Seoul, Japan (twice), Saigon, Phnom Penh, Taiwan, Washington and New York (U.S. Mission to UN). He earned Commendable, Meritorious and Superior Honor Awards; retired in 1982 at rank of counselor to become director of the Center for International Studies, University of St. Thomas, Houston, TX. He was named professor emeritus in April 1993 and is currently chairman of the Houston World Affairs Council.

Achievements in Cunningham's diplomatic career include directing rapid build-up of administrative services to support 20-fold expansion of the U.S. Embassy, Phnom Penh, in 1954-55; was the first career foreign service officer named director of the Department of States' Chinese Language Training Center in Taiwan; contributed to the success of Ping-Pong Diplomacy with China, for which Dr. Henry A. Kissinger commended him in his White House years; managing successfully, incorporation of the East-West Center, Honolulu; and realizing major savings in UN operating expenditures by directing U.S. initiative to have the General Assembly reduce use of first class accommodations for official travel by UN officers and staff.

At the University of St. Thomas, enrollments in the International Studies major quintupled under Cunningham's leadership, and the Center for International Studies became a prominent regional forum for serious discussion of world problems and international affairs.

JAMES H. CURTIN (LTJG) enlisted at 17 and was assigned to the V-12 Program at Holy Cross College in July 1943. He was commissioned at Fort Schuyler in January 1945, joined LCT-400 in New Guinea and proceeded to the Philippines by courtesy of a LST tow.

They provided back-up supply to Tacloban and Batangas and were working at Subic Bay when the war ended. He spent time on LCT-1205 at Davao, shipping badly needed hemp from recaptured plantations. He had the opportunity to work the China Coast wise trade as a civvy when "go-home" points were accumulated; he gave it serious thought, but decided to go home.

Graduated from Holy Cross in June 1948; married Trudy in June 1949; and was discharged in 1959. He worked 43 years with the Royal Dutch/Shell Group companies marketing oil and coal internationally. He retired in 1991 and resides in Midlothian, VA and Jupiter, FL. He devotes time to his five married children, 12 grandchildren, church, golf and fishing.

ROLLINS KING CUSHMAN (LTJG) was born in Leadville, CO on June 27, 1922. He enlisted Oct. 29, 1942 (Navy Day) and his military locations and schools included: V-12 Program, Peru State Teachers College, Peru, NE; Midshipman School at Notre Dame, IN; Harvard, San Diego, San Francisco; Leyte; Philippine Islands.

All of his duty was great. He received various medals and was discharged Sept. 16, 1946 in San Francisco, CA.

Cushman has been married 48 years (no children) and lives in Los Gatos, CA. He is retired from real estate; has condos in Hawaii and travels via cruise ships. He spent five months in 1990 in Australia.

ROBERT DOMINICK CUSUMANO (LCDR) was born on Staten Island, NYC and entered the V-12 Program at Yale University on Nov. 1, 1943. He was commissioned ensign in February 1946 and received a BS degree from Yale. After training cruise with 500 plus new ensigns aboard the USS *Cleveland* (CL-55), he was assigned to PCS-1387 at Key West, FL; next he served aboard the *Meredith* (DD-890), one of the first ships to fly the UN flag off Naifa, Palestine during the Israeli war of independence.

September 1949, he resigned his USN LTJG commission for same in the Naval Reserve. He changed his officer designator from line to civil engineer. He allowed the next 13 years from 1959, for billets in CB Div. 3-10 and public works officer at Freeport and Huntington, NY Reserve Training Centers, where he attained the rank of lieutenant commander before his discharge in November 1971.

Concurrent with Naval Reserve status, in 1950 he returned to Yale for a BCE degree and in 1952 to Johns Hopkins for Master of Sanitary Engineering degree. He was employed by Nassan County, NY as public health and air pollution control engineer and organized and directed NASSAU's first air pollution control program. After 28 years' service with NASSAU, he retired to Pequa Houseboat Charters and Adventures Cruises, Inc. He also operates two vans to transport furnishings from Florida to northern destinations. He resides in Punta Gorda, FL with his wife, Constance. He has a daughter, Linda, and son, Robert Jr. (Tampa), four grandchildren and one great-grandchild from a previous marriage. He has a son, Ronald, from his present marriage, who lives on Staten Island.

ROBERT D. DALLEN (CAPT) enlisted in the Naval Reserve on May 12, 1943, and entered Yale University's V-12 Program on July 1, 1943. He completed the NROTC Program at Yale; was

commissioned ensign on Oct. 24, 1945; and served as communications watch officer and CR Division officer aboard the USS *Providence* (CL-82), a flagship of the Mediterranean, from November 1945-August 1946.

Returned to Yale and received his MA degree (sociology) in June 1947, and continued studies there until 1948. He returned to active duty at the Naval Post Graduate School (Naval Intelligence), participated in the Naval Reserve for 37 years and took AcDuTra for 30 consecutive years. He served as CO, NR Intelligence Division 3-1 for three years, and later was Reserve Intelligence Area Coordinator (Area 19) and responsible for all Reserve Intelligence Units in the Washington, DC area (three years). He was awarded the Navy Commendation Medal upon completion of his tour.

In the 1950s he worked in school equipment sales, then as administrative assistant to the Director of Purchases and Purchasing Agent for Philip Morris, Inc. He joined IBM in sales in 1959 after receiving his MBA (management) from New York University Graduate School of Business Administration. Dallen worked in IBM's finance and insurance sales. In 1969 he was promoted to advisory marketing representative in the Special Programs Branch. He was responsible for sales to various federal classified agencies in Washington, and served in national federal marketing for over 20 years as national account representative on Navy and Air Force accounts, including Navy BuPers, Navy Supply Systems Command and Air Force Logistics Command.

Dallen retired in March 1990 as senior marketing representative. He attained 21 Hundred Percent Clubs and eight IBM Golden Circles during his career.

Capt. Dallen and his wife, Muriel, participate in community and church activities in Potomac, MD and Chatham, MA. Their son Robert Jr. (and his wife, Julie, reside in Rockville, MD with their son, Robert III) and their daughter, Carol (and her husband, David Keuch, live in Comus, MD).

PHILIP B. DALTON (LTJG) was born on July 21, 1923, in Brooklyn, NY, and graduated from the University of Illinois. He enlisted in the Navy in 1943 and attended the Purdue V-12 Program. He served aboard the USS LST-656 from March 1944-August 1945 as a navigator; and in the USS *Lancewood* as an XO from September 1945-February 1946.

Served with the rank of lieutenant (jg) in the European-African area, including the invasion of Southern France in 1944; the American Theater; and the Asiatic Pacific Area. He received his MA degree from Columbia in 1947, worked as a research chemist from 1947 to 1954, joined the GAF Corp in 1954 as a development engineer, and retired as president and chief operating officer in the 1979.

Dalton and his wife, Elaine, are residents of Sarasota, FL; they have two daughters and five grandchildren. He raises orchids and continues to serve as a consultant to companies in the States and abroad.

WILLIAM DURWOOD DANDO (AMMI3/c) was born on April 23, 1922, in Schenectady, NY. He enlisted in the USN on Jan. 1, 1943. Before V-12, he was an AMMI3/c and was discharged from the Navy (March 6, 1946) while still at Brown University, where he finished his BA soon after his discharge. He then went on to get his MA from the University of Houston.

His memorable experience happened at Union College; he was sitting on the dock when he heard a scream for help. Although fully clothed and possessing little knowledge of life-saving methods, Dando plunged into Mariaville Lake to save a man from drowning.

Dando is married and has three children and three grandchildren. He retired in 1984 from the Jacksonville School System as a school psychologist.

CHARLES W. DARNALL JR. (LTJG) was born on Aug. 22, 1923, in De Ridder, LA. He enlisted in the USNR in August 1942; attended Navy-7 (August 1942) Southwestern Louisiana Institute, Lafayette, LA; Navy V-12 (July 11, 1943) at SLI; Columbia Midshipmen's School (December 1943-April 1944) and made ensign.

Served in the USS *Olmsted* (APA-188) and participated in the invasion of Luzon, Lingayan Gulf (January 1945), invasion of Okinawa (April 1945), Occupation Honshu (September 1945) and Magic Carpet September 1945-June 1946).

His memorable experiences include the first time he saw a kamikaze hit a ship, the USS *Zeilin* (APA-1) on the West Coast of Luzon in January 1945.

Darnall was discharged on June 21, 1946. He received area ribbons and medals for the American Theater, Pacific Theater, Philippine Liberation and the WWII Victory Medal.

Graduated from LSU and practiced law in Franklin, LA for 38 years before retiring. He married Louise Braquet on Feb. 15, 1958, and they have three children: Charles III, Christine and Ann. They still live in Franklin, LA.

CALVIN C. DAUGHETEE (LCDR) was born on Jan. 9, 1925, in Salina, CO. He joined the V-1 Program in October 1942 while a freshman at Amarillo Junior College, Amarillo, TX. He was transferred to the V-12 Program at the University of Oklahoma in July 1943 and graduated in June 1945. He then attended Midshipman School at Cornell University and was commissioned ensign, USNR on Dec. 7, 1945.

Released to inactive duty in August 1946 after a short period of active duty as assistant communications officer on board the USS *Collett* (DD-730) in the South China Sea. He was recalled to active duty in April 1951 and assigned to the USS *Edmonds* (DE-406) as assistant engineering officer. After a two year tour, including six months in Korean waters, he was again released to inactive

duty. He then remained active in the Naval Reserve serving in Surface Divisions in Dallas, San Antonio and Ft. Worth. His last assignment was being the CO of Div. 8-85, Ft. Worth. LCDR Daughetee retired in July 1971 with over 27 years of qualifying service.

Married his long time sweetheart, Jo Ann Moore, in June 1948. They have three sons and a daughter, all blessed with families of their own, making them the proud grandparents of nine children. Their oldest son, David, served in the Vietnam War as a P-3 Orion crewman.

In June 1989 he retired from LTV Aerospace and Defense Co. Grand Pairie, TX. He worked for 34 years in various engineering assignments. He and Jo Ann are now enjoying retirement, living in a home they recently bought on the shores of Cedar Creek Lake, Mabank, TX.

WILLIAM W. DAVENPORT (LCDR) was born on April 2, 1926, in Jacksonville, FL. He entered the USNR in June 1944; attended Franklin and Marshall College in Lancaster, PA; joined the V-12 Program at Princeton University in Princeton, NJ; and was designated Naval Aviator, USN in 1950. He was stationed at the NAS, Pensacola, FL; Norfolk, VA; and Anacostia, DC.

His memorable experiences included flying SNJs, TBMs and AFs off straight deck carriers with LOS and flags on board at Tarawa, Saipan, Leyte, etc. and flying as instructor and check pilot in the Naval Reserve.

Davenport was discharged from active duty in October 1952. He went to work for American Airlines and retired as a captain after 33 years of service.

He and his wife, Jeannie, have two daughters, Caroll and Ginny, and one son, William. From 1986-94 he was FAA Air Carrier Inspector.

JESSE F. DAVIS was born on Jan. 23, 1927, in Yeagertown, PA. He enlisted in the V-12 Program at Bucknell University in July 1944 and transferred to the University of Rochester in November 1945. He was assigned to boot camp at Great Lakes after opting out of the program in March 1946. He served at the Brooklyn Navy Yard and

aboard the USS *Yosemite* (AD-19), which became the flagship of the Atlantic Destroyer Fleet at Casco Bay. He was discharged at Bainbridge, MD in 1946 as S2/c.

Graduated from Bucknell in 1948 with a BS degree in commerce and finance. He joined Ford New Holland in 1952 and served as a plant controller for 26 years before retiring in 1986. He was assigned as financial systems manager in Belgium for one year in 1982.

Currently active in local church work. He and his wife, Patricia, live in Lewistown, PA. They have a son, Richard, who served in the Army during the Vietnam era, and a daughter, Rhonda, who is a high school guidance counselor.

ROBERT D. DAY (LCDR) enlisted in the USN in June 1943 and entered the V-12 Program in March 1944. His tours of duty after Supply Corps School, Bayonne, NJ, included USS *Isbell* (DD-869) for 32 and one-half months (1946-1949); three cruises to China, where *Isbell* evacuated Embassy personnel from Tsingtao when the Red Chinese were taking over; Supply Annex, Stockton, CA (1949-1951); Naval Ship Yard, Pearl Harbor (1951-1953); Naval Supply Depot, Clearfield, UT (1953-1956); USS *Wasp* (CVS-18) (1956-1958); Navy Area Audit Office, Chicago (1958-1960); Naval Station Keflavik, Iceland (1960-1961); Chief, Naval Air Reserve Training Staff, Glenview, IL (1961-1963).

Day retired in November 1963. He worked for the Navy Federal Credit Union from 1963-1974 and entered the Civil Service with the Navy Accounting and Finance Center, Crystal City, VA in 1974 and retired in 1990.

He is enjoying retirement with more traveling, golfing, reading and spending more time with his four children and seven grandchildren.

ROBERT F. DELANEY (CAPT) was born on Aug. 2, 1925, in Fall River, MA. He enlisted with the regular Navy in 1943; rose to captain USNR and RADM, RI Naval Militia (RET); and commissioned with BNS, NROTC, Holy Cross College. He served first in destroyers, Pacific Fleet, then on the carrier *Ticonderoga* in the 6th Fleet.

Served in Intelligence for the balance of his 38 year reserve career including two tours as special assistant to CINCAF South (NATO/Naples) and assisted director psyops Vietnam. He was appointed to Milton Miles Chair of International Relations USN War College from 1971-1981.

Retired in January 1981 and received the Air Medal (Vietnam), Vietnamese Medal of Merit and all the usual WWII and Korean awards and medals.

Graduate education through Ph.D. at Harvard, Boston University, Catholic University, University of Vienna and USN Postgraduate School (Intelligence). He also holds a DHL from the University of Massachusetts.

He served in the U.S. Foreign Service for 20 years in west and east Europe, Latin America and Asia. He rose to assistant director, U.S. Information Agency. Upon retirement, he was appointed first director, Edward R. Murrow Center for Public Diplomacy, Fletcher School of Law and Diplomacy, Tufts University, then president American Graduate School of International Management at Thunderbird, AZ.

From 1981-1991 he served as CEO of Washington lobbying firms, Michael W. Moynihan Associates and the RFD Group. Retired, but writes a syndicated weekly column and at present is working on his ninth book.

Delaney married Patricia Ann Nestor and they have nine children and 14 grandchildren. They have alternate residences between Newport, RI and Cape Canaveral, FL.

E.D. DWIGHT DENNIS (CHIEF ENGR) was born on May 12, 1922, in Momence, IL. He enlisted on June 8, 1942, and entered the V-12 Program at Illinois State Normal University, Normal, IL from 1942-1943; went to USNR Engineering Midshipman School, Columbia University, NY and USNR Diesel Engineering School, Cornell University, Ithaca, NY, 1944. He served as chief engineer on LSM-200 and retired in 1959.

Graduated from the University of Illinois with a BS degree in mechanical engineering. He was employed by A.O. Smith Corp. from 1948-1962, then at Roper Corp., Kankakee, IL until his retirement in 1986. He served and chaired numerous national standards committees and organizations pertaining to home appliances.

He passed away on Feb. 20, 1992. He is survived by his widow, Ruth; seven children: William, Patrick, Timothy, Jon, Michael (LTC, AF), Kathleen (LTC, Army), Margaret (MAJ, AF); six grandchildren; and three great-grandchildren. *Compiled by widow, Ruth Dennis.*

M. GORDON DITTEMORE was born on Sept. 11, 1925, in Colorado Springs. He enlisted in the V-5, but was assigned to V-12 Program at St. Thomas, July 1, 1943; went to Tarmac School, Sandpoint NAS; Flight Prep at Cal Poly San Luis Obispo; St. Mary's Preflight; back to Occidental V-12; UCLA NROTC; and was called back to Chapel Hill Preflight at the University of North Carolina as an Aviation Cadet. He was discharged in October 1945.

Under the GI Bill he graduated from USC with a degree in civil engineering. He spent five years with the USBR working on Shasta Dam and siting of the Feather River Dam. He spent five years designing and building missile test facilities with North American Aviation. He then spent 31 years with TRW in design, development, and

deployment of the Minuteman and Peacekeeper missiles; the development of ASWFORPAC on Ford Island and Pearl Harbor; in Star Wars development; MILSTAR Satellite Program; and the start up program for the disposal of high level nuclear waste. He retired from TRW in 1989.

He and his wife, Helen, live in Redlands, CA. They have four daughters: an entrepreneur in villa rental, a nurse, a medical doctor, a teacher; a son in insurance; and 10 grandchildren. He enjoys woodworking, hiking, world travel, gardening and visiting their daughter in her villa in Italy.

BYRON F. DOENGES (LCDR) was born on June 18, 1922, in Ft. Wayne, IN. He was a member of the DePauw University Navy V-12 Program from 1943-1944 and division officer and commanding officer of the USN LSTs, 1944-1946.

Participated in the first wave landings and in casualty evacuation work on Iwo Jima and Okinawa; initial ground reconnaissance of Nakasaki after the atom bomb attack; and naval support to the Kuomintang during the Chinese Civil War. He held the rank of LCDR, USNR.

College degrees include the BA from Franklin College, 1946; MBA from Indiana University, 1948; Ph.D. from Indiana University, 1962; and an honorary degree, D.Ltrs, from Franklin College, granted in 1985.

Career Detail: (1948-1950) he was a teacher and administrator at Punahou Senior Academy, Honolulu; (1951-1972) administrator and teacher of economics at Indiana University and Willamette University, rising to the rank of Dean of the College of Arts and Sciences and professor of Economics; (1972-1993) senior officer of the Arms Control and Disarmament Agency, Department of State. In 1993 he retired from ACDA and became an independent writer and international economics consultant. He served as the chairman of the 1989 Navy V-12 National Colloquium and a member of the National Committee for the Norfolk reunion commemorating the 50th Anniversary of the Navy V-12. He is currently residing in Fearrington Village, NC.

HOWARD C. DOOLITTLE (CAPT) was born on Sept. 4, 1922, in Endicott, NY. He entered the USMCR on Oct. 6, 1942, and the V-12 Program at Dartmouth College in July 1943 directly from the USMC. He was commissioned 2nd lieutenant, Quantico, VA Officer's School.

Served in the Pacific Theater and China as a bomb disposal officer. He was released in May 1946, Brooklyn Navy Yard, NY and retired from the USMC with the rank of captain.

His memorable experience was taking 31 days to cross the Pacific and return to California from China in April 1946 on a LST (slow boat).

Returned to Dartmouth, graduated in June 1947 with his BA degree, immediately joined General Electric Company, and for 27 years worked in marketing management. He retired from General Electric and founded Active Realty as owner of a real estate and loan business in Simi

Valley, CA in 1974. He is still going strong and loving every day of activity.

He married Alice in 1950, and they have four children: Jean, Jerry, Jo Ann and James. For the last 28 years, they have resided in Woodland Hills, CA.

JOHN E. DORMAN (LTJG) attended Ft. Schuyler University, Miami, FL and UNC Chapel Hill. He entered the Navy V-12 Program from North Atlantic duty, USS *Yukon* (AF-9). He also attended Harvard University Encoding/Decoding School and Small Craft School at Hollywood and Miami, FL.

His memorable experiences include being instructed by his commanding officer of the Miami unit to take a group of transferees to the UNC Command. They arrived late at night on campus UNC and were informed that there was no room for them; however, that evening or the following day someone realized one of his group was Arnold Tucker, who later was assigned to West Point. Needless to say, quarters were found for them, probably at the expense of another group.

He is married to Ruth Mac Innes and they have five children, four of whom were boys who served in the Vietnam War. He is a sales manager for the Seafood Division of Sara Lee Corp.

THEODORE L. DORPAT (ENS) was born on March 25, 1925, in Miles City, MT. He enlisted in the USNR on July 1, 1943; attended the V-12 Program at State Teachers College, Valley City, ND; and Northwestern Midshipman School in Chicago.

His memorable experience was spending six weeks in Guam, Manila, Hong Kong and China trying to find his ship (AKA-106).

Dorpat was discharged on July 15, 1946, with the rank of ensign. He was awarded the Good Conduct Medal for serving overseas in the Pacific Theater.

Married and has one daughter, JoAnne and two granddaughters. He works as a physician, psychiatrist and psychoanalyst and has published 268 publications and two books.

THERON RUDOLPH DOSCH (RUDY) (QM1/c) was born on Oct. 15, 1923, at Auburn, IN. He enlisted in the USNR on June 7, 1942, and went to boot camp and QM School at Great Lakes, IL. A plankowner, he served on LST-345 from January 1943-October 1944 in North Africa, Sicily, Salerno and Normandy.

Assigned as Group 37 staff quartermaster aboard the LSM-360 at Galveston, TX in January

1945; saw action in Okinawa; and transferred from LSM-360 at Saipan for naval officer's training at the University of Utah, Salt Lake City, UT, during 1945-1946 academic year.

A memorable experience was going downtown to Salt Lake City on liberty, Oct. 27, 1945, Navy Day. He jumped into the pointer's seat of a 40mm Bofors gun on display, his former battle station in combat. In an instant, other gunners had manned their respective stations and those apprentice seamen with battle ribbons were ready for action once again.

Dosch was discharged as QM1/c in September 1948 to accept a commission in the USNR and was discharged from the USNR in August 1968.

Employed by the U.S. Geological Survey from 1950-1981 as a civil hydraulic engineer. He married Mary Dawn Stevens in Salt Lake City, UT in 1946. She passed away in May 1981. They have four children: Geraldine, Jeanne, William and Richard.

JOHN C. DOUTHIT (ENS) was born in Sioux Falls, SD in a Navy family (father, twin brother and brother-in-law). His father was an instructor at Midshipman School and eventually commanded V-12 Unit, Muhlenburg College, Allentown, PA.

Enlisted in the USNR in February 1944; entered the V-12 Program at St. Lawrence University, Canton, NY (March 44); transfered to NROTC Tufts College, Medford, MA (November 1944) and was commissioned ensign (June 1946).

Assigned to the USS *Tanner* (AGS-15) and was stationed at Norfolk, Vera Cruz, Mexico and New Orleans, LA. Memorable experiences were the Battle of Scollay SQ, downing of the schooner, *Takie Wicth*, and the taking of a tassle at the "Old Howard."

He formed many strong and lasting service friendships. He recalls fondly the Coke spot-light band appearances of Glenn Miller, Vaughn Monroe, Bennie Goodman and brief interludes with Kate Smith, Carole Landis and members of Jimmie Dorsey's orchestra. A member of the Tufts Unit was "Pat" Moynihan (now senior senator from New York).

Douthit was discharged in May 1947.

RICHARD C. DUBLE (LTJG) was born on Dec. 27, 1922, in Cincinnati, OH. He enlisted in the USN (tactical radar) on July 1, 1943. He entered the V-12 Program, Berea, KY (was a bugler) and University of Cincinnati where he earned his BS degree. Midshipman School was at Columbia University, NY.

He was stationed at Hollywood Beach, FL; Pacific Ocean; and participated in action at Iwo Jima and Okinawa. His memorable experiences include being jailed 75,000 miles in the Pacific Ocean aboard the USS *Lubbock* (APA-197).

Discharged on Sept. 6, 1946, with the rank of lieutenant (JG) USNR. His awards include the China Service Medal, Asiatic-Pacific Medal, American Campaign, Navy Occupation and WWII Victory Medal.

Duble has three daughters and three grandchildren. He is now retired. His brother, William, served aboard the LST-913 as QM in WWII.

JAMES T. DUFF (CDR) was born on Jan. 23, 1925, in Sandusky, OH. He entered the Colgate University V-12 Program on July 1, 1943; went to pre-midshipman School at Platlsburg, NY; and Cornell University Midshipman School, receiving his commission on Nov. 30, 1944.

Served with the amphibious units at Coronado and Oceanside, CA, he was officer-in-charge at the Officer Messenger Center on board the LST-505, Jinsen, Korea; and OinC Fleet Post Office, Tangku, China. Duff was discharged on Aug. 6, 1946.

Received his BS and MS degrees from the Ohio State University and his Ph.D. degree in microbiology from the University of Texas at Austin.

Employed by the Army Biological Laboratories in Frederick, MD from 1949-1965, and by the National Cancer Institute, National Institutes of Health at Bethesda from 1965-1983. He remained active with the Naval Reserve for 21 years in the Medical Service Corps, retiring with the rank of commander. He resides in Silver Spring, MD.

GEORGE R. DUNCAN was born on Sept. 10, 1923, in Inverurie, Scotland. He enlisted in the USN on July 8, 1941. Military stations/ships include: USS *Brooklyn* (CL-40) from September 1941-July 1944; North Atlantic troop convoys; invasions of North Africa, Sicily, Anzio and defense Port of Naples.

Entered the Navy V-12 Program at Dartmouth College in July 1944; NROTC and was discharged on June 26, 1946.

Duncan graduated from Dartmouth in June 1947 with his AB degree in government and attended St. John's University Law School, receiving his JD in February 1951. He spent a career in insurance and reinsurance and was vice-president of three different reinsurers. He is now retired and acts as a consultant.

Married Caroline on April 20, 1957, and they have two sons, James and Ross.

DUANE E. DUNKLE (CDR) was born on Sept. 16, 1923, in Johnstown, PA. He enlisted in the USNR on Sept. 18, 1942; entered V-12 Program at Muhlenberg College and Plattsburg Midshipman School.

Participated in battles at Iwo Jima, Okinawa and was with Task Force 77 in Korea. His memorable experiences include being boat group commander and taking the first group of marines into Yellow Beach at Iwo Jima.

Served five years active, 25 years in the Reserves and retired on Sept. 17, 1972. Dunkle was awarded the Asiatic-Pacific Medal with two ttle Stars and the Korean Medal with one star.

Worked for the Commonwealth of Pennsylvania as a public utility analyst and retired in 1984. He is currently residing in Camp Hill, PA. He is active in many civic and fraternal organizations, American Legion, VFW #4048, Navy League, Masons, Shrine, Elks, and the Lutheran Church.

KENZIE PARKS EASTER (LTJG) was born on June 12, 1924, in Lexington, NC. He enlisted in the USNR on July 1, 1943, and entered the V-12 Program at UNC, Chapel Hill, NC and Midshipman School at Harvard University.

Served in the U.S. hospital ship, *Repose* (AH-16) and was assigned to the 7th Fleet in the Pacific. His memorable experience was the typhoon in the Philippines; they were in the eye of the storm and many ships sank all around them.

Easter was discharged in October 1947. He received the Pacific Fleet Ribbon, Good Conduct Medal, etc.

He and his wife, Virginia, have two sons, Ken and Bob, and one daughter, Ellen. He retired in 1989 and enjoys traveling and tennis.

JACKSON EAVES (SN1/c) was born on July 21, 1925, in San Francisco, CA. He enlisted in the USN on July 1, 1943, and entered the V-12 Program at Colorado College for eight months and UCLA for 16 months.

Other stations include Port Chicago and Mane Island. He was discharged in May 1946 as seaman first class.

Eaves and his wife, Pat, have four children and seven grandchildren. He is a consultant for Wells Fargo Bank in San Francisco.

JOHN C. ECHERD (ENS) was born on Jan. 6, 1926, in Statesville, NC. He entered the Navy V-12 Program on July 1, 1943, at Duke University and was commissioned at Notre Dame in June 1945. He served in LST-287 until July 1946.

Received his BA degree from Carson-Newman in 1948; MA degree from Peabody/Vanderbilt in 1949; and taught in the Hamilton, TN schools until 1956. He served as specification engineer for the 3-M Co. until 1960 then worked for the Tennessee Valley Authority in the Power Program until his retirement in 1988.

Remained active in the Naval Reserve and retired as a commander in 1986. He was CO SD 6-75 and taught in the Reserve Officers School in Knoxville, TN. He is active in the National Management Assn. as local chapter president, national director and national area vice-chairman of the board.

He and his wife, Kathryn, live in Ooltewah, TN. They have three children: son, Hartly (served in Vietnam) and daughters, Amy and Cynda, and four grandchildren.

WILLIAM L. EDWARDS (LCDR) was born on Dec. 22, 1923, in Mankato, MN. He entered the USN V-12 Program in July 1943 at Gustavus Adolphus College, MN. One year later he was transferred to Notre Dame Midshipman School and commissioned in October 1944.

Assigned to new construction of USS *Hydrus* (AKA-28). He spent a number of months in the South Pacific in the initial Easter Sunday landings on Okinawa in April 1945; took Marines into Northern China at Singtao; returned Air Force personnel from Shanghi to Seattle; left Seattle for the Panama Canal and on to the Brooklyn Navy Yards for decommissioning. He was placed on inactive duty at Great Lakes in March 1946.

Graduated from the University of Minnesota Carlson Management School in June 1948. He is a member of the Navel Air Squadron VF811 as a guard officer, NAS, Minneapolis, MN. He retired as LCDR after 22 years of Reserve and active duty.

He is an agent of the Northwestern Mutual Life Ins. Co., 1948 to present. He currently resides in Mankato, MN with his wife, Ginny.

DELMAR W. EGGLESTON (YM2/c) was born on Feb. 11, 1923, in Parker, SD. He enlisted in the USN on Nov. 20, 1942, and his military stations/ships included: Bureau of Ships, USNATB, Ft. Pierce, FL and the USS *Catoctin*.

Entered the V-12 Program at Central Missouri State Teachers College in 1944 and NROTC at University of California, Berkeley in 1945. He was discharged on April 30, 1946, and received the Good Conduct Medal.

Eggleston is married and has two children. He is a retired housing manager.

JOE R. EISAMAN (SN2/c) was born in Pittsburgh, PA. He entered Dartmouth College in 1943 as a civilian, Class of 47, and transferred to Dartmouth V-12 Program in 1944. He attended Midshipman School at Cornell University and was medically discharged in August 1945 from Camp Shoemaker, CA and returned to Dartmouth as a civilian.

Graduated Class of 1947 and returned home to Pittsburgh, PA, where he and a boyhood friend decided to "go west, young man" and started an advertising agency in Los Angeles, CA in 1948.

In 1994 Eisaman, Johns and Laws Advertising bills in excess of $160,000,000 annually, representing such clients as Pennzoil Motor Oil, Neutrogena Skin Care Products, Giorgio Beverly Hills and General Motors dealer groups. Headquarters are in Los Angeles with branch offices in Chicago and Houston.

He looks back on friends made at Dartmouth and in the V-12 with great pleasure. He is a physical fitness fan. Eisaman has a wonderful wife and five great children.

RAMON J. ELIAS (LTJG) was born on May 16, 1925, in North Royalton, OH. He enlisted in the USNR on Nov. 1, 1943, and entered the V-12 Program at Brown University. He was stationed at

Newport, RI and the Atlantic. Elias was discharged in July 1946.

Married, has three children. He is the Director of the Dezign House.

HAROLD L. ELLIOTT followed the V-1, V-5, V-7 and V-12 service and received his discharge from the Navy at the age of 20 in 1946. He received his BS and Masters degree from Northern Arizona University, Flagstaff, AZ. He taught high school mathematics and coached all sports at Kingman and Yuma, AZ, prefaced administration and principalship at KOFA High School, Yuma, until 1975. Successful second career in investments continue to be the focus of business from 1975 to date.

His past and continuing service to community and state involves volunteer and professional efforts in all areas of education, school administration, Chamber of Commerce presidency, Rotary, Episcopal Church work, Arizona Academy and extensive service in NAU. He holds multiple honors in all of these areas.

He married Marion L. Dick in 1952 and they have two daughters, Diana and Linda. They enjoy world-wide travel and have home-hosted some 25 foreign students whom they very much enjoyed.

PERRY WILLIAM ENGLAND (A/S, (S), was born on Oct. 23, 1920, in Gastonia, NC. He enlisted Jan 7, 1942, USN V-12 Cadet. Attended Newberry College Aug. 7, 1943. Attended High Point College, BA, February 1946; graduate studies at Duke University; Schools of Arts, Science, Law and Medicine, 1946. Child psychologist, athletic director, Latin tutor (St. Michael's School), associate resident engineer SCSHD, principal St. John's (Corington, LA), import-export owner Old World Trade Entpr., civil engineer and published writer (autor and poet). Has many awards, certificates of recognition and merit with his poetry.

USNR Training Station, 1944; 6th Naval District, Charleston, SC USN Base; temporary assignment at Paris Island Marine Base. His memorable experience at Newberry was when he was told (while he was in the shower) that he had eight minutes to get ready for inspection. He showered, shaved, dressed in white uniform and raced to platoon to fall in place with a graceful swan dive on his belly over the dew drenched grass. Separated from active duty Jan. 31, 1946, and discharged March 18, 1948.

Active national rep. committee, NRSC, NRCC, Exec. Comm. NRCLC, charter and honorary life member RPTF. In October 1991 he received special recognition (along with five other Republicans) and honored at a special dinner and dance by Vice-President Quayle and comm. chairman for more than two decades of sacrifice and service to make a great Republican Party. Received gift from Vice-President. In February 1992 he was honored, full and complete presidential commission, Republic Party's highest most coveted award for selfless long-standing work, dedication, sacrifice, achievements for the Presidents and Republic Party to NRSC, NRC, NRCC. Was honored by Presidents Nixon, Ford and Reagan at special events with presentation of gifts.

His wife and son are deceased. He is home-confined and writes and signs publication releases when health permits.

WILLIAM B. ENRIGHT (LTJG) was born on July 12, 1925, in New York City, NY. He enlisted in the USNR on July 1, 1943, and attended Dartmouth College and Midshipman School at Ft. Schuyler, NY. He was stationed in the Pacific on the USS *Marcus Island* (CVE-77).

He was discharged in August 1946 and received several theater medals.

Enright married Betti in 1951, and they have three children: Kevin, Kim and Kerry. On June 30, 1972, he was appointed as U.S. District Judge in San Diego, CA.

THOMAS E. ENTENMANN (CDR) was born Oct. 15, 1922, in Philadelphia, PA. Attended V-12 School, Bloomsburg University (PA) then Bloomsburg STC, July 1, 1943; Midshipman School, Northwestern University, Tower Hall) and commissioned May 10, 1944. Entered the USNR Nov. 22, 1942 (V-7, DV (G). Upon commissioning was assigned to LST-77 operating in the ETO. They carried POWs from Anzio to Bizerte. Operated out of Salerno-Naples area and Palermo, Sicily carrying out missions incidental to LST.

Transferred to the British Navy in Bizerte to LST 551 and participated in the invasion of Southern France. Returned to the U.S. and was discharged in July 1946. Returned to Millersville University and graduated May 1947 with BS. Taught two years in Philadelphia public schools; earned his MA degree from NYU just prior to be recalled for Korean conflict.

Assigned to USS *Prosperpine* and served as 1LT and gunnery officer for two years. Returned to teaching in 1953, but remained active with Reserve Program in Philadelphia area. Commanded the 1st Fleet Div, then assigned as CO of Mobilization Team at Folsom, PA. Rounded out his 21+ years as a naval academy information officer. Awarded the American and European (one star) Medals, Naval Reserve Medal and National Defense Medal.

Married Margaret and has four children: Martha, Thomas and twins, Richard and David. They each have two children, making him grandfather of eight.

Enjoying the retired life by traveling, golfing, reading and knitting. He stays active in church work, Lions Club and drives for Meals-on-Wheels.

DAVID ANDREW EPTING (CDR) was born on Dec. 25, 1922, in Columbia, SC. He entered the Navy V-12 Program at Newberry College on July 1, 1943, and was commissioned as ensign at Notre Dame in January 1944.

Served with the Supply Corps, Armed Guard Center, Brooklyn, NY from January-August 1944; attended Harvard Supply Corps School in August 1944; was supply and dispersing officer APL-34 from January 1945-August 1946; served in the Reserves until 1973, then retired as commander.

His graduate work was done at the University of South Carolina. He worked as a letter carrier for the U.S. Postal Service and retired as treasury cashier of Southern Bell in South Carolina.

Epting is a life member of Ebenezer Lutheran Church. His community activities include Board of Governors of Town Theater, Executives' Club, Telephone Pioneers of America, Fair Association, President's Committee, Newberry College, and District Governor of South Carolina Lions.

Honors include the Algernor Sydney Sullivan Award from Newberry College, South Carolina Lions' Hall of Fame and Honorary Church Council member.

Epting married Julie Derrick and resides in the home built by his parents in 1927 in Columbia, SC.

DURWARD R. EVERETT JR. (LTJG) was born on Feb. 20, 1925, in Robersonville, NC. He entered the University of North Carolina Navy V-12 Program on July 1, 1943, and was commissioned ensign following graduation in 1944 from Columbia University Midshipman School, New York, NY.

Served in the Pacific as communications officer for Flotilla Staff, LCI rocket and mortar ships and Amphibious Forces in Okinawa Campaign. Following the end of WWII he served on CVE *Manila Bay,* ferrying troops home from the Pacific. He was released to inactive duty in May 1946 as lieutenant (jg) USNR.

Graduated from the University of North Carolina, Chapel Hill; received BS in commerce in June 1947; and joined Wachovia Bank of North Carolina in Raleigh, NC on July 7, 1947. He was Durham city executive in August 1970; Asheville in April 1974 as western regional executive and retired on Dec. 31, 1985, as regional vice-president. He is now living in Biltmore Forest, Asheville, NC.

He and his wife, Iris, have three daughters

living in Durham, Greenville and Raleigh, grandsons and one granddaughter.

CHARLES FARBER (LTJG) was born April 21, 1924, in Newark, NJ. He enlisted in the USNR on July 1, 1943 and entered the V-12 Program at Duke University. He was then assigned to the Midshipman School, OCS, at Notre Dame where he was commissioned.

Served in the PC-592, Pacific Fleet and participated in the planned amphibious invasion of Japan. His memorable experience was the typhoon off the coast of Japan in 1945.

LTJG Farber was discharged in July 1946. He is retired.

CLAYTON C. "BUD" FENTON JR. (COL) was born on June 14, 1923, in Beulah, MI. He enlisted Dec. 15, 1942, and entered the V-12 Program at the University of Michigan (1943-1944). He was commissioned on March 28, 1945, in Quantico, VA. He then earned his BS in industrial engineering at the Michigan State University (1948) and his MBA at the University of Michigan (1958).

Served in the USMC and was involved in supply, maintenance and logistics. He served overseas in Hawaii, Guam, Korea, England, Okinawa, and participated in 11 military operations in Vietnam. His memorable experience was being a captain and the commanding officer of the last Marine contingent to leave Korea.

Col. Fenton retired on Dec. 31, 1972. His awards include the Legion of Merit with combat V, Joint Services Commendation Medal, Combat Action Ribbon and 12 Commendations and campaign medals.

He and wife, Dottie, have two daughters, one son and twin granddaughters. They reside in Beavercreek, OH where he does volunteer work, gardening, fishing and traveling.

JOHN P. FINCH joined the V-12 Program at Stanford on Dec. 7, 1942. He was assigned to Cal Tech on July 1, 1943, and transferred to the USC in February 1944 to replace NROTC Class promoted to amphibs. He graduated as ensign in February 1945, qualified for Radar School at Hollywood Beach, FL and then Fighter Director School at St. Simons Island, GA.

Finch went to Oahu in June 1945 and, when the war was over, he attended Night Fighter School at Barbers Point, Oahu. He was on board the USS *Eaton* DD-510 for about a month at Shanghai during the occupation. Finch taught at the Fleet Training Center, Oahu, until he was released from active duty.

Attended Stanford for a BS, then worked 39 years at General Electric in Erie, PA. As a manufacturing engineer, he was involved in the manufacture of locomotives.

Finch and his wife, Dorothy (deceased 1992), had five children, a foster daughter and three foreign exchange students. Their son, Mark, is a commander in the Navy.

RICHARD T. FISHER (LCDR) attended Millersville University in Pennsylvania for one year prior to entering the V-12 Program at the University of Pennsylvania. Then, he was assigned to the Midshipman School at Notre Dame where he was subsequently commissioned.

After completing several DD service schools, he was assigned to the USS *Borie* DD-704, remaining on board until the ship was decommissioned in 1946. From 1950 until 1970, he was an active member of various NR drilling units and retired as a lieutenant commander.

Returning to Millersville, LCDR Fisher satisfied the requirements for his BS degree in education. He then earned his MS at Penn and taught in the Philadelphia School System for 10 years. He was next employed by the Radnor Township School District in Wayne, PA, from which he retired in 1984. In 1958 he married Alta Acheson, and they reside in Devon, PA. Their daughter, Judy, also a Penn graduate, is a computer science engineer.

REV. DONALD B. FITZSIMMONS (LCDR) was born Jan. 21, 1925, in Boston, MA. He entered the V-12 Program in 1943 at Dartmouth and, upon graduation in November 1944, continued in the V-12 Program at Yale Divinity School. He was released to inactive status in January 1946 and honorably discharged in June 1947.

Graduated from the Divinity School in 1947 and, after ordination, he became a prep. school chaplain. Later, he was a minister to the youth at one parish and an associate minister at another parish. In 1957 he returned to the USN and served in the Chaplain Corps for 20 years.

After naval retirement in 1977, he has enjoyed supply preaching and volunteer denominational work. He and his wife, Mary Louise, live in Hingham, MA. Their son is an attorney.

RAYMOND J. FLOOD (LTJG) was born in Chicago, IL on Aug. 25, 1923. He enlisted in the USN on Dec. 15, 1942 and entered the Navy V-12 Program at the Illinois Institute of Technology, Chicago, IL on July 1, 1943. He graduated with a BSCE in February 1944, then attended Midship-

man School at Camp Peary, VA and was commissioned ensign CEC, USNR, on May 6, 1944.

Attended OTS, Camp Endicott, RI; joined the 70th USNCB (Amphibious Forces) on Aug. 15, 1944; served at Camp Parks, CA; Oahu; Russell Islands; Okinawa; Guam; and Nagasaki, Japan. He was assigned to the 53rd USNCB on Guam, then CBMU 503 at Peleliu, Palau Islands. He was released from active duty at Great Lakes, IL on July 10, 1946.

Served as construction engineer superintendent with contracting & material company, Evanston, IL (1946-1956); and as vice-president and chief geotechnical engineer with Walter H. Flood & Co., Inc., Chicago and Hillside, IL (1956-1983). Flood is now retired. He married Barbara Burnett in 1951 and they have three children: Raymond Jr., Jean M. and Gregory G.

EDWARD J. FRIEDERICH (LT) enlisted in the USMC, OCS Class 1942, in St. Louis, MO. He entered the V-12 Program at Miami University in 1943 and was discharged from the USMC in 1944 to enter Navy Flight Training at the University of Pennsylvania. After training at Atlanta, GA, he was assigned as an air controller at the NAS in Corpus Christi, TX. In 1948 he was appointed ensign and assigned as aircraft maintenance officer to the USNAS, St. Louis, MO.

Graduated from St. Louis University with BS and MS degrees and was certified in secondary education. He taught in secondary schools and the St. Louis Community College; owned and operated a vending business for 25 years; and retired from the Human Development Corp. as a community services' supervisor.

Married, he and his wife, Kathleen, have two sons and two daughters. They make their home in beautiful Lake Sherwood, MO, where he participates in local and state horseshoe tournaments and in daily walks which keep him physically fit.

NORMAN FRIEDMAN (LTJG) was born in Boston, MA on April 10, 1925. He enlisted in the USNR in June 1943; attended the V-12 Program at MIT and Harvard; and Midshipman School. His military locations/stations included Cambridge, MA; Throg's Neck, NY; Norfolk, VA; Key West, FL; and Guantanamo, Cuba.

His memorable experience was serving as sonar officer on board the USS *Stribling* (DD-867) and commended by the hero of *They were Expendable* (book and movie). At one point they were deployed with their squadron to steam from Guantanamo to Port-au-Prince, Haiti to calm potential violence.

Friedman was discharged from active duty in the Spring of 1946, with occasional active duty recall in the following years. He received the American Theater and Good Conduct Medals.

He has two children and two grandchildren. Friedman is a retired English professor and active psychotherapist, etc.

HERBERT L. FRITZ (HERB) (LTJG) joined the USN as an apprentice seaman in May 1942. He was a first class petty officer with duty in the South Pacific until March 1944 when he entered the V-12 Program at the University of Washington. He stayed in the program until January 1946, at which time he was discharged on points and returned to civilian life.

Finishing his schooling at the University of Washington, he obtained his degree in electrical engineering, and was subsequently commissioned in the USN. He stayed in the Reserve until his work took him to Canada in 1955. He resigned his commission as lieutenant junior grade in 1956.

After working as an engineer for Tacoma City Light, and later as an engineer for a major construction firm in Tacoma, that company opened a branch office in Vancouver, British Columbia. He went north to manage Tide Bay Construction Co. and Tide Bay Dredging Co., later purchasing those companies, which he later sold to the Dillingham Corp., a world-wide construction organization. He retired from Dillingham Canada as vice-president and director of Operations for Canada in 1973. He is a lifetime professional engineer in British Columbia, extremely active in amateur golf, winning several provincial titles and serving at different times as president of the Pacific Northwest Golf Assoc., British Columbia Golf Assoc., and as a governor of the Royal Canadian Golf Assoc.

After his retirement, he developed Nico-Wynd Estates, a luxury housing (recreational) development in South Surrey, B.C., which has its own golf course, marina, tennis courts, indoor swimming pool, and recreation building. He lives there now (with a view overlooking the golf course with his wife of 47 years, Mary Anne. Their four children, all professional people, live close by with their families.

FRANK R. FULTZ was born on Feb. 9, 1922, in Louisville, KY. He was a student at the University of Louisville when he enlisted in the USN on Sept. 28, 1942. He was assigned to Class V-1 and deferred for two calendar years from date of enrollment as a college freshman. Active status, July 1, 1943, in the V-12 Program at the University of ouisville. On March 6, 1944, he left the University of Louisville for USNR Pre-Midshipmen Training at Asbury Park, NJ, followed by Midshipman School, Columbia University, New York City.

Next came ATB at Camp Bradford, Norfolk, VA. After training he was assigned to USS LST-1097. The 1097 was commissioned on Feb. 9, 1945. He was assigned to the Asiatic-Pacific Theater and participated in the Okinawa Campaign from May 21 to June 30, 1945. LST-1097 received one Battle Star for WWII service.

Graduated from the University of Louisville on June 14, 1947, with a Bachelor of Arts degree in sociology. He went to work for the Boy Scouts of America in June 1947 and retired on March 1, 1982, with 35 years of professional service.

He married Jean Louise Thielen of Owensboro, KY and has three children: Marsha, John and David. At the time of this writing he resides in Wayne, PA.

ROBERT H. FUST was born in Omaha, NE on Aug. 8, 1924. He entered the V-12 Program on July 1, 1943, at Doane College, Crete, NE; Notre Dame Midshipman School, June 1, 1944; and was commissioned ensign in October 1944. He was assigned to the USS *Dorothea L. Dix*, troop transport, in December 1944; sailed from New York through the Panama Canal; and spent one and a half years transporting troops throughout the Pacific area.

Made the landing at Okinawa in April 1945; decommissioned the *Dix* at New York in April 1946; was discharged in June 1946; and resigned from the Reserve in the Spring of 1949.

Attended the University of Missouri for his BS in public administration (1948) and the University of Denver for his MS in government management (1949). He served with the New Jersey State League of Municipalities for 31 years, doing municipal research. He served as executive director of the League for 22 years, retiring in 1980.

Moved to Denver and currently resides there with his wife, Margaret. They have a married daughter, Heidi, who lives in Tampa, FL.

HERMAAN GADON (LTJG) was born Oct. 17, 1924, in Worchester, MA. He enlisted in the USN in July 1943 and entered the V-12 Program at Dartmouth College. He volunteered for naval gunfire liaison while attending Small Boat Training School in Miami, FL. He served in the Pacific Theater on the USS *Morris* (DD-417) during the invasion of Okinawa where the *Morris* was hit by a Japanese kamikaze (*Kate* torpedo bomber) on April 6, 1945, and put out of commission.

Completed service on the USS *Baltimore* in

Japanese waters and on LST-1012 in Chinese waters. He was discharged as LTJG on July 31, 1946.

Graduated from Dartmouth College with an AB degree in 1947, and from Massachusetts Institute of Technology with Ph.D. in economics in 1953. As professor of management, he helped found the Whittemore School of Business and Economics, University of New Hampshire in 1964; The Indian Institute of Management in Calcutta, India in 1967; the MBA Program, IMEDE, Lausanne, Switzerland in 1971; and the Executive Program for Scientists and Engineers at The University of California in San Diego in 1983.

Retired from UCSD in 1991 and now lives in La Jolla, CA. He is married and has four children.

EARL B. GARDNER (ENS) was born on Sept. 18, 1923, in Ladoga, IN. He joined the service in May 1943, entered the V-12 Program at DePauw University and Midshipman School at Notre Dame University.

Other duty stations included Tsingtao, China; Ft. Schuyler, NY; USS *Cassin* (DD-372); Mare Island; and Great Lakes. His memorable experience was riding the subway to the wrong end of New York City looking for Ft. Schuyler. Also memorable was being a member of the China area champion basketball team.

He was discharged in May 1946. Gardner is married and they have one daughter and three grandchildren. He is now retired.

EDGAR DAY GATES (ENS) entered the Navy V-12 Program in May 1943 at Baldwin Wallace College; Miami University; Asbury Park and Columbia MS Program. He was primarily a Navy musician (trombonist); attended the USN School of Music, 1945; and was with the USS *Pennsylvania* (BB-38) Pacific Flagship Band, 1945.

He engaged in the battle at Wake; torpedoed Okinawa (Aug. 12, 1945); and was discharged to the Reserve in May 1946. He graduated as a certified psychologist, APA, Miami University, AB, 1948; Ohio University, MA, 1949; and Ph.D. work at Tulane, University of Maryland, 1951;

Gates was director of personnel and held corporate positions in Baltimore, New York and Connecticut: General Foods Corp., director of Personnel, 1960-70; vice-president of Personnel for Graybar Electric, 1970-80; vice-president and chief AO of American Maize Products, Stamford, CT, 1980-87.

He is retired and has homes in Greenwich, CT and Palm Beach, FL, where he lives with his wife, Patricia. They have a married son, Philip; a daugh-

ter, Karen; and four grandchildren. He is listed in various *Who's Who* directories and in the *USS Pennsylvania Directory*.

RICHARD I. GAVIN (ENS) entered the V-12 Program on July 1, 1943, at Hampden Sydney College, VA. After two semesters there, he was transferred to Cornell University in March 1944 to pursue a mechanical engineering course. He was commissioned an ensign in February 1946 and, with hundreds of other V-12ers, was sent on a training cruise on the USS *Montpelier* (CL-57). He was released to inactive duty on July 9, 1946, returned to Cornell University in the Fall of 1946, and received his BSME degree in February 1947.

He commenced work with the firm of Sargent & Lundy Engineers in Chicago, on March 3, 1947, as a junior mechanical engineer. He was appointed an associate of the firm in July 1964 and was elected a partner in January 1968. As a project director for major electric power plant projects, he was involved in many of these projects in the United States as well as Indonesia and Mainland China. He retired from the firm on Dec. 31, 1989.

He was married on Sept. 16, 1950, to Jean Macferran; they have one daughter, five sons and 14 grandchildren. The Gavins winter in Scottsdale, AZ and spend the rest of the time in Northbrook, IL.

ROBERT D. GEE (LCDR) was born in Elkins Park, PA on Sept. 3, 1925. He enlisted in the USN in 1943 and, after participating in the V-12 Program at UNC for two years, he entered the V-7 Program at Notre Dame. After being commissioned, he served aboard the USS *Healy* and participated in the occupation of Japan.

Served on the East Coast in Minewarfare School and on minesweeps. He attended General Line School at Newport, RI and had duty aboard the USS *Badoeing Strait* at the start of the Korean War. He served in the defense of Pusan, support of the withdrawal of the Chosin Reservoir, and in the Inchon Landing.

Received awards and medals for the American Theater Campaign, Pacific Theater Campaign, WWII Victory, Presidential Unit Citation, China Service, National Defense, Japan Occupation Service, Korean Campaign, five stars for UN Service and authorized Command at Sea designation.

After retirement in 1963, he earned an MBA at the University of Hawaii; he stayed there and taught accounting in addition to serving as undergraduate advisor at its College of Business.

He and his wife, Robbie, live in Honolulu, HI. Their daughter, Katherine, and granddaughters,

Puakaliloa and Kahaowaiolu, visit often from their home on Molokai. His hobbies include volunteering at Sea Life Park and Hanauma Bay, his Macintosh computer, bridge and crossword puzzles.

ROBERT N. GIFFORD (ETM3/c) entered the V-12 Program at the University of Southern California in July 1943. He was transferred to the University of New Mexico in March 1944, transferred (because of eyesight) to the radio tech program in mid-1945 and was discharged as an ETM3/c in May 1946.

His degrees are a BA in psychology (1948); BSEE (1951); MS in engineering biotechnology (1963); and Ph.D. in instructional technology (1980). Most of his career has been in aerospace in the Los Angeles area, initially as an EE, but after 1958 as a human factors and training systems engineer. He also taught electronics and industrial psychology courses evenings at local colleges.

Gifford and his wife, Ramona, a public school music teacher, have four children. He retired from Northrop and went to consulting in February 1994. His interests are in equipment design, real estate, genealogy, walking and revisiting some of his great friendships, such as those formed in the V-12.

J. RAY GILMER (ENS) was born in Wolfe City, TX on Jan. 6, 1926. He entered Millsaps College for the V-12 Program in March 1944; commissioned ensign in July 1945 at Notre Dame Midshipman School; and attended Gunnery Training School at Anacostia Naval Base, Washington, DC. He was assigned as 2nd Div. officer on AA cruiser, USS *Tucson*, a gunnery training ship for the Navy Boot Camp, San Diego, CA, from January 1946 until his discharge in 1947.

Graduated from Southern Methodist University in Dallas, TX with a BS degree in electrical engineering in 1948. He joined Varo, Inc., an electronics firm in the defense and space industry, and resigned as vice-president of Corporate Development in 1971 to form Neuro Systems, Inc. Currently, he is founder and president of Neurotone Systems, Inc. which manufactures medical electronic devices for treatment of stress.

Gilmer and June Womack were married on Aug. 4, 1945. They have two daughters, Jamie (married to Dr. Craig Williams) and Gayann (married to Jim Snow). They also have five grandsons and one granddaughter. The Gilmers reside in Garland, a suburb of Dallas.

WILLIAM F. GIROUARD (MAJ) was born in Massachusetts on May 10, 1921. He enlisted on

94

Jan. 23, 1942, and served in the Pacific 3rd Def. Bn., 1942-43. He entered Oberlin College V-12 Program in November 1943 and was commissioned 2nd Lt. in September 1945 at Quantico, VA. He was released to inactive duty and commissioned in the Marine Reserves in November 1945.

Returned to Oberlin and graduated in 1947. He taught public and private schools in Boston and North Dartmouth, MA until recalled to active duty in June 1951. He was assigned to Drill Instructor's School (OinC) MCRD, San Diego and released to inactive duty in November 1952. In 1971 he resigned from the Reserve as major.

In 1955 he entered the School of Engineering, University of Southern California, and received his BSIE in 1958 and MSIE in 1959. He was director/manager Ind. Engr. and Facilities, Singer Co. from 1959-86; adjunct professor, USC from 1959-86; Who's Who in the West, Fellow, Institute of Ind. Engrs. and Institute for Adv. of Engrs., and is presently professor IE and Mfg. Eng. Cal Poly, Pomona.

Girouard is married and has two children.

NEIL GORCHOW (ENS) was born in Sioux City, IA on June 23, 1925. He entered the Navy V-12 Program at St. Ambrose College in July 1943 and continued in the V-12 Program at Depauw University. He was commissioned ensign at Northwestern Midshipman School in May 1945; attended Harvard Navy Communication School in June 1945; and served in the Atlantic aboard the destroyer, *Eugene A. Green* (DD-711), as communication officer until his discharge in July 1946.

He graduated from Iowa University in June 1948, continued graduate studies, then joined the family business. In April 1956 he joined Sperry Univac. After assignments in St. Paul, Dayton and Washington, DC, he moved to world headquarters in Bluebell, PA in December 1965 as vice-president of Systems Programming. He served as chairman of the executive committee and held vice-presidential positions both in marketing and engineering. Gorchow was cited by NASA for contributions to the Manned Space Flight Program.

Following his retirement in 1984, he served as a computer consultant; was appointed as a member of National Defense Executive Reserve; and is past-president of Beth Sholom Congregation, Elkins Park, PA.

He resides in Rydal, PA and Sarasota, FL with his wife, Roslyn (Wein). They have four married children and five grandchildren.

EDWARD C. GOSS entered the V-12 Program at Cornell on July 1, 1943. He was commissioned

at Notre Dame on March 7, 1945. His first assignment was to deliver harbor tug, USS *Chahao* (YTB-496), from Brooklyn, NY to Leyte Gulf for the invasion of Japan (almost made it). Returned fleet rescue tug, ATR-86, from the Pacific to Boston for decommissioning. He was discharged in July 1946.

Memorable experience was when the *Chaha* was strafed on its first day underway off the Ne Jersey shore by USMC "Corsair." Their aim wa bad, no hits were registered, but 20mm fire def nitely was unfriendly. They had considered th Marine Corps at least an Ally. He still number many Marines among friends in his local Nav League Chapter. Sixteen swabbies are foreve grateful for the lousy markmanship of the Marin Corps and Harry Truman's decision-making.

Goss graduated from Clark University i 1951. He was employed by Citibank; establishe his Armank (NY) office in 1965; relocated Vero Beach in 1983; and is a self-employe publishers rep.

JOSEPH WILLIAM GOSS JR. (LTJG) a tended Washburn University and UCLA; appren tice seaman through lieutenant, USNR Suppl Corps. He was born in Terre Haute, IN on Nov. 1 1925. He was dispatcher for Aircraft Component Van Nuys, CA in 1943; enlisted on May 14, 194 (age 17); was assigned to V-12, Washburn Unive sity, Topeka, KS; temporary assignment at Wint General Hospital, USA from Dec. 10, 1943-Jan ary 1944 with a broken leg.

Assigned to the University of California, L Angeles V-12 Supply from February 1944-Ju 1945; Naval Supply Operational Training Crs Bayonne, NJ from July-September 1945; was o Times Square V-J night; and released to inactiv duty at Lido Beach, Long Island, Sept. 17, 194 He was commissioned ensign at age 19; Nav Supply Operational Training Center, Bayonn August 1947; USS *Carpalotti,* destroyer escor Naples, Sardinia, Marseilles, December 1952; ar discharged as lieutenant junior grade on April 2 1962.

Married Aug. 11, 1958, to Patricia N. Man San Francisco; she passed away in 1989. His so Walter Woodbridge Goss, CAPT, USA INF, r tired from Berlin duty and is with Averitt Truc ing, Mobile, AL. He has a granddaughter, Sara Megan Goss, and a daughter, Sarah Hope Goss, French teacher in Magnet School, Long Beac CA.

Worked at the confirmation desk, E.F. Hutto 1945-46; UCLA 1946, BS in business administr tion (class speaker); Stanford Law School, 194 LLB (redesignated JD), and member of first Mo Court Board; HQ, European Cmd. and USA E rope Communications Zone, Fontainebleau ar Orleans, France (G-4 Supply and General Pu chasing Divs.) 1950-53. He was city attorney f Oxnard, CA, 1953-59; the California Stat

Controller's attorney, Sacramento, 1959-61; Municipal Court Judge, Oxnard-Port Hueneme Dist., 1961-65; Associate Justice, High Court, Trust Territory of the Pacific Islands (Micronesia), 1965-68; District Court of Guam, 1965-67; Associate Justice, High Court of American Somoa from 1967; Acting Chief Justice, 1969; member, Board of Land Appeals (U.S.), 1972-81; Virginia gentleman (pseudo), Hatcher's Run, Lincoln, Loudoun County, VA.

JAMES S. GRATTON (LTJG) was born in Palisades Park, NJ on April 5, 1925. He joined the service in June 1943 and entered the V-12 Program at Yale College.

Other duty stations included New Haven, CT; Bunker Hill NAS, Peru, IN; University of Pennsylvania, Philadelphia, PA; and New Haven, CT.

He was discharged in the Summer of 1946. He is married to Carole S. Stevens and they have two children, John Duncan and Jeffrey Nielsen. Gratton is a retired attorney.

WILLIAM H. GRIFFY (ENS) was born in Anderson, IN on Aug. 30, 1924. He began his second year at the University of Notre Dame in July 1943 in the Navy V-12 Program, then entered Pre-Midshipman School at Asbury Park, NJ in July 1944. He was there on Sept. 1, 1944, when a hurricane came up the coast and destroyed the boardwalk and concessions; he stood guard duty against the looters.

Commissioned as ensign in January 1945 at Northwestern Midshipman School in Chicago, IL, he then served as assistant communications officer at the Naval Operating Base, Kodiak, AK. He was released from active duty in Seattle in June 1946.

Graduated from the University of Notre Dame in 1948 with a BS in business administration and spent the next 32 years in technical sales. He retired from Tenneco, Inc. in 1980.

Along with his wife, Valerie (Williams), he raised five children. Daughter, one son and three grandchildren live locally, and two sons live in the Pacific Northwest. After being widowed, he married Eleanor (nee Ash) in 1984. He has resided in Pasadena, CA since 1959 and is now a genealogy library volunteer researching family history.

JOHN L. GROVES was born in Cow Lake (Beedeville), AR on Sept. 14, 1921 and graduated from high school in Senath, MO in 1940. He worked in war production, 1940-41. As a Gen. Motors tech., he joined the USN on March 14, 1942, and entered the V-12 Program at the University of Notre Dame.

Other duty stations included USNTS, San Diego; USS *MacCawley,* USS *Rendova,* USS *President Jackson,* CASU #32, VU-1 USNAS Barbers Point. He participated in all the naval battles for Guadalcanal; Rendova, New Georgia; Bougainville; Cape Gloucester and Solomons, earning eight Battle Stars.

His memorable experiences include the sink-

ing of the USS *MacCawley* and being rescued during the battle by the USS *Talbot* onto the USS *President Jackson;* and being a participant in invasions of Bougainville and Cape Gloucester as a member of medical "beach party" to evacuate the wounded.

Recommended for OCS in May 1944 and selected the University of Michigan, Notre Dame, Washington Med School in St. Louis and all were full. He went four semesters to Notre Dame; the war ended and he returned to naval duty aboard USS *Rendova,* CASU #32 Maui, Ford Island, USNAS Barbers Point. Mustered out CHPHM (temp) on Dec. 31, 1947.

Retired from his own Super K Industries, Flint, MI, where he presently resides with his wife, Gloria. They raised two sons and two daughters, all college graduates. They are proud Papa and Mama. He feels great pride in having participated in the V-12 Program and of his fellow V-12ers who went on to become officers in all ranks in OUR NAVY. He would like to hear from anyone he served with.

JAMES W. GUERIN (CDR) was born in Chicago on Feb. 14, 1923. He enlisted on May 27, 1942, as apprentice seaman and was commissioned to ensign at NROTC at Notre Dame. He was assigned to USS PC-797 as gunnery, ASN officer and OOD. When the 797 was put in mothballs, he went to LSMs.

Assigned to the USS *Prinz Eugen,* an old German heavy cruiser, he was 1st Div. officer and OOD. He went to Bikini for Operation Crossroad (first postwar nuclear explosion experiment) and was the first person to board the *Prinz Eugen* after Test A. For Test B (water explosion) he was possibly the first person to see it. He became a plankowner, submarine division, in San Francisco and qualified as submariner.

His best duty was in the PC and escorting ships taking trucks to Russia (later found out that those trucks were taken to Japanese islands). He was sent to inactive duty in October 1946 and retired as commander. His medals include the WWII Victory Medal, American Campaign, Asiatic-Pacific Campaign, Pistol Markmanship, Armed Forces Reserve and U.S. Naval Reserve Medal.

EDWARD R. GUNION (LCDR), enlisted in the Naval Construction Bn. in 1942 after two years with Pacific Telephone Co. He entered the V-12 Program at Chapel Hill, NC in November 1943 and was commissioned at Harvard University in 1946.

Served at Guam, Marianas Islands until released to inactive duty in August 1946. Then he entered the University of California, Berkeley and graduated in 1947. He returned to Pacific Telephone; retired from the Navy Reserve in 1971 and from Pacific Bell in 1983 after 41 years.

Married, he has two sons, a daughter and two granddaughters. For the past 10 years, he has been enjoying the seashore and golf at Rio del Mar, Aptos, CA.

ROBERT G. HALE, M.D. (LTJG) was born in Philadelphia, PA on April 24, 1923. He enlisted in the USNR on July 1, 1943; entered the V-12 Program at Muhlenberg College, PA, receiving his BS in 1947. Midshipman School was at Columbia University, NY; NTC at Miami, FL; AA (Gunnery) at Great Lakes, IL and CIC School at Pearl Harbor. He  qualified as drill instructor, gunnery officer and 1st lieutenant.

Ships and stations served on included Third ND, New York City, NY; Pre-Midshipman School, Asbury Park; USN Frontier Base, Tompkinsville; and USS PCE R 855. He participated in the American and Pacific Theaters (1944-47), ComServ4Pac. His memorable experience was action in Okinawa in 1945.

Hale was discharged on July 3, 1946. His awards include the WWII Victory Medal, American Theater Ribbon, Asiatic-Pacific Ribbon with one star and the Philippine Liberation Ribbon.

A graduate of Jefferson Medical College in 1951, he interned at Montgomery Hospital, Norristown, PA. Other hospital affiliations include Sacred Heart Hospital, Chestnut Hill Hospital Staff (1952-70), Thomas Jefferson Med. College and Temple University Med. College. He was active in many professional organizations and an Eagle Scout in The Boy Scouts.

He and his wife, Edna (RN), have three children: Robert, Stephen and David. He is a retired physician.

JOHN LYMAN HALL (AKC) was born in Danbury, NC on July 1, 1920. He enlisted in the Navy Seabees in Washington, DC on April 24, 1942. He joined the 5th NCB when it was commissioned and was sent to Midway Island to rebuild Navy facilities damaged in the Battle of Midway. Other stations included Camp Bradford, NAS Quonset, RI; San Diego NTC; V-12 Program at Gustavus Adolphus College for his BA in sociol-

ogy; Pre-Midshipman School, Asbury Park, NJ; USNMS Cornell Univ., Ithaca, NY and NCTC, Davisville, RI.

His memorable experience was when Secretary of Navy, Frank Knox, inspected Seabee fortifications; boat bulk heads, warehouses and airfield were all constructed in 11 months.

Discharged Sept. 8, 1980. His decorations include the WWII Victory, Armed Forces Reserve, USNR Medal, American and Asiatic-Pacific Campaigns and the Good Conduct.

In 1946 he married Elizabeth Schiener, who passed away in 1983. He has two children and five grandchildren. He married Charlotte Palmer in 1990. Hall plays tennis, volunteers at retiree affairs at Pope AFB, participates in senior games and is a director of Sandhills Area Land Trust.

IVAN D. HALSEY (SK2/c) was born in Rockview, WV on July 19, 1922. He joined the USN on Sept. 25, 1942, and entered the V-12 Program at Newberry. Other stations include Port Everglades and Hollywood, FL.

His memorable experience was being the only one in his class to solve a chemistry problem and he received 100% on his semester exam. Also memorable was the mock invasion exercise at Port Everglades.

Halsey was discharged on points March 5, 1946. He received the Good Conduct Medal.

Graduated from National Business College in 1947 with an accounting degree. Worked first as a salesman, changed to insurance in 1955 and received CLU designation in 1973. He retired Oct. 1, 1984, taught a pre-licensing course in Virginia, 1984-90, and received a National Management Award in 1980 from GAMC.

Currently resides in Bluefield, WV with his wife, Thelma. They have two daughters (both with Master's degrees), Nancy Moore and Belenda Bell, and three grandchildren.

GERALD L. HALTERMAN (LTJG) was born in Carbondale, IL on March 13, 1921. He enlisted in the USN on Sept. 8, 1939; went to recruit training with Co. 39-26, San Diego; and served aboard the USS *Oklahoma* (BB-37) for 18 months (Nov. 10, 1939-April 15, 1941) in 4th and C Div. He transferred to NYD Pearl Harbor 14th Nav. Dist. Comm. Office, as yeoman 3/c and witnessed the Dec. 7, 1941, Japanese attack from the top floor of the USN Receiving Station at Merrys Point Fleet Landing.

As yeoman 1/c, he was selected for officer training (V-12) and sent to Ambrose College, Davenport, IA for their first class, commencing April 23, 1943. He spent four semesters there and was battalion commander the last two semesters. He played 3rd base and pitched for their baseball team, was chosen for the tri-city all-star team and appeared on local radio to relate his experiences of the Japanese attack on Pearl Harbor.

Commissioned ensign, USNR, on March 6, 1945, from Ft. Schuyler, NY Midshipman's School. He chose Supply Corps training at Harvard College. Halterman met Rosalie Sharpe of Wellesley Hills and they were married when WWII ended. They have three sons: David, Ronnie and Glenn, and six grandchildren. He retired April 1, 1986, from Missile Systems Div., Raytheon Co. after 35 years in various administrative and engineering positions.

He is a member of Pearl Harbor Survivors Assn. (national and state) and attends USS Oklahoma reunions. He is class secretary of Amherst College, Class of 1947. His ship, the USS *Oklahoma*, took five or more Japanese torpedoes, capsized and suffered 429 killed. Their names are now engraved on plaques at the embarkation point for the USS *Arizona*. His three buddies in the Communications Office all perished in the attack.

DOYLE ROACH HAMILTON JR. (ENS) entered the V-12 Program at Southwestern Louisiana University, Lafayette, LA on July 1, 1943. After a short stay at Asbury Park, NJ Pre-Midshipman's School, he became "an officer and a gentleman" at Fort Schyler, Bronx, NY.

Ensign Hamilton served aboard the USS *Louisville*, followed by Pearl Harbor, Guam, and a tour of duty with the port director, Naha, Okinawa, where he received a Commendation Ribbon for having successfully headed a rescue team that removed eight men from a sunken ship. Finally came duty on an LST and an LCI where he served at Tsingtao, China and finally Shanghai.

Returned to the States in July 1946 and immediately entered Tulane Medical School, graduating in 1950. After two years of post-graduate study, Dr. Hamilton opened his family practice office where he continues to serve the people of Monroe, LA. Dr. and Mrs. Hamilton have three children and six grandchildren.

Hamilton says, "I am indebted to the USN and particularly the V-12 College Training Program for its part in helping me become a man and educating me at the same time."

ROBERT B. HAMILTON JR. was born Oct. 8, 1924. He entered the V-12 Program at the University of Pennsylvania and graduated from Notre Dame Midshipman School in October 1944. He served with Underwater Demolition Team 22 in the Pacific from November 1944-December 1945.

Returning to Coronado, he was assigned to the Shore Patrol at the Naval Gun Factor in Washington, DC and 5th Naval District until placed on inactive duty in July 1946. Returning to Penn in the Fall of 1946, he graduated with the Class of 1948 and was employed by the Atlantic Refining Co. and the Atlantic Richfield Co. for 41 years. After serving as a full-time consultant to Royal Dutch Shell for three years, they have since functioned in that capacity to other major oil companies.

After 20 years he retired from the Reserve, and after 20 different addresses, he settled near his hometown (Williamsport, PA) in Montoursville, PA. His son, Robert B. III, and daughter, Cynthia Hamilton Phillippe, and three grandchildren, reside in the Capitol District in Upstate New York.

JOHN C. HAMMETT (LT) was born in Washington, DC. He joined the USNR on July 1, 1943; entered the V-12 Program with two semesters at the UVA, 1943-1944, and two semesters at the University of Richmond in 1944. Hammett received his AB in 1948 and his MBA in 1952 at GWU.

Other stations include the USS *Graylab*, YMS-120 and Ft. Schuyler, USNRMS, in March 1945. His memorable experiences were ports of call in Hawaii and the Panama Canal.

On July 13, 1946, he was released to inactive duty. Hammett and his wife, Dolores, have four daughters and nine grandchildren. He retired from Addressograph-Multigraph Corp. in 1972 and from HHS-PHS-FDA in 1988.

CHARLES BROOKS HANDY entered Westminster College (Missouri) V-12 Program on July 1, 1943. He was commissioned ensign in January 1945, Northwestern Midshipman School, Chicago. He served in the Pacific aboard LST-1103 and was communications officer and executive officer when the ship was decommissioned in June 1946.

The 1103 carried LCT and pontoons to Okinawa. While en route to Pearl Harbor for cargo to invade Japan, WWII ended, and they transported the port director's unit to Japan and repatriated Japanese colonists from the Pacific Islands to Japan. He was released to inactive duty in June 1946.

Education: BA from Westminster College in 1947; MA from Iowa University in 1956; Ph.D. from Iowa State University in 1970; on Iowa State faculty for 34 years, retiring as emeritus professor of accounting and emeritus dean, College of Business, in 1992. He is a CPA.

Handy resides in Independence, IA with his wife, Mary Catherine. He and deceased wife, Donna, have two children: son, Mark, with Armstrong World Industries and lives in Covington, LA with his wife and son; and daughter, Karen, is a student at Iowa University.

ROBERT A. HANLEY JR. (LTJG) was born in New York City on Jan. 11, 1925. He completed the V-12 requirements in September 1944; was commissioned at Columbia University; assigned to Ammunition Task Force and was sent to Leyte, Philippines out of Hawaii. Upon successful completion of the mission to Leyte, Task Force vessels were used to ferry first occupation troops to Japan at the war's end.

Thereafter, he was assigned as Island Commander, Marianas, to assist in rehabilitation of the island of Rota. His last active billet was as communications officer, Pearl Harbor.

After discharge in September 1946, he returned to Union and graduated in 1948. He then entered the U.S. Foreign Service with assignments at embassies in Paris and Baghdad. He joined the Port of New York Authority (PA) in 1954. Career highlights include: Brussels (1957-58) representing P.A. at World's Fair; Cleveland (1961-62) to obtain Midwest shipping tonnage; (1969) assigned to World Trade Center project; and London (1992) as "Owner's Representative-Europe." He retired in 1993.

He married his Schenectady sweetheart in Paris, and they have four married children and nine grandchildren. His only son, Timothy, served aboard USN vessels in Vietnam waters during the Vietnam conflict.

B.J. HANSEN (CAPT) entered the V-12 Program at Kansas State in Pittsburg, KS as a V-5 Naval Aviation Cadet (V-12a) in Los Angeles in December 1943. After 18 months in V-12, he transferred to the University of Oklahoma NROTC and graduated as ensign, USNR. He was then ordered to Commander Amphibious Group I aboard the Flagship, USS *Mt. McKinley* (AGC-7), as Admiral's Staff Officer performing duties as communication watch officer, cryptographer for Bikini atomic tests, beachmaster's communicator and assistant division officer.

During the Korean War, as one of 84 Ready Reservists qualified for Air Intelligence, he was ordered to the original Air Intelligence Team that established Air Intelligence in the 2nd and 6th Fleets with NATO navies in 1951-53, ranked as senior lieutenant, USNR, operating throughout Europe, Scandinavia and the Mid-East with Air Group 17 and the USS *F.D. Roosevelt* (CVA-42).

Three years after his Reserve retirement in 1963, he was sent to Vietnam by the Navy and his Company to perform service for the Navy as a senior vice-president with an assimilated Navy rank of Captain, USN (1966-67).

His last business association was five years in the Kingdom of Saudi Arabia. He now manages his investments in the U.S. For the Navy and Marine V-12 50th Anniversary Reunion in 1993, he recreated a skit he wrote, produced, directed and acted in 50 years before at his Kansas V-12 Unit. It was called, *V-12, We Love You.*

He has five daughters and seven grandchildren living in the States and England. He is listed in numerous *Who's Who* publications, including:

Who's Who in the World, America, World's Who's Who in Commerce and Industry and the *Royal Blue Book in U.K.* Hansen is a published magazine, newspaper and book author and holds patents for his inventions, as well as enshrinement at the International Swimming Hall of Fame for swimming and waterpolo achievements.

STANLEY G. HARDY (LCDR) joined the V-1 Program at Montclair State, NJ in November 1942 and entered V-12 at Drew University, NJ on July 1, 1943. He was commissioned ensign at USNRMS Notre Dame on Jan. 20, 1944; additional training at NTS Cornell, completing in April 1944; assigned to USS *Shangri-La* (CV-38) and commissioned at Norfolk Navy Yard on Sept. 15, 1944. He served in *Shangri-La's* Gunnery and Engineering Departments until May 1946, participating in the Okinawa and Japanese homeland campaigns. *Shangri-La* was the flagship of fast carrier task force commander, Adm. John S. McCain. He was released to inactive duty in June 1946.

Completing his education at Montclair State in 1947 with a BA in science education, he began a career in the chemical industry. He eventually took a position in scientific intelligence with the Dept. of the Army in Washington, the for 16 years he worked for the U.S. International Trade Commission, retiring in 1976 as a senior analyst and assistant chief of the USITC Chemical Div.

Joined the Naval Reserve Div. at Port Newark, NJ in the early 50s, later changing his designator to Naval Intelligence (1635) as a member of the Washington USNR Intelligence Div. 5-2 from which he retired in 1966 as a lieutenant commander after 20 years service. His decorations include the American Theater, Asiatic-Pacific with two Battle Stars, WWII Victory Medal and the USNR Medal.

After 27 years in the Washington area, he and his wife, Louise (a retired Dept. of Defense analyst), relocated to Punta Gorda, FL, where they have resided for more than 10 years.

EDGAR B. HARGER (SN2/c) was born in Noblesville, IN on Jan. 24, 1926. He enlisted in the USN on July 1, 1944, and entered the V-12 Program, University of Illinois State Normal University, IL and Depauw University, Greencastle, IN.

He was discharged on June 4, 1946. His decorations include the WWII Victory Medal and the American Area Campaign Medal.

Following discharge, he joined the Indiana State Police, served 23 years and retired as lieutenant district commander. In 1979 he was elected to the office of county sheriff, serving eight years. He was then elected to the County Council, serving four years and three months before passing away unexpectedly from cardiac arrest on Feb. 24, 1993.

He leaves his wife, Mavis, and two sons,

Mike and Tom. A daughter, Kathy is also deceased. *Submitted by his wife, Mavis (an RN), who says he spoke often of experiences in the V-12 Program and was thankful for the education he derived from those years.*

DONALD J. HARRINGTON (LT) entered the V-12 Program and graduated on Feb. 23, 1946, from Michigan with a BSE degree in aeronautical engineering. Commissioned as ensign on the day he graduated, he was immediately assigned to short tours of duty at the Navy Depot in Washington and at a missile test facility at Wright-Patterson AB in Dayton.

Permanent assignment was as Navy Bureau of Aeronautics representative, engineering officer, at McDonnell Aircraft Co. in St. Louis. While at McDonnell he worked on the design of the first carrier-based jet fighter plane called Phantom-1.

Released to inactive duty, effective Oct. 1, 1946, but remained active in the Naval Reserve. He now is in the retired Reserve with a rank of lieutenant.

Harrington attended the University of Michigan Graduate Engineering School and then worked as a research engineer and as a math instructor, including three years in the automatic transmission design group at Ford Motor Co., which designed the first Ford automatic transmission. He then returned to school again and earned his LL.B degree from the University of Detroit. He then attended the Graduate Law School at George Washington University and earned a Master's degree in law (LL.M) in 1953.

After working as a patent examiner at the U.S. Patent Office, Harrington was a patent attorney for Chrysler Corp. for three years. He then was a patent attorney for Ford Motor Co. for 34 years and retired in 1989. He is presently a partner in a patent law firm in Southfield, MI.

Don and Monica Harrington have been married 45 years and have four children.

NOBLE HARRIS (SPIKE) (LCDR) was born Aug. 18, 1922, in Meadow, TX, the son of D.J. and Clara Copeland Harris. He entered the Navy V-12 Program on July 1, 1943, at Texas Christian University, Ft. Worth, TX. He was commissioned ensign, Navy Supply Corps School, Harvard University on May 31, 1944.

His duty stations included asst. disbursing officer, Navy Amphibious Trng. Ctr., Solomons, MD; overseas, he became officer-in-charge, Purvis Bay Fuel Tank Farm; became the fuel and supply officer for Naval Base, Tulagi, Br. Solomon Islands; and served briefly at the Disbursing Office, Pearl Harbor.

Released from active duty in March 1946 was in the inactive Reserve until June 30, 1969. He received his BS degree in business administration from Abilene Christian Univ. in June 1947, joined the South Texas Lumber Co. until 1950, then worked on his MBA at the University of Texas. In 1952 he began a career in the real estate brokerage business in Abilene which has lasted over 41 years, to date. He has served in the following capacities: president, Abilene Board of Realtors; member, then vice-chairman, city of Abilene Planning & Zoning Commission; member, city of Abilene Landmarks Commission; president, Abilene Navy League; president, Abilene Rotary Club and governor of Rotary International District 5790.

He married the former Bette Joyce of Snyder, TX and they have two children, Paul Joyce Harris and Camille Harris Tomlinson.

RICHARD EARL HARRIS (CAPT) was born in Florence, SC on Dec. 22, 1922. He gradu-

ated, as valedictorian, from Graceville, FL HS in 1940; attended the University of Florida, 1940-43; joined the USMCR in 1942 and entered the Duke University V-12 Program in July 1943. Boot camp was at Parris Island and OCA and SOCS at Camp Lejeune.

Commissioned 2nd lieutenant and served in the 6th Mar. Div. on Guadalcanal, Okinawa, Guam and Tsingtao, China. He was decorated with the Bronze Star in 1945; Purple Heart with Gold Star in 1945; and Presidential Unit Citation for Okinawa in 1945. He is now a captain in the retired USMCR.

Returned to the UF, graduated with a BS in pharmacy in 1947 and worked as a pharmacist from 1947-94 in Clermont, FL. He is active in the Methodist Church, Lions, Merchants & Community; is author of several books, *Yesterday's Remedies For Today's Ills, My Dad's Sayings, Fishin' Tips 'N Tales* and *Much Ado About Many Things*. He enjoys fishing, camping, hiking and writing.

He married Audrey Miner on July 7, 1946, and they have five children: Terry Robert, Sara Dee Kerr, Paula Jean Fleming, Mary Kathryn Jones and Richard Kirk, and nine grandchildren.

DANIEL E. HARRISON (CAPT) entered the Navy V-12 Program at Rensselaer Polytechnic Institute in June 1940. He majored in aeronautical engineering and received a BAE degree with the Class of 1946. He received his commission as ensign in 1947 directly from the Navy Dept., Washington DC; was a weekend warrior in the Navy Reserve, primarily at Anacostic and Andrews AFB outside of Washington DC; department head of maintenance for several S2F Squadrons (ASW); and was weapons effect commanding officer for NASRU units. He retired as captain at the age of 60.

Civilian work included aeronautical research at NACA (now NASA, Longley Field, VA), Johns Hopkins Applied Physics Laboratory as staff engineer, Navy Missile programs. He joined the GE Missile and Space Dept. as systems engineer and later as program manager of missile projects.

Harrison married Margaret Knott from Dinwiddie, VA in 1951 at the NAS, Breezy Point, Norfolk, VA. They have two children, Ann and Dan, both married. He retired to New Bern, NC and is enjoying the many activities the area offers, and his two grandchildren, Joseph and Bailey.

PHILAS JACKSON HARTSELL (JACK) was born in Charlotte, NC on Aug. 24, 1925. He entered the Navy V-12 Program in November 1943, while attending the University of North Carolina at Chapel Hill, and continued the program at the University of South Carolina at Columbia.

He began Naval Training School at Great Lakes, MI; Hospital Corps School in San Diego; served at the USNH in Newport, RI and Okinawa and was discharged from Camp Shelton, VA in 1946.

Hartsell graduated from the School of Mortuary Science in Nashville, TN. He has since owned and operated a family funeral home business in Midland, NC, where he currently resides with his wife, Ramelle. They have one son, two daughters and three grandchildren who live nearby.

JAMES L. HARTSOCK (AVCAD) was born in Clearfield, PA on May 27, 1925. He enlisted in

the Navy V-5 Program in April 1943 and was assigned to the V-12a Program on July 1, 1943, at Pennsylvania State University.

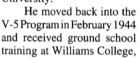

He moved back into the V-5 Program in February 1944 and received ground school training at Williams College, pre-flight at the University of Georgia, flight training at Memphis and Pensacola NAS, and he was released from active duty in September 1945 as an aviation cadet.

Completed undergraduate work at Penn State and received his degree in industrial engineering in June 1949. He accepted a position at Eli Lilly and Co., Indianapolis and remained with Lilly for 40 and a half years, retiring in 1989 as a systems associate.

Currently serving as a church business administrator in Indianapolis. He is married and has one son, now a student at the University of California in San Diego.

JOBE E. HARVEY (SN2/c) was born in Sheridan, AR on Jan. 29, 1927. He joined the USNR on March 2, 1944, and entered the V-12 Program at Arkansas A&M University, Monticello. Other duty stations and schools included Great Lakes Naval Training Center; Shumaker, CA; University of Louisville, KY; Treasure Island, San

Francisco; Kaizer Shipyard, Portland, OR; an Midshipman School/Marine OCS.

Harvey was discharged from the service o July 15, 1946, and he graduated from Texas Chris tian University in 1948 with a degree in busines administration. He was self-employed for 42 year and retired in 1990.

He is married and has three children, thre grandchildren and one great-grandchild. Currently he is an investor and a member of the Board O Directors. First National Bank.

HARRY HASEKIAN (SGT) was born in Lo Angeles on April 6, 1923. He entered the V-1 Program at Redlands University in 1942. He serve in the USMC as a sergeant in the South Pacific an headed a scouting team on the invasion of Okinawa and mop-up on Guadalcanal and Guam. He partici pated in the Japanese surrender and occupation i Tsingtao, China. He was discharged in April 1946

Witnessed the press taking pictures of Simo B. Buckner, commander general of the 10th Arm (including the USMC), as he stood high on a roc and was killed when a Japanese soldier dropped 30mm shell. His souvenirs of the war include riddled blanket from action on Okinawa when h was practically buried alive (burst eardrums) an a samurai sword taken from a Japanese officer. H also received various decorations and medals fo action.

Graduate of the University of Southern Cali fornia, he then attended Law School. He has bee with Prudential Ins. for 40 years and is active i church and community. He and Dorothy hav been married 43 years and have four sons: Charles Harry, John and Stephen III.

H. HARLOW HAYES (CDR) was born i Leona, OR on Nov. 23, 1922. He enlisted in the V 1 Program in 1942; entered V-12 in 1943 at Par College; was commissioned in the Chaplain Corp in 1947 and served combat duty with the 1st Ma Div. in Korea. He served aboard ship and variou Navy and Marine installations in the States an overseas until retirement in 1971.

Awarded the Bronze Star and Navy Com mendation Medals with combat "Vs" for service i

Korea. Hayes received a BA from Park College; MA from Union Theological Seminary; MA from Boston University; Ed.D. from Seattle University; and post graduate, Harvard.

After Navy retirement, he served as dean, Whidbey Campus, Skagit Valley College, Oak Harbor, WA and retired in 1989. The new classroom and library building was dedicated as Hayes Hall in 1993. He and his wife, Harriette, have three children; they presently operate a farm near Oak Harbor where they raise registered polled Hereford cattle.

DALE C. HAYNES (LCDR) was born in Newark, OH on May 29, 1925. He entered the USN in August 1943; NTC Great Lakes, IL; Navy Radio School, University of Wisconsin; Navy Radio Tech School, Texas A&M; radioman 2/c. He entered the Navy V-12 Program at Denison University from July 1944-June 1945 and NROTC at the University of Kansas, July 1945-June 1946. He was released to inactive duty in June 1946 and commissioned ensign in April 1947.

Graduated from Denison University with a BS in physics in June 1947. He was recalled to active duty from USNR Organized Reserve in September 1950 during the Korean War. He graduated from Naval General Line Post Graduate School, Monterey, CA, in Advanced Communications, in December 1950.

Assigned to the USS *Rockbridge* (APA-228), attack transport, where he served as communications officer and assistant operations officer in both the Pacific and Atlantic until released to inactive duty in October 1952. He was discharged from the USNR in 1966 as lieutenant commander.

Worked for Westinghouse Electric Corp. for 23 years as merchandise manager of Major Appliance Div., and 18 years as district manager of the Commercial Div. for Sears Roebuck & Co. He retired in 1990 and currently resides with his wife, Burneal, in Westerville, OH.

RALPH C. HEATH went into the V-12 Program (after high school) at the University of North Carolina on July 1, 1943. He was commissioned at Columbia University Midshipman School in April 1945. Heath attended Advance Line Officers School at Miami, FL where he met his only wife, Martha Sandidge. He served on the YMS-218 in mine sweeping operations at Pusan, Korea and Bikini, and was discharged in June 1946

Graduated from UNC, Chapel Hill, with a BS in geology and joined the U.S. Geological Survey in 1948. He worked in Florida, New York, New England, North Carolina, and served as district chief in both New York and North Carolina.

Since retirement from the USGS in 1982, he has taught hydrogeology courses at NC State, Duke, UNC, and he operates his own consulting business. He served as first chairman of the Board of Registration, American Institute of Hydrology, 1983-84, and in 1990, as the Darcy Distinguished Lecturer of the National Ground Water Association.

JAMES B. HELME (1LT) was born in Rye, NY on April 27, 1924. He joined the USMC Intelligence on Nov. 9, 1942, entered the V-12 Program at Cornell, and graduated from Princeton in 1947.

Other duty stations and schools included: 9th Plt. Commander's Class; graduate of Naval Air Combat Intelligence School; Philippine Islands; Quantico; and Newport, RI.

Transferred to inactive duty on Dec. 7, 1945 and discharged on July 4, 1948. Helme has four children and seven grandchildren. He is a retired pediatrician and does Philatlic Research.

JACK F. HELMS (LTJG) was born Oct. 7, 1923, in Tacoma, WA. He enlisted in the service in November 1942; entered the V-12 Program at Willamette, Salem, OR; and Midshipman School at Harvard, Boston, MA.

Duty stations include the USS *Riddle* (DE-185) and the South Pacific. He participated in action at Iwo Jima, Okinawa and twice in the Philippines.

He was discharged in June 1946. Retired from Occidental Chemical, but stays active in the Shriners and Elks. He and his wife, Jean, have a daughter, Teri Lynn Carey, and a grandson, Daniel Carey.

W.L. HEMEYER (WIL) (LCDR) enlisted in the USN in March 1942. He served in the Armed Guard (Pacific) on convoy duty after "boots" at San Diego and Signal School at Treasure Island.

Entered the V-12 Program in November 1943; transferred to NROTC in June 1944; was commissioned in February 1946, followed by subsequent duty aboard the USS *Sicily* (CVE-118).

Attended the University of Illinois for his MA degree in history. For 34 years he taught history and economics in Illinois high schools. He is now retired and is a lobbyist for retiree causes. He enjoys sailing and is a volunteer sailing instructor at Great Lakes. He lives with his wife, Jeanne, in Beach Park, IL. They have three children and five grandchildren.

EUGENE L. HENRIOULLE (SN1/c) was born in Chicago, IL on Sept. 5, 1925. He enlisted in the USN on Feb. 2, 1944; entered the V-12 Program at the University of Notre Dame, Central Missouri State University; and Midshipman

School at the University of California in Berkeley.

His memorable experience is the men he served with who came from all walks of life and made him realize how people have to be able to work together to achieve their goals. The V-12 helped him to receive a college education and become a mature and responsible citizen. He was discharged Nov. 21, 1945.

Henrioulle is a retired sales representative. His family includes wife, Virginia; son, Ron; daughter, Denise; step-son, Tim; and step-daughter, Colleen.

HENRY CURTIS HERGE SR. (LCDR) played an active leadership role in American education and internationally over a 48-year span. He began his professional career as a secondary school teacher of English. WWII interrupted his tenure as the district-wide principal of schools on Long Island. He enlisted in the USN and subsequently became commanding officer of the Naval War College Training Program, V-12 Unit, at Wesleyan University, Middletown, CT.

Just prior to V-J Day, he was tapped to be associate director of the American Council of Education study of Armed Services Training Programs where, in 1948, he authored the study, *Wartime College Training Programs of the Armed Services*, followed by appointment as state director of Higher Education in Connecticut.

From 1953-1965, Dr. Herge served as dean and professor of the Rutgers Graduate School of Education; accepted assignment as associate director of the Rutgers Center for International Studies; served as a consultant for the International Cooperation Administration and the Agency for International Development in the Rhodesias and Nyasaland, Paraguay, Jamaica; and in South America for the Organization of American States as senior research fellow.

Dr. Herge is the recipient of three earned degrees at the New York University, a Ph.D. at Yale University and an Honorary degree at Wesleyan University. He is now retired and resides in Florida.

NELSON O. HEYER (RADM) enlisted in the USN in 1943 and entered the V-12 Program at Union College. He was commissioned in 1945 and assigned to the commissioning crew of the USS *Cadmus* (AR-14) as assistant chief engineer. He received a BSEE degree from Union College in 1947 and a MS in industrial management in 1951 from Stevens Institute.

Admiral Heyer transferred to the Naval Reserve Security Group Communications. He was promoted to rear admiral in 1979 as the first Reserve flag officer of the Reserve Cryptologic Force and retired in 1983 after 40 years of service.

Retired in 1986 from IBM Corp., Armonk,

NY as director of Human Resource Planning and held positions in design engineering, manufacturing and personnel management. In 1989 he was visiting professor of Labor & Industrial Relations, University of Illinois and in 1987 he formed Heyer Associates, Human Resource Consultants, having blue chip and government clients.

Heyer holds office in the Naval Order of the U.S., American Legion and Naval Reserve Association; Distinguished Fellow of the Human Resource Policy Institute, Boston University; vice-commander general of the Naval Order of the U.S. and is active in community and civic organizations.

His spouse is the former LTJG Nora Frain, USNR. They have four children: Jane, Anne, Susan and Thomas (Union 1978), and four grandchildren: Matthew, Michael, Laura and Douglas.

RAYMOND HIATT (LTJG) was born in Kalamazoo, MI on Feb. 2, 1925. He enlisted on May 27, 1943, and entered the V-12 Program on July 1, 1943. He was commissioned ensign on Feb. 20, 1946, and trained at Newport NTS and on the USS *Cleveland*. He was Air and CIC off. and asst. nav. on the USS *Gardiners Bay;* sea duty included Tsingtao, China; Sasebo and Yokosoka, Japan; and the Marshall Islands.

His memorable experiences include: the close call at NTS hosing back a gasoline fire; sailing through a hurricane on the *Cleveland*; serving on summary court-marshal; experiencing ship crash at the fuel dock at Pearl Harbor; and two weeks of sea sickness. He was released from active duty on April 7, 1947, and returned to Yale for his BS in electrical engineering, Class of 47N.

Married 41 years to Margaret, they have four children and eight grandchildren. After 37 years in elec. construction and design at Gilbert/Commonwealth, he retired in Jackson, MI. He also worked in W. Pakistan and at NASA Nuc. React. Test Fac. He enjoys his family, golfing, skiing, sailing, travel and financial investing.

Attended the Navy V-12 50th at Norfolk in 1993 with his wife. He is grateful for the V-12 opportunity and Navy experience.

DONALD J. HICKEY (ENS) entered the V-12 Program at Berea College, Berea, KY in March 1944; transferred to the University of North Carolina, Chapel Hill, NC in November 1944 for Naval Supply Corps training and, after one semester, entered the NROTC Program. He was discharged in May 1946, graduated from the University of Dayton, OH, and was commissioned ensign in June 1949.

Retired in 1989 from the Dayco Corp., Dayton, OH as vice-president of a sales division after 37 years in engineering and technical sales assignments. He continues to pursue interests in farming and raising purebred cattle.

Resides in Miamisburg, OH (suburb of Dayton) with his wife of 28 years, Sunny. They have two sons, both first lieutenants in the USAF. Steve, a pilot, is based at Moody AFB, Valdosta, GA and Paul, a 1992 graduate of the Air Force Academy, is a cost analyst with the Brilliant Eyes Program, stationed at Los Angeles AFB.

NORMAN W. HICKS (LT) was born in San Antonio, TX on Sept. 22, 1925. He joined the USMC on Feb. 20, 1943; entered the V-12 Program at Southwestern Louisiana Institute (now University of Southwestern Louisiana); and OCS at Quantico, VA.

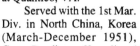

Served with the 1st Mar. Div. in North China, Korea (March-December 1951), Camp Pendleton, Hawaii and Washington, DC.

Hicks left the USMC on Oct. 1, 1970, after 27 and a half years of service. His awards include the Silver Star and Purple Heart for the Battle of Horseshoe Ridge (April 1951), Army Letter of Commendation Medal and the Good Conduct Medal.

Earned BS and MA degrees from the University of Maryland. He is married to the former Elaine Prentiss of San Antonio, TX and they had one son, Norman Jr. (deceased); one daughter, Linda; and three grandchildren. Hicks is retired, but owns a wild game ranch as a hunting preserve for family and friends.

KENDALL A. HINMAN (LTJG) was born in Madison, WI on March 31, 1924. He entered Illinois State Normal University V-12 Program on July 1, 1943. He was commissioned ensign in October 1944 at the Harvard Graduate School of Business Administration, Boston, MA.

Served in the Philippines as ship's store officer, Naval Air Base, Buyuan, Island of Samar. He supervised 75 people in the ship's store, including laundry and tailor shop.

Graduated from the Principia College with a BA in economics in 1947; from Harvard Business School in 1948; and worked in accounting and computing in aerospace with Systems Development Corp., Santa Monica, CA and earlier in industry with Monsanto Chemical Corp. He retired as lieutenant jg in the Supply Corps, but is still in the USNR Honorary.

Hinman is married and has one child.

ALLEN HODGES (LTJG) was born in Greenville, SC on Feb. 16, 1925. He graduated from Lees McRae College and entered the Navy V-12 Program on July 1, 1943 at Newberry College.

He received his commission at Plattsburg,

NY on June 27, 1944, and married Elizabeth L. Swanson on the same day.

Following training in attack boats and UDT at Ft. Pierce, FL, he was assigned to the USS *Horace A. Bass* (APD-124) as boat officer and top watch officer. The *Bass* earned two Battle Stars, including the Okinawa Campaign where the *Bass* served on picket duty. Eleven kamikazes were downed and a submarine was sunk by the *Bass*. Hodges was the boarding officer of the HMS *Nagato* prior to the Japanese occupation.

Discharged as a lieutenant jg in March 1946, he earned a BA, MA and Ph.D. in psychology. He was chief psychologist, at State of Minnesota (1956-59), assistant professor, University of Minnesota (1959-61) and joined NIMH in 1961. He retired from DHEW in 1983 as assistant regional director, Region VIII.

Hodges and his wife have four daughters and five grandchildren. They reside in Ocala, FL where he practices as a clinical psychologist.

CHARLES LOUIS HOFFMAN (LCDR) was born in Dayton, OH on May 10, 1925. He entered the V-1 Program at the University of Dayton in 1943; received his AB degree from Oberlin College in 1945; MD from St. Louis University in 1949. He served in the USPHS, Regular Corps.

Medical internship was at U.S. Marine Hosp., Baltimore, 1949-50. From 1950-1993 he served in various positions with duty at Marine Hospital, Kirkwood, MO; 2nd Coast Guard Dist., St. Louis; Marine Hospital, San Francisco; Marin County, CA; Neumiller Hosp., Tamal, CA; active staff, Marin Gen. Hosp., 1955-93; chief of med. staff, Ross Gen. Hosp., CA.

He is co-founder of Med. Ins. Exchange of California, 1975; med. dir. of Aldersly Danish Home, San Rafael, 1987-92; and med. dir. of Rafael Conv. Hosp., 1987. He is recipient of numerous professional honors and awards. His military awards include the Navy Good Conduct Medal, WWII Victory Medal, American Theater and Korean Service Medal.

Hoffman and Nancy Adele Fahrendorf were married on June 14, 1947. They have four children: Thomas, Mary Lynne Lamb, Lori Brustkern and William Edward.

THOMAS F. HOGAN (S1/c GM Striker) was born in Bronx, NYC on Nov. 29, 1926. He joined the USN on Feb. 7, 1944, and entered the V-5, V-12A and V-6 Program; attended Dartmouth College, Hanover, NH. His stations included Hanover; Great Lakes, IL; Norfolk, VA; and USS *Duncan* (DD-874).

Served at Wake Island on picket duty, Aug. 1, 1945; Eniwetok, Aug. 14, 1945; Seoul, Korea occupation, destroying Japanese mines; then to Russia and down the China coast before going to Yokosuka and Sasebo, Japan, then back to Pearl Harbor.

Hogan was discharged on June 5, 1946. His awards include the American Theater, Asiatic-Pacific and Victory Medal.

He is retired from the NYC Fire Dept.

THOMAS C. HOLLEMAN JR. (1LT) was born in McAlester, OK on Nov. 17, 1923. He enlisted in December 1942 and attended Southwestern Louisiana Institute (now University of Southwestern Louisiana), in July 1943, in the V-12 Program.

Served at Parris Island, Camp Lejeune, Quantico OCS, Charleston Navy Yard, two years in the Panama Canal Zone, Camp Pendleton and one year in Korea, 1951-52.

Participated in the Battle of the Hwachon Reservoir and the Punchbowl Battle with the 3rd Bn., 1st Mar. Regt. He resigned his commission in October 1957. His awards include the Bronze Star with Combat V and the Navy Commendation Medal with Combat V.

Worked on offshore drilling rigs in the Gulf of Mexico and started an oil field service company in 1959. He sold the company in 1990. He resides with his wife, the former Eleanor Landry, in New Iberia, LA. They have three sons, two daughters and 12 grandchildren.

LEROY C. HORPEDAHL was born in Glyndon, MN on April 13, 1924. He enlisted in the Navy on Dec. 1, 1942, while a freshman at North Dakota State University, Fargo, ND in the V-12 Program. He was called to active duty on July 1, 1943, to the University of Minnesota and graduated as BME in June 1945

Attended Midshipmen's School at Notre Dame University and was commissioned Nov. 2, 1945; sent to Amphibious Training Center, Coronado, CA and assigned duty aboard the LC(FF) 790 until June 21, 1946. He was discharged from the Reserve on April 14, 1954.

Taught in the Mechanical Engr. Dept. at Tri-State College, Angola, IN; Montana State Univ. in Bozeman, where he received his MSME; and in the Mechanical Engr. Dept. at North Dakota State University, Fargo, ND. He joined the staff of the Los Alamos National Laboratory on May 31, 1951.

Participated in the above/below ground nuclear weapon testing; was present for arming,

detonating, and observing the largest yield thermonuclear device ever tested at Eniwetok Atoll in the South Pacific. He retired March 2, 1986, and continues to serve the laboratory as a consultant.

He and his wife, Alice, have been married over 45 years and live in Los Alamos, NM. They have four children: Kristine, Gary, David and Paul, and nine grandchildren.

GEORGE E. HOTZ (LCDR) enlisted in the Navy V-7 Program while attending the University of Toledo. Called to active duty July 1, 1943, in the V-12 Program at Bowling Green State University, he was commissioned May 10, 1944, with the 18th Class at Tower Hall, Northwestern University Midshipmen's School.

After completion of SCTC School, Miami, he was ordered aboard USS PCS-1459 and served in the North Pacific and Aleutian Islands until February 1946. He assumed command in Seattle and the ship was decommissioned in April 1946. He was released from active duty in May 1946.

Recalled to active duty in October 1950, he served on the staff of COMPHIBPAC, with flag, aboard the USS *Mount McKinley* (AGC-7). He served six months in the Korean Theater of Operations, during which time he gave logistic support to VADM Joy at the onset of peace negotiations in Kaesong, Korea.

LCDR Hotz was released from active duty in October 1952, remained in the inactive Reserve, serving in Military Sea Transportation Svc. Co. 4-14, Toledo, OH until his retirement in January 1969.

In civilian life, he worked for 32 years in engineering at Haughton Elevator Co. and later Schindler Elevator Corp., Toledo, OH, retiring in 1980.

C. RANDOLPH HUDGINS JR. (LT) was born in Norfolk, VA on July 15, 1924. He enlisted, July 1, 1942, in the USNR; entered the V-12 Program at Hampden-Sydney College; SCTC at Miami, FL; U.S. Naval Base, Manus, Admiralty Islands; Midshipman School, Northwestern University Class-22 and graduated in January 1945.

Assigned to NAS Los Negros, Admiralty Islands with One AVN; decommissioned 13 AVN at war's end; participated in V-J Day in San Francisco; and was discharged June 30, 1946. Hudgins was recalled to active duty during the Korean War and assigned to CINCLANTFLT Intelligence Div. in Norfolk, VA. He was discharged on June 1, 1953.

Hudgins was formerly a trust officer with the National Bank of Comm., Norfolk; since 1956 he has been an investment advisor, Commonwealth Investment Counsel.

Married Anna Black and they have four children: Jane Frazier, Randy, Anna Brooke Wallace and Bill.

WILLIAM A. HUDSON was born in New York City on May 19, 1925. He joined the Corps in August 1943; went to Underwater Demolition School at Ft. Pierce, FL; was assigned to 3-K-25 and was wounded on Iwo Jima on March 15.

Assigned to Cornell University's V-12 Officer Training Program and was discharged in July 1946. His awards include the Bronze Star and Purple Heart.

Graduated from New York University with BS and MA degrees, then taught physical education in the Los Alamos, NM School. He retired in 1982, but worked as a financial planner until 1987. He is now working as a fitness consultant at the Los Alamos National Laboratory Wellness Center.

He is a fitness buff and ran 50 miles on his 50th birthday, 60 miles on his 60th and, at age 65, was on a six-man relay team that swam across the English Channel. He has competed in marathons and triathlons.

Hudson and his wife, Maureen, live in Los Alamos, NM. They have four children: Bill, Art, Ty and Jan. Their youngest son, Ty, served in the USMC for five years and Bill, the oldest son, served in the USN during the Vietnam years.

ARTHUR H. HUEBNER was born in Clifton, NJ in December 1926. He spent his boyhood in Lake Grove, OR and joined the USN V-5 Program in January 1944. He was assigned to V-12A at Willamette University, then to V-12 NROTC at the University of Washington and was commissioned in June 1946.

Volunteered for an extra year of duty and was assigned to the USS *Furse* (DD-882) at the Bikini Bomb Tests. The *Furse* helped pick up drone planes after the airburst and he was assigned to help the tugs beach sinking ships after the underwater burst. None were saved, but the bomb beached a few. The *Furse* then served mostly as a carrier plane guard out of Pearl Harbor, San Diego and San Francisco until she went in for overhaul at Mare Island. He was discharged from there in June 1947.

Returned to Washington and married Marty in December 1947, obtained his BSME in August 1948 and started his career as a design engineer for construction/logging/industrial (forklift) equipment at Hyster Co. in Portland, OR and retired as vice-president of engineering in 1986. He and Marty now split their time between golf courses and homes in Vancouver, WA and Cannon Beach, OR.

HERBERT HUMPHREY (LCDR) entered the Navy V-12 Program at Yale in March 1944;

transferred to NROTC in July 1945; graduated from Yale and was commissioned ensign USNR in June 1946. He served as XO, then CO, USS *Saluda*, IX87, doing sonar experiments in New England waters until released from active duty in May 1947.

Recalled to active duty in December 1950 and served as XO, USS *Umpqua* (ATA-209) doing ocean-going towing off SE United States; XO, USS *Allegany* (ATA-179), Hydrographic Survey Group One in the Persian Gulf collecting data for charts; sailing master, U.S. Naval Academy, Annapolis, MD. He was released from active duty in August 1953.

After completing 20 years of Naval Reserve satisfactory service in April 1966, he transferred to the Naval Reserve retired list as lieutenant commander in August 1986. As a civilian he was employed in the Administration Division of Kodak Research Laboratories, Rochester, NY, retiring in 1988, after 35 years.

Resides with his wife, Vandy (who teaches high school science), in Rochester, NY. He has one daughter, Beth (North Carolina); one son, Herb (Massachusetts); and step-sons, Barry (California) and Jay (Massachusetts). No grandchildren.

WILLIAM CRAIG HUNLEY (BILL) (LT) was born in Chesterfield, SC on Jan. 3, 1922. He enlisted in the USNR in September 1941; entered the V-12 Program and Midshipman School, NROTC, at the University of South Carolina.

Served in the USS *Greer* (DD-145) and USS *Lang* (DD-399) and hunted subs in the Atlantic. Discharged, he was called back during the Korean War and served on the USS *Isherwood*. He received a Letter of Commendation for life-saving.

After the Korean War, he returned to civilian life and worked for Belk stores; was collector of revenue for the city of Monroe, NC and a real estate developer.

In 1952 he married Lois Crawford and they have two children. He continues to play volley ball in the church league.

YORICK G. HURD (S1/C) was born in Santee, NE on June 30, 1922. He attended Tufts College and Cornell University and entered the V-1 Program as a freshman in the fall of the year, went on active duty in July and graduated from the V-12 Program at Great Lakes.

Ordered to Cornell Midshipman School, where he was a student, on V-J Day. He went to Great Lakes and ended his active duty at Corpus Christi NAS where he taught physics at the station

and at the college. He was discharged on points, May 3, 1946.

His father and grandfather served in WWI and his great-great uncle (for whom he was named) was a surgeon in the Civil War. Since 1946 Hurd has spent two years teaching physics at Simmons College; 12 years at 20th Century Fox Films Research Dept.; 32 years at L.E. Carpenter Co. The name was changed several times through the years to Vicrtex Inc., Dayco Corp. and Forbo Industries. At present he is retired.

JOHN SPENCER HYNDMAN was born in Wichita, KS on Oct. 20, 1922. He enlisted March 19, 1942, and attended the V-12 School at Notre Dame in 1943-44. His basic training was at Parris Island and OCS at Camp Lejeune. He was commissioned on Sept. 30, 1944.

Landed at Iwo Jima, Feb. 19, 1945, as part of a replacement draft. Two weeks later he became the sixth platoon leader of 1st Plt., Co. B, I-28 and received a critical head wound from a shell fragment after only four days, on March 9. He spent one year in various Navy hospitals and was medically retired on April 1, 1946.

Graduated from Wichita State and married the former Lois Williams. He was employed by Amoco Oil in Corpus Christi, TX, where two sons and a daughter were born. Both his sons and his brother were Marines.

In 1960 he opened his own consulting geology firm until his death on Sept. 11, 1990. He was buried at Arlington National Cemetery. He had fond memories of his year at Notre Dame and was an avid fan of their football team. He often said that Notre Dame's spirit almost matched the USMC "esprit de corps."

Hyndman was active in the USMC League and MOWW and served as its representative on the Mayor's Council for Veteran's Affairs and he also belonged to several geological professional organizations.

WILLIAM G. HYZER was born in Janesville, WI on March 25, 1925. He entered the V-12 Program at the University of Wisconsin in 1943; transferred to the University of Minnesota in 1944; graduated in February 1946 with a BS degree in electrical engineering.

Served as engineering officer on PCE-902 in the Pacific Theater and was discharged in 1947. He received a BS degree in physics from the University of Wisconsin and joined the Parker Pen Co. as a research physicist in 1948.

Entered private practice in 1953 as a consulting engineer specializing in imaging technology, a

professional career that continues to demand his full time and attention. He has received numerous awards and citations for his scientific contributions to this field.

He is the author of two scientific books and more than 500 scientific articles and patents, including *Photographic Instrumentation Science and Engineering*, written under contract to the USN.

In 1989, as a consultant to the National Geographic Society, he and Thomas Davies, RADM, USN (RET), performed photogrammetric analyses of photographs taken by Adm. Robert E. Peary during his 1909 polar expedition, confirming Peary's long-contested claim to the North Pole. He and his wife, Mary, live in Janesville, WI in the house where he was born. They have three sons: David, John and James.

KARL H. IRWIN (XO) was born in Columbus, OH on June 30, 1925. He entered the Navy V-12 Program at Middlebury College, Middlebury, VT on July 1, 1943. He was commissioned as ensign on Mar. 6, 1945, at the USNR Midshipmen's School, Ft. Schuyler, The Bronx.

Served as XO on LCT-918 in the South Pacific and was discharged to the USNR on July 24, 1946.

Irwin received a BS degree from Ohio State University in September 1949. In June 1951 he received his LLB degree from the Ohio State University College of Law. He devoted his legal career to public service and worked as an attorney for the Dept. of Taxation, state of Ohio. Then he served the Finance Dept. of the city of Upper Arlington, retiring as deputy finance director in the fall of 1985.

He was the loving husband of Marian Irwin for over 40 years and a devoted father to his eight daughters and his son. Sadly, he died of colon cancer on Aug. 28, 1992. *Submitted by his daughter, Joan Fishel.*

DALTON M. IVINS (CDR) is a native of Memphis, TN. He enlisted in May 1943 and entered the V-12 Program at Tulane University, New Orleans, LA in July 1943. He completed the full college program and received his commission as ensign and his Bachelor's degree in mechanical engineering in February 1946.

Completed Officers Steam Engineering School in Newport, RI and was assigned to the Pacific Reserve Fleet at Tongue Point and Astoria, OR. He was released to inactive duty in July 1947 and remained active in the Reserve for 27 years. He commanded drilling units in Memphis, TN and Natchez, MS and a training center in Natchez until retiring as commander in 1978.

Worked in various facility engineering and manufacturing engineering management positions for four major industrial concerns and retired after 41 years. He was a member of engineering organizations and held professional engineering licenses in several states. He is now engaged in private consulting practice.

With his wife, Mary Ann, he lives in Columbia, SC. They have two daughters and a son who served as a Navy fighter pilot.

WILLIAM H. JACOBSON (LT) was born in Chicago, IL on April 28, 1925. He enlisted in the USNR on July 1, 1943, and entered the V-12 Program at Middlebury College. Jacobson attended Midshipman School at Notre Dame and was commissioned ensign in March 1945.

Served in the USS LSM-8 and USS LST-772 and participated in the movement of North Korean POWs to island of Koje Do.

Jacobson graduated from the University of Illinois in February 1949 with a BS in DSSWV and majored in industrial engineering. He worked as a steel sales representative for Youngstown Sheet and Tube Brainard Steel Div. and retired from Wisconsin Steel Div. as Eastern Regional Sales Manager.

CDR Jacobson retired from the USNR on April 28, 1985, after 20 years. His awards include the Armed Forces Reserve Medal, Navy Reserve Medal, American Campaign Medal, Asiatic-Pacific Campaign, WWII Victory, Asia Occupation, National Defense, Korean Service and United Nations Service medals.

His family consists of wife, four daughters and three grandchildren. He makes his home in Indianapolis, IN.

JOSEPH R. JAMES (LTJG) was born in Seattle, WA on July 15, 1924. He joined the USNR on Dec. 8, 1942, serving as a signal officer in communications. He entered the V-12 Program at the University of Washington NROTC.

Served in the USS *West Virginia* (BB-48) in Okinawa and the USS *Goodhue* (APA) after the war. His memorable experiences included: participating in the victory parade and surrender ceremony in Tokyo Bay; and visiting the British battleship HMS *Duke of York* in Tokyo Bay.

Discharged on Feb. 6, 1946, he received the Asiatic-Pacific with one star, Philippine Liberation and the American Theater.

Married with two grown children in family business. President/CEO of Ye Olde Curiosity Shop, Inc. He is retired.

RAYMOND W. JASEK (LTJG) was born in Chicago, IL on Nov. 21, 1925. He entered the V-12 Program in July 1943 at St. Lawrence University, Canton, NY and was commissioned an ensign after attending Midshipman School at Cornell University in May 1945.

Attended General Line Officers School in Miami and Amphibious School in San Diego. He served as skipper of an LCT and later as division commander of an LCT Div. in Samor, P.I. The LCT Div. included 11 boats and 110 men and in July 1946, he oversaw the de-commissioning and moth balling of the division. He had the distinction

of signing his own orders to send the division personnel, as well as himself, home. His final rank was lieutenant junior grade.

After graduation from the University of Illinois, he joined IBM selling and leasing computers for eight years, after which he left to set up his own company leasing small computers and systems. He is still in the leasing business in his own company.

He lives with his wife, June, in Tiburon, CA where they have lived for the last 30 years. Their three children and three grandchildren live nearby. He remains ever grateful for the wonderful opportunity he received from the V-12 Program and the USN.

WILLIAM J. JASPER (LTJG) was born in Wilkinsburg, PA in 1925. He entered the Navy V-12 Unit in July 1943 at Bucknell University. In 1945 he transferred to University of Pittsburgh Dental School. He returned to inactive duty in January 1946, was ordered to active duty in 1948 and upon graduation, he was commissioned LTJG, Dental Corps, USN.

His active duty was from 1949-1966; served in carriers, *Antietam* and *Forrestal*, and in nine shore stations including Taiwan. He retired as commander.

A Jewish lay leader throughout his service career, he proposed establishment of first permanent Navy Jewish Chapel and suggested name, Commodore Levy, Norfolk Naval Station (Levy Chapel dedicated 1959). His medals include the American Campaign Service, WWII Victory, National Defense Service, Korean Service with two Battle Stars, Korean Presidential Unit Citation and the United Nations Service.

Earned Masters of public health from University of North Carolina in 1967. After residency he was employed by the District of Columbia Health Dept.; was head of dental auxiliary programs, Northern Virginia Community College, 1969; was acting head, Dept. of Public Health Dentistry, Hebrew University School of Dental Medicine, Israel, 1971; and joined Baylor College of Dentistry faculty, Dallas, TX in 1974 until retirement in 1980.

His naval and veterans activities include FRA representative, County Veterans Council; Navy's Raleigh Recruiting District Advisory Council; Board of Directors Navy League's Triangle Council; and Durham VA Medical Center volunteer. His honors include: Fellow, American and International Colleges of Dentistry, Certificate of Merit, Alpha Omega Dental Fraternity, Letter of Commendation, Fifth Naval District.

His family consists of wife, Retha, and children, Noreen Berger and Dr. Warren Jasper.

JACK B. JOEL entered the Navy V-1 and later the V-12 Program, Wabash College in January 1943. Attended Northwestern Midshipman School and was commissioned ensign in 1945. He served in the Pacific and Japan aboard various ships until released to inactive duty from acting commanding officer, PCS(D) 1391 in 1946.

Graduated from Wabash College in 1947 and Indiana University Law School with JD degree in 1951. He was an attorney for Pure Oil Co. for two years; Cities Service Oil Co. for five years; senior vice president, director and general counsel, U.S. Testing Co. and numerous subsidiaries for 23 years. Following retirement, he has been in private practice.

Joel is listed in *Who's Who in the World;* was honored in 1987 at the House of Lords (London); elected chairman emeritus of Products Liability Committee of International Bar Association (chm. 1983-87); chm. of Consumer Protection Committee of International Bar Assoc. (1988-92); and currently vice chairman of International Law Committee of American Bar Assoc. and Class Agent for Wabash Class of 1947.

Resides in Wyckoff, NJ with his wife, Patricia Henchie Joel.

BILLY F. JOHNSON was born in San Marcial, NM on Jan. 24, 1922. Home schooled on the family cattle ranch, he graduated at Albuquerque High School in 1938. Johnson was a cowboy until he enlisted in the NROTC in 1942; was commissioned ensign in 1945; OinC of LCT-467, an interesting trip until sunk in Okinawa; and he was a Flotilla repair officer until reassigned to Seattle, WA in 1946.

Married in 1945, he divorced in 1952. He was a building contractor until 1952; bouncer, Anchorage and Whittier, AK; taught science Aztec HS; mathematics NMMI; moved to Seattle and was plaster patternmaker; later an industrial engineer at Boeing Airplane Co. Married Seattle police detective.

He taught Jr. HS Spanish, French and Latin; taught every grade kindergarten through college plus administration; and he has Bachelor's degree UNM, Master's and Doctorate, University of Washington.

A master of Exemplar Lodge #284 F&AM, 1967; he is a lifetime Mason and life member of VFW. He married a Seattle police detective and has four children and six grandchildren. He makes NW style steambent boxes, carving tools, writes novels, swims, fishes, clams, crabs, etc. His "heroine" considers them all a pestilence, but thankfully, she goes with him.

DANIEL THOMAS JOHNSON (LCDR)

entered the V-12 Program at Washburn University on July 1, 1943. He transferred to Central Missouri State in February 1944; was commissioned ensign at Cornell University; and attended Sonar School in Miami. He served on the USS *Washington* until he was released to inactive duty on July 3, 1946.

AB, Manhattan (KS) Bible College (1948); MA, church history, Butler University (1949); BD, Central Baptist Seminary (1952); MA, history, Kansas State University (1962); and Ph.D., University of Wisconsin (1974).

Transferred to the Chaplain Corps in 1954 and retired as a lieutenant commander; he served as a chaplain in the Reserve for 20 years and on two 90 day summer duties. In addition to several ministries in the Christian Church, he taught at Kentucky Christian College (1954-56), Manhattan Bible College (1956-62) and Western Illinois Univ. (1965-90). During the last 12 years at WIU, he was the Director of Scheduling and of the Kaskaskia Program (a self-designed degree program).

He married June Randall of Omaha, NE on Aug. 1, 1947. They have two sons and three grandchildren. They retired in 1990 and moved to Bella Vista, AR where they now reside.

ROBERT H. JOHNSON (CAPT),

Arizona State Teachers College and Louisiana Tech, 2nd Plt. Commanders Class, Quantico. He was a platoon leader at Camp Lejeune, then combat cargo officer of the USS *Bexar* (APA-237). He joined the Associated Press in Dallas in 1946 and completed degree at SMU in 1947.

Recalled as infantry officer and 3rd Bde. embarkation officer at Camp Pendleton, then was combat cargo officer, Transport Div. 14, in Korea, and discharged as captain.

As Dallas AP bureau chief, he wrote the first bulletins and directed coverage of President Kennedy's assassination. As AP sports editor, he covered Israeli massacre at Munich Olympics in 1972. And as managing editor in 1975, he oversaw coverage of President Nixon's resignation.

Named assistant to AP president in 1977; moved to Albuquerque as bureau chief in 1984 to return to southwest; and retired from AP in June 1988.

He taught advanced reporting at New Mexico State University and the University of New Mexico, and is now executive director of the New Mexico Foundation for Open Government.

JAMES PAUL JOHNSTON (LTC)

was born in East Moline, IL on May 1, 1926, and attended public elementary and high school there. He enlisted in the USNR on April 27, 1944, and was sent to the V-12 unit at Miami University, Oxford, OH. He was released to inactive duty in October 1945, finished his pre-med studies at Miami, graduating cum laude

and in Phi Beta Kappa. He attended the University of Illinois Medical School, graduated in 1950, interned in Indianapolis and served a year of surgical residency in Chicago.

Called to active duty on July 1, 1952, and served two years as the OIC of Mobile Photofluorographic Unit #9 in the 9th Naval District. After a four year surgical residency at Iowa Methodist Hospital in Des Moines, IA, he was in the private practice of general surgery in Moline, IL for 31 years. During this time he was attached to the local USMCR unit as its MD until he retired as lieutenant commander in 1967. In 1984, he enlisted as lieutenant colonel (MC) in the 123rd FA ILANG and served until 1989.

He was solo practitioner his entire time in practice and was active on the local hospital staff and the county medical society being president of both.

Since retirement in 1989, he has done volunteer surgical work in Africa, St. Lucia and Jamaica and has worked at multiple Indian Health Service units. At home he is active as one of the curators at the county historical society.

Married Joyce in 1953 and they have four children: Martha, Susan, Carolyn and David.

ERNEST FOX JONES (CWO4)

was born in Los Angeles, CA on Nov. 3, 1924. He joined the USN on Jan. 14, 1943, and entered the V-12 Program at Millsaps College, Jackson, MS. He was stationed at Pearl Harbor, T.H.

NROTC, University of South Carolina; graduated with MS, BS, and MBA. He retired on Nov. 5, 1984. His awards include the Good Conduct Medal, American Theater, Asiatic-Pacific, WWII Victory, National Defense, Naval Reserve Medal and the Armed Forces Reserve Medal.

Jones is married and has two children and two grandchildren. The former owner of Jones Garden Center, he is now retired.

PHILIP B. JONES (AVCAD)

was born in Corning, NY on Nov. 19, 1925. He joined the USN on Nov. 8, 1943, and entered the Navy V-5 Program at St. Lawrence University from March 1944-March 1945.

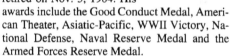

He was stationed at Ft. Worth, TX; went to pre-flight at University of North Carolina; and received BS degree in mechanical engineering from Bucknell University in 1948.

Jones spent 37 years with Corning Inc. in various engineering management positions. He retired in June 1984 as senior associate of Corning Inc.

Married Frances in 1948 and they have five children and 12 grandchildren. He now lives on Keuka Lake, Hammondsport, NY

THOMAS RAY JONES (LTJG)

was born in Madison County, TN on July 22, 1922. He attended Union University; the V-12 Program at the University of the South; Harvard Graduate School of Business (Naval Supply Corps); the University of Tennessee (business administration); and Northwestern University (Master's in hospital administration).

LTJG Jones served in the Naval Supply Corps in Espiritu Santo, New Hebrides Islands, Samar, Philippine Islands. He was discharged in 1946.

He was assistant to Dean of Students, The University of Tennessee (4 years); administrator North Mississippi Community Hospital (6 years); administrator Jackson-Madison County Gen. Hosp., (6 years); consultant in hospital administration, U.S. Public Health Service and Health Care Financing Administration, Rocky Mountain States, Denver (17 years) until retirement in 1984.

Community and International Service: Rotary clubs of Tupelo, Jackson and Lakewood by Denver, CO; volunteer organizer and coordinator of highly successful U.S. "Rotary for Mexico" project for sending donated hospital, medical and dental equipment for use throughout that country; audio-visual presentations from travels for Colorado Mountain Club and other community organizations; and a volunteer at the Denver Museum of Natural History.

He is single and lives in Denver, CO.

WILBUR C. JONES (LTJG)

was born in Oakland, MD on Dec. 4, 1925. He joined the service on March 1, 1944 and reported to the V-12 Program at Worcester Polytechnic Institute, Massachusetts. Commissioned an ensign in February 1946, and he went to Newport, RI in March for a three month training cruise.

Released to inactive duty in the USNR as ensign in July 1946 and retired from the federal government after 31 years of continuous service in July 1994 as lieutenant junior grade.

Single, never married, he lives in Chevy Chase, MD.

VIRGIL ALEXANDER JULIAN JR. (S2/c)

was born in Independence, MO on May 17, 1928. He joined the service on July 2, 1945, and entered the V-5 Program at Kansas State, Pittsburg, KS and V-12 at Dartmouth College, Hanover, NH.

Stationed at USNTC Great Lakes, IL. He was discharged Dec. 25, 1946, and received the Good Conduct Medal and WWII Victory Medal.

Graduated from UMKC in June 1951, economics and taught accounting at UMKC from September 1962 to June 1968. Julian has been a certified public accountant in Missouri and Kansas City since 1962. He and his wife, Bettie Jean, have four children: Gregory, Wayne, Sylvia and Cynthia.

ROBERT C. KADRI (LTJG)

was born in Summit, NJ. He enlisted in the Stevens V-12 Unit n July 1943, attended Midshipman School in Chicago and was commissioned as ensign in January 1945.

Served as radar officer aboard the *Greer* (DE-23) in the Pacific Theater and was present at the surrender of Wake Island. He received the WWII Victory Ribbon.

After the service he entered New York University and received both a BA and MPA degrees. He was employed for many years by Prudential Insurance Co. and was associate manager, Group Credit Ins., prior to early retirement in September 1986. As a retiree, he helped restore Prudential's historic archives.

Kadri and his wife, Virginia, reside in New Providence, NJ. Their daughter, Barbara, is a paralegal with AT&T. Their son, Ronald, is a scenic designer with Papermill Playhouse. Their grandson, Ryan, was born in July 1993.

ALAN J. KAPLAN (1LT) entered the Navy V-12 Program at Dartmouth College on July 1, 1943. He was commissioned ensign in January 1945 at Northwestern Midshipman School, Chicago, IL. He served as OinC in LCT-813, 1945-1946 and was released to inactive duty .

Returned to Dartmouth College, was recalled to active duty and served aboard the LST-973 and LST-1083 in the Korean Theater; first lieutenant USS *Bayfield* (APA-33); Bureau of Naval Personnel, XO; USS *Taloga* (AO-62); CO USS *Jamestown* (ACT-3) and CO of USS *Virgo* (AE-30) in Vietnam Theater. After shore duty he retired in 1972.

Received his Ph.D. in professional psychology from the U.S. International University in 1978. He is now retired on Orcas Island, Eastsound, WA with his wife, Maggie. Their son, Brad, is a captain in the USN and their daughter, Pamela, is married and resides in California.

MICHAEL G. KAPNAS (CDR) enlisted in the Navy V-12 Program when he was 16. Enlistment changed to 17th birthday, June 24, 1943, and he entered active duty on July 1, one week later. He was commissioned ensign on March 29, 1945, at age 18 from Abbott Hall, 23rd Class.

Served as skipper of LCT-602, LCT-538, YF-864 and LCT Type Commander, 16th Fleet. He returned to Northwestern (1947-48) and received his BS degree. He served in the Naval Reserve for 28 years before retiring in 1971.

His billets included CO, Fleet Div. (-15; XO

Surface Div. 9-37 (Forrestal Awd.); CO Surface Div. 9-38; Battalion Commander 9-10 and Group Commander 9-75. Civilian occupations included 25 years with steel industry (USS and Inland); director of Public Relations, Whiteco Ind. (1972-80); president Kapnas Public Relations/Advertising (1980-88).

Retired to La Costa, CA in North San Diego County with his wife, Helen. Their daughter, Irene, was killed in a plane crash in New York in 1976 while a student at City College. Their son, George has BSE from Purdue and MBA from Indiana. George is a program manager at Lear Astronics, Santa Monica and is married to Susie Marie of Cypress, CA; they have two daughters, Kara and Kelsi. Michael Kapnas is currently working two days a week as pharmacy courier at Scripps Hospital in Encinitas after having volunteered for over a year.

DUNCAN E. KARCHER JR. entered the Navy V-5/V-12A Program at Duke University on Nov. 1, 1943. Attended Princeton University Pre-Midshipman School then Northwestern Midshipman School and was discharged in 1945.

Graduated Polytechnic Institute in Brooklyn with Batchelor of electrical engineering degree in June 1949. He continued graduate studies in servomechanisms and automatic control system applications. Karcher specialized in automatic control systems for the Arma Corp., Sperry Gyroscope Co., Bell Aircraft Corp., Republic Avn. Corp. and the Grumman Aerospace Corp.

Projects at Grumman included the OAO Satellite, A-6A aircraft, Lunar Landing Module, E-2C aircraft, X-29 aircraft and the NASA Space Station "Freedom." After 42 years in the aerospace industry, he is now with the commercial real estate department of Roosevelt Savings Bank in Garden City, NY.

He and his wife and four sons reside in Garden City, NY.

GEORGE S. KARIOTIS (ENS) was born in Boston, MA on Feb. 22, 1923. He joined the service in November 1942, entered the V-12 Program at Tufts College and received BSEE from Northeastern University.

Military locations include the USN Academy and he served in the USS *Cogswell* in the Pacific. He was discharged in March 1946.

Kariotis founded Alpha Industries, Inc. in 1962; and he was Massachusetts Secretary of Economic Affairs, 1979-1982. Married since 1945 to Ellen Arvanites, they have two children, Kathryn and Stephen, and three grandchildren. Semi-retired, he is chairman of Alpha Ind.

CARL L. KASEY (ENS) was born in McPherson, KS on Nov. 28, 1922. He entered College of St. Thomas V-12 Program in St. Paul, MN on July 1, 1943. He was commissioned ensign on Feb. 24, 1944, at Columbia University Midshipman School, New York, NY.

Served as officer in the Amphibious Forces at

San Diego, on the island of Maui and on the Amphibious Operating Base near Pearl Harbor, HI. He was discharged to the Naval Reserve on July 15, 1946.

Graduated from McPherson College in May 1947 with BS in science. From 1957 to 1985 he taught science in McPherson schools. After his retirement in 1985, he and his wife, Betty, have resided in McPherson. He does volunteer work daily with Churches United in Ministry, helping the disadvantaged. They are the parents of two children, Karen and Dennis. Karen lives in Hesston, KS and has two sons in college and Dennis lives in Lincoln, KS.

NELSON H. KASTEN was born in Cape Girardeau, MO on Aug. 6, 1924. He entered the Southeast Missouri State (SEMO) University V-12 Program on July 1, 1943, and received his ensign commission in June 1944 from Midshipman School at Columbia University in New York City.

After amphibious training in San Diego, he was assigned to the USS *Magoffin* (APA-199) which participated in the April 1, 1945, invasion of Okinawa. Subsequent tour was as navigator on the USS *Nemasket* (AOG-10) with Western Pacific duty until released to inactive duty in July 1946.

Remained in the Naval Reserve until his 20 year retirement as lieutenant in 1964. He received his BS in chemistry in 1947 from SEMO and a certificate in chemical technology from Washington University in 1952. Following 38 years of employment at Monsanto Co., he retired in 1986 as senior research specialist.

Currently he is a member of The Retired Officers Association (TROA) national organization. He served as president of the Greater St. Louis Chapter during 1992-93, and he was the treasurer of the Missouri Council of Chapters, TROA in 1994.

Kasten resides in the St. Louis area with his wife, Alice. They are the parents of a son living in Montana and a daughter and three grandchildren in the St. Louis area.

ROBERT I. KATZ was born in Springfield, MA on Dec. 16, 1924. He entered the V-12 Program at Tufts College on March 1, 1944, and was discharged while in the V-12 Program at Tufts Medical School on Nov. 7, 1945. Several months later he was offered, and accepted, a commission as ensign in the USNR.

Upon graduation from Medical School in 1948, he went to New Orleans, LA for internship and subsequent training. He was recalled to active duty soon after the outbreak of the Korean War and was assigned to the 2nd Mar. Div. at Camp Lejeune, NC. He was released from active duty in early 1953, resumed his training in surgery and became a drilling reservist which he did until his retirement from the Naval Reserve in 1985.

One year later, he took a sabbatical from his practice in Los Angeles, CA and was recalled for active duty for special work as surgeon for a WestPac deployment aboard the USS *New Jersey*. He enjoyed it so much, he closed his office and continued doing ADSW for the USN at various hospitals. He had a second deployment with the USS *New Jersey* and topped it all off by being on active duty during Desert Shield.

When he retired for good, he had a naval career from WWII to Desert Shield. His awards include the WWII Victory Medal, American Theater of Operations Medal, Naval Reserve Medal, Armed Forces Reserve Medal with two HGs, National Service Medal with Bronze Star, Naval Reserve Sea Service Ribbon and the Unit Commendation Ribbon.

He still practices medicine part-time and plays tennis.

LAWRENCE C. KEATON (LT) was born in Gainesville, TX on Nov. 24, 1924. He joined the service in May 1943; entered the V-12 Program at the Southwestern Louisiana Inst. and University of Oklahoma; and received his BS degree in mechanical engineering from the University of Oklahoma in 1945.

Military locations include Pearl Harbor; Samar, Philippine Islands; and Davisville, RI. He participated in the Philippine Liberation. Keaton graduated USNR Midshipman School in Quonset Point, RI and received his discharge in 1960 as lieutenant. His awards include the American Campaign, Asiatic-Pacific Campaign, Victory Medal and the Philippine Liberation.

From 1963-1976 he had various engineering positions in construction, maintenance, process and design, maintenance planning, quality control, plant operations, project development engineer and world-wide director, all for Phillips Petroleum Co. He was managing director for Nordisk Philblack AB, Malmo, Sweden (1965-73); (1976-81) Sevalco Ltd., Bristol, England; (1981-85) consulting engineer and (1985-present) managing partner SEAL of Northwest Texas.

He belonged to SAME, Society of American Military Engineers and numerous professional societies.

RALPH M. KEENAN (LCDR) was born in Oakdale, PA on May 17, 1921. He joined the

106

service on July 1, 1943 and entered the V-12 Program at Bucknell University and the Midshipman School at Northwestern.

Served in the USS LCI(L)-872 and particpated in the neutralization and occupation of Palau Islands.

Awards include the WWII Victory Medal, Naval Reserve Medal, Atlantic Theater and Asiatic-Pacific Theater.

He and his wife, Betty (Cotts), have three daughters and six grandchildren. Keenan is retired from the Cyclops Corp. in Pittsburgh, PA.

HARRY WARREN KELLEY (ENS) was born in Butte, MT on July 6, 1922. He joined the service on Sept. 7, 1941, and entered the V-12 Program at the University of the South, Sewanee, TN.

Military stations and schools include the Aviation Machinists Mates School, Jacksonville, FL and Great lakes Training Station, Chicago, IL. His memorable experience was in 1941 when he won an award for receiving the top grade (97) in the Air Trade School, Jacksonville, FL.

Kelley was discharged on Oct. 4, 1945. He and wife, Mary, had one son, Larry, who lives in Houston, TX. Harry Warren Kelley passed away Oct. 21, 1993.

FREDERICK R. KENNEDY (CAPT) was born Nov. 19, 1922, in Waltham, MA. He joined the service in August 1942 and entered the V-12 Program at Dartmouth, receiving his BA degree. He was stationed at Parris Island and attended OCS at Camp Lejeune from July to November 1944.

Joined the 6th Mar. Div. in January 1945 and participated in the Okinawa Operation from April 1 to May 30, 1945. He was wounded on May 30 and evacuated. His memorable experience was regaining consciousness on June 10 aboard a hospital ship.

Kennedy was released from active duty in April 1946. He was recalled to active duty and served with the 2nd Div., Camp Lejune in March 1951. Discharged in August 1952, he received the Purple Heart.

Married his high school sweetheart in 1947. They have one married daughter, one son in California who is single and two grandsons. Kennedy was in the banking profession for 36 years and retired, Jan. 1, 1983, as vice-president of operations. He is now a golf bum in Bedford, MA.

PATRICK F. KENNEDY (CAPT) of Fall River, MA enlisted in the USN in February 1944. He entered the V-12 Program at Brown University; transferred in the fall of 1944 to Dartmouth College; graduated with Class of 1946 with BA degree and commissioned ensign, Supply Corps.

Served as supply officer, USS *Joseph P. Kennedy Jr.* (DD-850) in the Atlantic and USS *Cape Esperance* (TCVU-88 in the Pacific and Naples, Italy with VR-25 and FASRON 77. On shore duty, he specialized in financial management.

Kennedy graduated from the Naval Postgraduate School in 1959; received an MBA from the George Washington University in 1964; served on the staff, Navy Comptroller; OIC, Navy Regional Finance Office, Pearl Harbor; director, Financial Control, SPCC Mechanicsburg, PA and Comptroller, NAVELEX. He retired as captain, Supply Corps in 1973. He was awarded the Legion of Merit.

From 1974 to 1980, he was comptroller Naval Research Laboratory, Washington for which he received the Navy Superior Civilian Service Award. In retirement, he lives in Alexandria, VA with his wife, Grace.

WALTER S. KENTON (SGT MAJ) Louisiana Polytechnical Institute, entered the V-12 Program on July 1, 1943, as a Marine. He served as a gunner in a scout bomber squadron in the 3rd Mar. Air Wing and was discharged as a corporal on July 1, 1946.

Employed by the Colorado Interstate Gas Co. for seven years before joining the professional ranks of the Boy Scouts of America. He served the BSA in the Midwest in both local scout councils and on the National Staff. He retired on Jan. 1, 1986, after 32 years.

Joined the Naval Reserve Program in January 1957 and served until April 1975 when he went into the Retired Reserve. Kenton attained the rank of sergeant major, serving in the 12th Spec. Forces, Airborne, a MP Co., a training battalion in the 89 Div., a USAR School and an infantry company in the 35th Div. of the KSNG.

Kenton was retired from the Reserves in August 1985 with 31 years active and reserve time. His awards include the Expert Infantry Badge, Master Jumps Wings. Canadian Jump Wings, Air Crew Wings (Marine), the Army Reserve, Army Achiemvement, Army Good Conduct, WWII Victory, Marine Good Conduct, and the American Defense Medal.

Resides in Carol Stream, IL, near Chicago and with his wife, Agnes, they sail the Great Lakes aboard their sailing vessel *ScotFree*.

ROY F. KERSCHNER (LTJG) was born in Pine Grove, PA. He entered the Navy V-12 Program at Muhlenberg College, Allentown, PA on July 1, 1943. He was assigned to Pre-Midshipman School at Asbury Park, NJ in November 1944 and was commissioned ensign at Cornell University in June 1945.

Ordered to Communications School at Harvard University and on completion was assigned to the 4th Naval Dist. HQ, Philadelphia in

October 1945; ordered to CINCPAC Communications Pool, Pearl Harbor in January 1946 and was assigned to Hawaiian Sea Frontier Merchant Shipping Control (MERCO PEARL).

Separated to inactive duty in July 1946 and served in the organized Reserve, Communications Supplementary Activities (CSA) Group, Pittsburg from 1950 to 1952. He was discharged as lieutenant jg in May 1958. He graduated from Muhlenberg College in 1948 and was employed by Aluminum Co. of America (Alcoa) in Pittsburgh, PA for 41 years.

He retired in 1989 and resides in Gibsonia, PA with his wife, Alda. They are the parents of David (employed by McClaren Hart in Pittsburg) and Barbara (employed by Lockheed at the Kennedy Space Center, FL).

BURK KETCHAM (LTJG) entered the V-12 Program at Princeton and Union. He was commissioned at Columbia in 1945; was communications officer of AE-1 in the Pacific and XO of PYC-37 in the Atlantic. He received his BA degree from Union in 1948 and MS from Columbia.

He was married to Helen Schmid (Wellesley College) for 38 years until her death in 1989. He is the father of two sons.

His business career was devoted to city planning consulting work as a principal of firms in Newark, Boston, Los Angeles and San Francisco. He was consultant on the redevelopment of the Newport, RI waterfront and now operates his own consulting firm.

Recent activities include solo backpacking in New Zealand and competitive rowing. He finished second for his age group in Alden Ocean Shell event at the Head-of-the Charles races in Boston last October.

JERALD KIRSTEN (CAPT) was born in Portsmouth, OH on Sept. 13, 1924. He joined the service Dec. 10, 1942 and entered the V-12 Program at the University of California, Berkeley; and USN Pre-Flight School at St. Mary's College.

He was stationed at NAS Norman, OK. His memorable experience was Command of FIRSTPAC 1387. He retired on Feb. 19, 1981. His awards include the American Defense, WWII Victory, Expert Marksman and the 20 year Naval Reserve.

Married to Claudine and they have three children: Claudia, Kenneth, Susan, and six grandchildren. He is a retired certified public accountant and still flying a Cessna 182M. He also enjoys golfing and traveling. (His mother, YN3(F) Frieda Hardin, was in the USN in WWI and still wears her uniform with pride.)

WARREN A. KITCHEN was born in Chico, CA on Oct. 22, 1925. He joined the USN on Nov. 3, 1943 and was assigned to Colorado College V-12 Program in February 1944. He was told, "the Navy needs doctors, you're a pre-med." He was discharged in November 1945, joined the Re-

serves, was recalled during the Korean War and served with the 1st and 3rd Mar. Div.

Graduated from UC Berkeley with AB in zoology and UCSF with a DDS degree and specialized in Orthodontics. He practicied in San Francisco for 36 years and still practices one day per week. He has been active in professional organizations, community activities and his church. He gets his exercise growing Christmas trees.

He and his wife, Nancy, live in Healdsburg, CA. They have two daughters, Anne and Susan, and five grandchildren.

EDWARD L. KITTS (ENS) was born in Houston, TX on Oct. 10, 1925. He joined the service on June 8, 1943, and was called to Bucknell University along with 35 other V-5 enlistees. All were placed in one platoon and called V-12(A). After eight months at Bucknell, he transferred to Ohio Wesleyan University, and after eight months at Ohio  Wesleyan, he was sent to Pre-Midshipman School at Princeton University for five weeks, then onto Cornell University where he was commissioned as an ensign on Aug. 1, 1945.

Attended Torpedo Officers School at Destroyer Base San Diego, CA. He was assigned to the USS *DeHaven* (DD-727) in November 1945 and stayed with the *DeHaven* until August 1946 in the Northern China Destroyer Patrol. Kitts was discharged from active duty in 1946.

A Reservist from 1947-48, he was gunnery officer of PBY Sqdn., Port Columbus. He graduated from Ohio Wesleyan University in June 1948; did post-graduate work at the University of Texas in 1957; and graduated from the University of Texas Medical Branch, Galveston, TX in 1962. He has been in family practice in Selma, CA for the past 30 years.

WALTER E. KNOUSE JR. entered the V-12 Program at Trinity College, Hartford, CT on July 1, 1943 (one week after leaving high school). He transferred to V-5 in January 1944, received his wings in April 1946 and served on four carriers: USS *Ranger, Saipan, Roosevelt* and *Midway*. He flew nine different planes including the F-H-2 (first Navy pure jet).

Left the Navy in 1948, entered the University of Pennsylvania and graduated from Dental School in 1953. He has been involved in surgical implantology for the last 28 years. He served in dental organizations in various capacities includ-

ing chairing the Admissions and Credentials Board of the American Academy of Implantology. Besides his private practice, he has lectured nationally and internationally since 1970.

Flies his own twin engine plane to give lectures and examinations through-out the U.S., Canada and the Caribbean. He attended the V-12 50th Reunion and flew his own plane to NAS Norfolk (the only attendee to do so).

Inducted into the International Forest of Friendship in Atchison, KS (Amelia Earhart's birthplace) in 1990, a memorial to aviation and aerospace donated to the U.S. on the 200th Anniversary in 1976. He served as an elder in two different churches, presently the first Presbyterian of Lambertville, NJ.

He has three children and five grandchildren.

HAROLD CAMPBELL KNOX was was born in Trenton, TN on May 9, 1926. He attended Milligan College V-5 from July 1944 to June 1945; and Central Michigan College from June-October 1945.

Fulbright professor University of Honduras and Escuela Superior del Profesorado, 1969 calendar year; lecturer and professor of English and journalism at Michigan State University and Humboldt State University (Arcata, CA); and at present, he is professor of English at Sacramento City College.

Knox was married and divorced. He has a son Carlton who lives in San Francisco. Knox and Lori were married in August 1993.

WALTER H. KOEHLER JR. (LTJG) was born in North Little Rock, AR on Jan. 13, 1924. He entered the V-12 Program on July 1, 1943, at Washburn University, Topeka, KS. After two semesters, he was assigned to Notre Dame Specialists Midshipman School, then to Harvard Supply Corps School. He was commissioned in June of 1944 and joined the *George F. Elliot* (AP-105) as disbursing and supply officer in May 1945. He served in the Pacific and later decommissioned the ship in Norfolk, VA.

Presently resides in North Little Rock, AR, and is active as chairman of Koehler companies, a bakery serving supermarkets and restaurants throughout the United States. He has served as chairman of the Arkansas Red Cross; the district governor of Rotary; director, of One National Bank, Little Rock; and he is president of two other family businesses.

He and his wife, Evelyn, have three sons, seven grandsons and two granddaughters.

ROBERT W. KOENIG (ENS) was born in Freelandville, IN on April 12, 1926. He entered Denison University, Granville, OH on July 1, 1944; was commissioned ensign in June 1946 at the University of Kanasa; was a graduate of the University of Kansas in June 1947; Bonebrake Theological Seminary in 1950 and the Christian Theological Seminary in 1960.

Ordained a minister in the Evangelical United

Brethren (now United Methodist) Church in August 1950. He was awarded the Outstanding Alumnus Award by the Christian Theological Seminary, the Honorary Doctor of Divinity degree by the University of Indianapolis, and the Outstanding Alumnus Award by Freelandville High School.

Pastor in the United Methodist Church for 43 years before his retirement in 1990. He served as district superintendent and as executive director of the Church Federation of Greater Indianapolis.

He lives with his wife, Kathryn (Deal), at Plainville, IN. They are the parents of two sons, Joseph and Mark, and one daughter, Callie.

ALBERT M. KOGA was born in San Francisco, CA on Nov. 10, 1915. He was an instructor of mathematics and engineering in the V-12 Program at Doane College, Crete, NE; a graduate of the College of Engineering, University of California, Berkeley; Doane College invited him to serve on the faculty despite the fact that he had been detained in a  War Relocation Center due to his Japanese ancestry even though a loyal American-born citizen.

He has always appreciated and never forgotten the trust and confidence the college, the V-12 Program and the students had in him during those trying, uncertain times and making it a wonderful, memorable experience for him.

After the end of the V-12 Program, he worked as a chief engineer for 30 years at Hub Electric Co. (later a subsidiary of Westinghouse Electric Corp.), a firm specializing in custom designed lighting and control systems for theaters, public and private buildings. He worked in close collaboration with numerous architects and engineers nationally and internationally.

Currently, he is an independent consulting lighting engineer.

JOHN H. KREKELBERG, III (LTJG) was born in Brainerd, MN on Jan. 30, 1925. He entered the V-1 Program on Jan. 29, 1943, and the V-12 Program on July 1, 1943. He received his commission on March 8, 1945, at Notre Dame and was assigned to the USS PC-781. He was released from active duty on Aug. 12, 1946.

His tour of duty was to Midway Island and Pearl Harbor. He decommissioned and recommissioned the ship twice, Green Cove Springs, FL and Charleston, SC. He sailed the ship to Chicago, IL for use as a reserve training ship. On Jan. 15,

1951, he received reserve promotion to lieutenant junior grade.

Returned to school and received his BS in education at the University of Minnesota. Then he joined the Honeywell Inc. in 1951 and spent 36 years in various positions of management. Krekelberg retired in 1986 and he resides in Minneapolis, MN with his wife. They have two children, Julie Zinniel (lives in Richfield, MN) and 1st Lt. Stephen (presently serving in Army Aviation in Hanau, Germany).

RICHARD J. KRUGER (CAPT) was born in Willoughby, OH on Sept. 4, 1924. He enlisted on Sept. 8, 1943, and was MM3/c prior to entering V-12 Program. He was commissioned ensign, USNR on June 5, 1946. He served on the USS *Estes* (AGC-12), Tsingtao, China.

Recalled to active duty for the Korean War as engineering officer in the USS *Grady* (DE-445). He returned to Naval Reserve inactive duty assignments in July 1953 in the San Francisco area, serving as commanding officer of four different units and retired on Sept. 4, 1984.

A registered professional engineer, and a registered environmental assessor, state of California. He completed a successful career in the engineering/construction business as an officer in the Thorpe Group. He is now performing technical consulting work in energy related fields.

Kruger and his wife, Bernice currently reside at Star Ranch, Hat Creek, CA. Their oldest son, Capt. Paul F. Lorence, USAF was killed in action over Libya on April 15, 1986. Their daughter, Pamela Lorence Glass, is a designer and manufacturer of fine jewelry. Captain Jefferey M. Kruger, USMCR served with the 1st Bn., 2nd Marines during Desert Shield/Desert Storm. His special memory was great "liberty" in the Twin Cities of St. Paul and Minneapolis.

JEROME H. KRUPP (LTC) was born in Belleville, IL on Sept. 16, 1926. He was salutorian, Vandalia Community High School, 1944; BS in naval science, University of Illinois, June 1947; DVM, University of California, June 1958; and MS in radiation biology, University of Arkansas Medical Center, January 1972.

Enlisted in the USNR in January 1944, active duty with Navy V-5 Unit, Mercer Univ., Macon, GA, 1944-45; Navy V-12, Univ. of South Carolina, 1945-46; NROTC, Univ. of Illinois, 1946-47; commissioned ENS, USNR in June 1947 and was promoted to LTJG in April 1950. Attended Comm. Off. Short Crs., Monterey, CA, 1951; comm. watch off., CTF 95, 1951-52; TAD, 1st Mar. Cas. Co.,

1952; comm watch off, COMAIRPAC, 1952-53 RAD June 1953. Promoted LT, USNR, June 1953 and resigned commission January 1958. Appointed CAPT, COANG, January 1960, resigned August 1965; appointed CAPT, VC, 123rd TAC Hosp KYANG, September 1965; recalled to active duty USAF, January 1968. USAMRIID, Ft. Detrick MD 1968-70; promoted MAJ, 1969, Univ. of AK Med. Ctr., 1970-72; USAFSAM, 1972-75.

Traffic Dept. trainee, SW Bell, Oklahoma City, OK, 1947-49; traffic engr., Illinois Bell Chicago, IL, 1949-51; county vet, Clinton Co., IL 1958-59; general practice, Flora, IL, 1959, 1960 instructor, Dept. of Surgery, Univ. of Colorado Med. Ctr., Denver, 1960-65; asst. professor and dir. Animal Care Center, Univ. of Louisville, 1965-68; GM-13, radiation sciences, USAFSAM, San Antonio, TX, 1976-86; chief scientist, SRL, Inc. Brooks AFB, TX, 1989-92.

Married Mary Frances Edwards on Sept. 1 1947, Hillsboro, IL. They have two children, Terry Krupp (Huntington Beach, CA, born Oklahoma City) and Patricia Susan Laffin (The Woodlands TX, born Hillsboro, IL), and five grandchildren.

DONALD G. KUBLER was born in Easton. MD on April 4, 1923. He entered the USN as a seaman in December 1941 and was assigned to the USS *Sands* from March 1942-August 1944, FC. 2nd. He served in Guadalcanal, New Guinea, New Britain, Admiralty Islands and Hollandia operations. He was assigned to the V-12 prep program and then to Newberry College in October 1944.

Transferred to the University of South Carolina from November 1945-June 1946. He was discharged in June 1946 with BS degree from University of South Carolina in August 1947; Ph.D. in chemistry from University of Maryland in June 1952. Kubler was a chemist at Union Carbide, 1952-58; professor of chemistry at the Univ. of South Carolina; Hampden Sydney College and Furman University, 1958-85.

Retired in 1985 but taught one year each at Grinnell College in 1986 and at King College in 1988. He lives in Marietta, SC with his wife Rose B. Kubler. They are the parents of three sons and a daughter and grandparents of three children. He is active in the Greenville Literacy Program, wood carving and storytelling.

His most memorable experience in the Navy was with the sea in all of its beauty whether angry or calm from the California Coast to the Bearing Sea to the South Pacific. It would scare him to death today, but the storms really are deeply engraved in his mind as great experiences.

FREDERICK W. KUHL (LCOL) enlisted in the USN in June 1942 and entered the V-12 Program at Park College. He was commissioned at Camp McDonough, Plattsburg, NY in June 1944 and assigned to subchaser training schools at Naval Base Charleston and NTC Miami.

Served aboard barracks ship USS APL-9 in the Admiralty Islands, attached to USS ABSD-2 and USS ABSD-4, Svc. Force, 7th Fleet, providng berthing, messing and medical services to crews performing battle damage repairs to various Pacific Fleet vessels. Was selected for change of designator to intelligence during Navy's 1950s reserve recruitment efforts. He retired from the Reserves in 1975, after nearly 34 years.

Graduated from the University of Oregon with a BS (1947) and MS (1961) in journalism, with minors in English and history, being awarded membership in Kappa Alpha, national journalism scholastic honorary. He received his Ed.D. (1981) from Nova University. He worked 10 years in corporate public relations, advertising, and newspapers in Washington and Oregon before beginning a teaching career in high schools and higher education. Currently professor emeritus at American River College, Sacramento, CA, where he has headed journalism sequence and taught English composition and technical writing.

After retiring from the Navy Reserve, he has served more than 10 years in the all-volunteer California State Military Reserve as lieutenant colonel, public affairs, headquarters unit and as adviser to the Guard's *Grizzly* magazine, which has won numerous area and national honors for 'journalistic excellence." The state's Military Dept. awarded him two Medals of Merit and the Commendation medal, the Department's second and third highest service awards, respectively.

HOWARD L. KUHL (LTJG) was born in Witt, IL on March 31, 1925. He enlisted in the USNR V-1 Program in February 1943; active duty, USNR V-12 Program at University of Illinois, Urbana, IL from July 1943 to June 1945. He received his BS degreee in chemical engineering and attended USNR Midshipman School, Notre Dame, IN from July to November 1945, was commissioned ensign and served aboard the USS *F.D. Roosevelt* (CVB-42) from November 1945 to June 1946.

Attended the University of Illinois School of Medicine; internship at USNH, San Diego, CA; general surgical residency, Veterans Administration Center, Wood, WI; surgical service, VAH, Oakland, CA; and private surgical practice from April 1958-March 1987, Modesto, Stanislaus County, CA.

Certification: Diplomate, American Board of Surgery, 1957 and Fellow, American College of Surgeons, 1958. Field rep., part-time, Joint Commission for Accreditation of Health Care Organizations, Chicago, IL from September 1987 to February 1989. Kuhl is now retired.

He married Betty Olivier and they have two daughters, Carol and Mary, and three granddaughters.

DONALD F. KUHLMAN (LTJG) was born in Pemberville, OH on Oct. 25, 1924. He joined the USN (line officer) on July 1, 1943, and entered the V-12 Program at Bowling Green State University. He trained at Little Creek, VA and had sea duty on USS *Cornell* (LSM-462).

His memorable experiences include: the end of the Okinawa Campaign; the awesome typhoons; trip to Nagasaki and Hiroshima after the A-bomb; UN formation (1945) in San Francisco; and Gen. Marshall Mao Chiang peace attempt (Peking). He was discharged in May 1946 and received awards for Okinawa Campaign, Pacific Theater and Japanese occupation.

Graduated from BGSU in 1947 with BS in education. Married Mary Hirschy and they have three children: Robert, Craig and Katherine Wilhelm. He worked in the auto business from 1949-1983 and is now semi-retired on a farm.

JEROME J. KUTZEN (1/LT) was born in Detroit, MI on Jan. 21, 1923. He joined the service on Dec. 15, 1942 and entered the V-12 Program at Oberlin College.

He served primarily as senior TQM and landing officer in the Pacific. He was stationed aboard the USS *Fergus* and was responsible for landing troops and equipment. He received several commendations from Troops Cos., both Army and Marine, and from Ship's Divisions Commanders. He also served as the ship's legal officer.

During service, he was stationed at Okinawa, Guam, Saipan, Philippines and Aomori in northern Japan before the surrender. His memorable experience was riding out the typhoons in the Okinawa area and the kamikazes attacks.

He married an Oberlin College student, Carol Tedoff, in 1946. Carol passed away. He has two children and three grandchildren. He is president of Kutzen & Co., Inc., an investment banking merger firm and has been with them for 30 years.

DONALD M. LADD JR. (CAPT) enlisted in 1942 in the V-12 Program at Denison Univ., OH and received BA degree after the war. Other duty stations include Parris Island, Lejeune, Quantico, Johnson Island and Maui. He served with the 4th Mar. Div. and was discharged in 1946.

Ladd served in the Korean War from 1951 to 1952 with the 1st Mar. Div. on the east and west coasts of Korea. He was discharged in 1952 and received two Battle Stars and the Good Conduct Medal. His memorable experience was traversing 200 yards of a Korean road selected at the time for artillery registration by the enemy.

Graduated from Stanford Law School and

was admitted to the California Bar in 1950; the legal staff of Union Pacific, 1953-1956; senior deputy prosecutor, 1956-1958, for city of Pasadena; deputy DA from 1958-1971 and assistant DA from 1971-1988, retiring after 30 years in the district attorney's office, Santa Clara County, CA. A certified criminal law specialist; Phi Alpha Delta law fraternity; and listed in *Who's Who in the West*.

Married to June Martin, he has three sons and one grandson.

BRUCE LAINGEN president of the American Academy of Diplomacy, a non-profit, limited membership society of 100 men and women, now retired from government service but who have held senior positions in the conduct of American foreign policy, whether as career or political appointees. The Academy is dedicated to fostering the highest standards in the conduct of diplomacy, particularly those nominated by the president as ambassadors.

Ambassador Laingen was born in Minnesota in 1922. He served in the USN in WWII and in the Foreign Service from 1949-87, including tours in Germany, Iran, Pakistan and Afghanistan. After serving as deputy assistant secretary for European Affairs, he was Ambassador to Malta from 1977 to 1979. Later that year he returned to Iran for a second tour, as Charge D'Affaires of the Embassy, before being taken hostage in the Iran hostage crisis, 1979-81.

Following his release, he served as vice president of the National Defense University in Washington, DC until his retirement from the service in 1987. He was executive director of the National Commission on the Public Service (the Volcker Commission) from 1987 until it completed its work in 1990.

In addition to his work with the Academy of Diplomacy, Ambassador Laingen serves on the boards of A Presidential Classroom for Young Americans, the Mercersburg Academy in Pennsylvania, No Greater Love, and the National Defense University Foundation. Currently he is also a member of the National Commission on the State and Local Public Service.

Among his awards and citations, Ambassador Laingen holds the Award for Valor from the Dept. of State, the Distinguished Public Service Medal from the Dept. of Defense, the Distinguished Alumnus Award from St. Olaf College, the Golden Plate Award from the American Academy of Achievement, and a Presidential Meritorious Award.

He is a graduate (cum laude) of St. Olaf College in Minnesota and the National War College and has an MA in international relations from the University of Minnesota.

Ambassador Laingen has honorary degrees from Columbia College in Missouri, Hahneman University in Philadelphia and the College of Osteopathic Medicine of the Pacific in Los Angeles. He is the author of *Yellow Ribbon: the Secret Journal of Bruce Laingen (1992)*. He and his wife,

Penne, live in Bethesda, MD. They have three sons, each of whom chose naval officer careers.

JOHN WALTER LALONDE (LTJG) was born in Dallas, OR on May 15, 1921. He enlisted in the service on June 1, 1942 and entered the V-12 Program at NTS Purdue University, West Lafayette, IN; EE&RM Oklahoma, A&M Stillwater, OK; ARM Corpus Christi, TX; USNRMS at Cornell University, Ithaca, NY; Officers Steam Engineering School, NTS, Newport, RI; and records officer (prospective) USS *Leyte* (CV-32).

He participated in the Aleutian Campaign, Asiatic-Pacific Theater, NAS Kodiak, Territory of Alaska from March-August 1943. His memorable experience was listening to Tokyo Rose broadcast of bombing Kodiak NAS (when, in fact, it never occurred).

Released to inactive duty on April 11, 1946, he transferred to retired Reserve on Jan. 16, 1967. His awards include the WWII Victory Medal, American Campaign Medal, Asiatic-Pacific Medal and the Good Conduct Medal.

He and his wife, Mary Louise, have three children: Jeanine, Bruce and Beth. He retired Sept. 1, 1985, from the Boeing Co. in Seattle as safety manager of the Fabrication Div. and BCAC Engineering Div.

ALBERT W. LAMPTON (SK3/c) was born in Scammon, KS on Aug. 9, 1925. He entered the University of the South (Sewanee, TN) V-12A Program on Nov. 1, 1943, went to pre-Midshipman School at Princeton, NJ in March 1945 and Midshipman School in Chicago, IL in April 1945. He was reassigned to general duty at Great Lakes NTC in May 1945; assigned to the Officer Separation Center, Great Lakes in October 1945; and was discharged as SK3/c (disb) on May 17, 1946.

Lampton graduated from Pittsburg (KS) State University in 1948 with a BS in chemistry. He was employed by the Spencer Chemical Co. from January 1949 to December 1956; worked for the USAEC from January 1957-July 1978 at the Hanford Plant in Richland, WA; and retired from Federal Service on July 26, 1978.

He is presently residing in Richland with his wife of 42 years and they have two sons, one daughter, two granddaughters and four grandsons.

LEE E. LANDES of New York City, NY entered the V-12 Program on July 1, 1943, at Union College and received an AB degree in February 1944. He was commissioned ensign in the Supply Corps on Aug. 16, 1944, at Harvard. He served in WWII and in Korea at NAS Pensacola; LCI Group 61 Pacific; Fleet Activities Sasebo; NAS New York & Squantum, MA; COM 3 and COM 9.

Landes received his MBA degree from Harvard in June 1949, was promoted to rear admiral in November 1972 and retired from active Reserve status in November 1979. He was awarded the Legion of Merit.

He retired from Ford Motor Co. in February 1980 as a world-wide technical planning manager for advance cars and trucks; he worked in finance and product planning and has been a full-time volunteer since 1980; he founded MADD in Michigan in 1982; and he is active with the Sea Cadets, Navy Recruiting, ROTC and the Reading Program in Detroit.

Landes resides in Livonia, MI with his wife, Suzanne. They had two sons, Stephen and George. Stephen and his wife and children, Benjamin and Gwenyth, live in Ann Arbor. Their younger son, George, was killed by a drunk driver in October 1981. Both sons graduated from the University of Michigan.

H. PARKER LANSDALE (2LT) was born in Worcester, MA on March 18, 1923. He enlisted in August 1942, entered the V-12 Program at Oberlin College. He went to Yale Divinity School and received his BD in 1950, he attended Graduate School and received his MA in 1953 and Ph.D in 1956. He made a career with the YMCA.

His military stations include Parris Island, Lejeune, Quantico OCS and FMF Pacific with the 4th Mar. Div. His memorable experience was the friendships formed. Lansdale was discharged in August 1946. He is a member of Croghan's Heroes.

Married, he has seven children and seven grandchildren. He is retired.

CHARLES B. LARKIN (COL) was born in Madison, WI on June 6, 1924. He entered the service July 22, 1943, and attended V-12 program at Willamette University, Salem, OR from March 1944-June 1945 and the University of Wisconsin from September 1945-February 1946.

His military assignments in the USN include: USNTS, Farragut, ID; USN hospitals at Great Lakes, IL; Long Beach, CA; and Bremerton, WA. He served in the USAF as a medical officer from 1950-52, active duty; in the USAF active Reserves from 1952-1984 when he retired as a colonel. He served as medical commander, neuropsychiatrist and consultant to the Surgeon General USAF.

Discharged from the USN in July 1947 and from the USAFR in 1984. He practiced medicine in Wisconsin and California from 1949-87 as an internist and neuropsychiatrist.

His family includes his wife, Irene (deceased since 1989), seven children and eight grandchil-

dren. In 1991 he married Mary M. Conners. He likes to travel, read, attend sporting events, walking and church activities.

ERLING LARSON JR. entered the V-12 Program in July 1943 at Miami of Ohio. He attended College of Medicine, University of Iowa from 1944-1948; interned at City Hospital, St Louis; Naval Operating Base, Norfolk and USS *Philippine Sea* (CV-47) from 1949-1950; residency in internal medicine at Veterans Hospital in Des Moines from 1950-1953; and was a medical officer at Naval Ordnance Plant, Indianapolis from 1953-1954.

Larson was in private practice in Davenport IA from 1954 to 1992; was clinical associate professor of internal medicine, 1976-1992 AOA was Iowa internist of the year 1984; teacher of the year in 1989; received the Laureate Award, American College of Physicians in 1990; president of Iowa Medical Society in 1983; and was a delegate of the AMA from 1973-1981.

From 1978-1981 he was president of the University of Iowa Foundation; from 1984-1988 was chairman of President's Club; and from 1979 1993 he served on the board of directors of Citizens Federal Savings Bank. He received the Distinguished Alumni Achievement Award in 1983 and the Finkbine Medal in 1979.

Larson is now semi-retired in Chesapeake VA with his wife Jan M. Larson, M.D. They have two sons, Erling III (an M.D.) and David (engineer), two daughters, Karen and Melissa (both teachers), and nine grandchildren.

ROBERT M. LAURION was called to active duty on July 1, 1943, at Northwestern after a civilian year in pre-Journalism at Northwestern University. Three semesters later he was offered the opportunity to join a group of V-12 physics majors at the University of Illinois. They celebrated V-E Day at Illinois before going to Midshipman School at Notre Dame. This eleventh Midshipman class became the last because they celebrated V-J Day just before being graduated and getting their commissions.

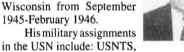

There was a fast year of duty at Pearl Harbor on PCEs before separation to inactive duty. He spent the rest of 26 Navy years in various Reserve units and on annual active duty training tours as he moved around in his civilian career.

Laurion has been in communication work generally and telephone work specifically during this time and retired a short time ago as president of American Communications Consultants, an engineering consulting firm. He still does some independent consulting work in this field.

He and his wife enjoy living in Madison, WI and spend time at their farm 65 miles north when not visiting their four children and families in various parts of the country or traveling to interesting places in the world.

JOHN M. LAW (CAPT, JAGC) was born in Chicago, IL on Dec. 5, 1927. He enlisted as V-5 Aviation Cadet in 1945 and attended V-12 units, St. Ambrose College and Colorado College from July 1945 to February 1946. After V-12, Tarmac, NAS Moffett, he entered selective flight training at NAS Livermore and preflight at NAS Ottumwa.

Discharged to the USNR in July 1946, he was commissioned ENS DL USNR, 1948. He served in Active Reserve billets for 32 years including Air Intelligence Officer, NAS Denver, 1950-1960 and in 1961 member, executive officer and commanding officer, Naval Reserve Law Co. until his retirement in 1977.

Graduated from Colorado College in 1948; the University of Colorado School of Law in 1951; and entered private practice in 1951. His practice in recent decades has been primarily in litigation in the field of professional liability insurance defense of attorneys, health care and other professionals as founding and name partner in the two firms he has been associated with since 1957, until retiring in 1993.

Law and his wife, Carol, reside in Denver. Their children: John, Lucy, Fritz and Beth, are graduates of the University of Colorado, are married and have presented them with nine grandchildren. He is a member of the Denver Country Club and the University Club of Denver and enjoys golf, travel and photography.

GEORGE ELLETT LAWRENCE (SN2/c) was born in Grenada, MS on July 22, 1927. On June 6, 1945 he joined the service and entered the V-5 Program at Millsaps College, Jackson, MS and Tulane University, New Orleans, LA. He was stationed at Great Lakes NTC/Naval Bureau of Personnel, Arlington, VA.

He was discharged on Aug. 15, 1946. Married to the same girl for 40 years, they have two daughters and four grandchildren. He joined the family business, Lawrence Printing Co., in Greenwood, MS after his graduation from the University of Mississippi in 1949. He was president of the company from 1971-1993 and chairman of the board from July 1993 to present.

DONALD L. LEDBETTER (CDR) completed his engineering degree in V-12; received his BS in civil engineering from the University of Oklahoma; and remained in the CEC Reserve. He had active duty in the Pacific during the cleanup after WWII and at NavSchools Construction during Korea.

Entered the nuclear engineering industry at Argonne Labs and spent most of his career as project manager/project engineer on nuclear and high tech facilities including reactor installation for the USS *Long Beach,* nuclear rocket program, defense nuclear waste repository and on nuclear power plants. He had assignments as manager and vice president of business development for Bechtel and Kaiser Engineers focused on new technology for space and defense programs. His Navy assignments and Reserve training provided valuable experience for an exciting career managing first of a kind projects.

Received his MBA with construction emphasis from Golden Gate University, San Francisco. Currently he is a consultant on Project Management and on nuclear waste projects and is chairman of the National Society of Professional Engineers Task Force on career transition. He is a registered nuclear engineer in California and a registered professional engineer in Oklahoma, Kansas and New Mexico.

JOHN WESLEY LEE JR. (LCDR) was born in Indianapolis, IN on Feb. 13, 1924. He graduated in three years from Crispus Attucks HS at the age of 16. He attended Indiana University from 1940 until entering the USN in April 1943 as an officer's stewards mate. In July 1944 he entered the Navy V-12 Program at DePauw University. He was commissioned ensign in July 1945 at Northwestern University Reserve Midshipman School, Chicago, IL.

Served in Pacific Fleet ships *Ramapo* (AO-12), *Sepulga* (AO-20) and *Capricornus* (AKA-57) as gunnery officer prior to being released to inactive duty in August 1946. He was recalled to active duty in March 1947 as the first Black American in the Regular USN at Great Lakes, NTC, IL. He attended the General Line School, Newport, RI and graduated from the Navy Postgraduate School, Monterey, CA earning a BS in aerological engineering and a MS in physical oceanography from Scripps Institute.

Returning to the fleet, he served in *Kearsarge* (CV-33), *Wright* (CVL-48), *Constellation* (CVA-63), *Toledo* (CA-133), *Storms* (DD-780), and *Cotten* (DD-669) as commanding officer Oceanographic Detach. 2 and XO of *Aeolus* (ARC-3). His final tour was on staff Naval Deputy CinC Allied Forces Central Europe in Fontainebleau, France. Lee retired June 30, 1966, as a lieutenant com-

mander. He worked for RCA as a production engineer and at the Naval Avionics Center, Indianapolis as a logistics engineer, retiring in 1989.

Married Geraldine V. Bridgewater on July 16, 1947; they are the parents of Deborah Lee Fleary, a physical therapist; Cheryl Ann Lee, a school teacher; and John W. Lee III, a marketing rep. The family moved 29 times before returning to Indianapolis.

WILLIAM J. LEGG was born in Enid, OK, in 1925 into a pioneer Cherokee Strip family. After passing the 1943 V-12 national examination, he enlisted instead at age 17 in the Naval Aviation Program, V-5. Transferred in 1944 from the Hutchinson, Kansas NAS to the V-12A Program at Pittsburg State University, he thereafter completed the V-12 Supply Corps Program at the University of Texas at Austin, receiving a BBA degree and a commission as ensign. In 1950 he married Imogene Hill of Bartlesville, OK and they have three daughters and currently four grandchildren. While working full time, he earned the Juris Doctor degree from the University of Tulsa in 1954, with the highest over-all law school grade average.

Entering the private practice of law in 1962, he is now senior counsel in the 100-person firm in Oklahoma City, Andrews Davis Legg Bixler Milsten & Price, P.C. having practiced internationally as an oil and gas specialist. He has been published extensively on legal subjects, and has served as an adjunct professor of law, and as an instructor in many legal seminars throughout the United States.

An ordained Evangelist in the Reorganized Church of Jesus Christ of Latter Day Saints, he has participated in outreach ministry in many domestic and foreign locations.

In addition to membership in state, national and international bar associations, he has served as an officer and director of several oil companies and other commercial corporations, is a trustee of Graceland College in Lamoni, IA, a Research Fellow of the Southwestern Legal Foundation in Dallas, and a trustee, officer or active member of several similar national and local professional, civic and charitable organizations.

He is listed in *The Best Lawyers in America, Who's Who in American Law, Who's Who in America,* and *Who's Who in the World.*

C. FREDERICK LEYDIG (LTJG) was born in Denver, CO on Jan. 24, 1925. He entered the V-12 Program at the Illinois Institute of Technology; graduated in October 1945 with a BS in chemical engineering; and from DePaul University College of Law in May 1950 with Juris Doctor degree. He is a fellow, American College of Trial Lawyers.

Duty stations include Anacostia Fire Control School and the USS *New Jersey.* He was in the Reserves from July 15, 1946-March 24, 1958.

Retired as senior partner in Leydig, Voit & Mayer, LTD with offices in Chicago, Washington, DC, St. Louis and Rockford, IL. Married to Patricia

Schwefer on July 2, 1949, they are the parents of four children: Gregory (Sarasota, FL), Deborah Pfaff (Barrington Hills, IL), Gary (LaGrange, IL) and Suzanne Schubring (Hanover, Germany).

WILLIAM A. LIDDELL (DREW) (LTJG)

was born in Shreveport, LA on Jan. 6, 1925. He enlisted in January 1942; entered the V-12 Program at Georgia Tech and the University of North Carolina; and attended Midshipman School at Harvard.

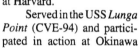
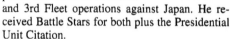

Served in the USS *Lunga Point* (CVE-94) and participated in action at Okinawa and 3rd Fleet operations against Japan. He received Battle Stars for both plus the Presidential Unit Citation.

Liddell was discharged in July 1946. He became an executive assistant of National Technology, Inc.

RONALD LIGHT (LT) was born in Cleveland, OH on Aug. 5, 1925. He enlisted in June 1943

and entered the V-12 Program at Oberlin College. Other stations include Great Lakes Naval Station, University of California and Alameda Naval Air Station.

His final discharge was in June 1952. Light went into the practice of Orthodontics in Concord, CA. He has two sons, one daughter and three grandchildren.

WALEN F. LILLY (ENS) was born in Manhattan, MT on Aug. 13, 1925. He entered the Navy

V-12 Program on July 1, 1943, immediately after graduating from high school at Manhattan, MT. He attended Montana School of Mines four semesters and Ft. Schyler Midshipman School on Throggs Neck Long Island and was commissioned ensign in February 1945.

Served in both the Atlantic and Pacific on the USS *General R.M. Blatchford* AP as deck officer and advanced to ships 1st lieutenant. He was discharged in July 1946. The *Gen. Blatchford* transported thousands of troops to and from various war zones. In August 1945 the *Blatchford* was underway in the Pacific with veteran troops recently engaged in the European Theater, to invade Japan. The atomic bomb changed their plans.

Lilly feels fortunate for the opportunity to get part of his university training in the V-12 Program and experience the war as a naval officer.

He spent 25 years as a high school science teacher and also developed an international reputation as a fly fishing outfitter retiring in 1982. He has written several books on western fly fishing and has a reputation among the best.

O. VICTOR LINDELOW (LTSG) was born

in Minot, ND on June 10, 1924. He entered the V-

12 Program on July 1, 1943, at Minot State Teacher's College, Minot, ND. In October 1943, he transferred to the V-12, University of Minnesota in Minneapolis, receiving his Bachelor of chemical engineering degree. He attended Notre Dame Midshipman School, July 12, 1945, and the war ended Aug. 15, 1945.

Lindelow transferred to Great Lakes, IL in October 1945, completed his service at Cleveland, OH and was discharged in May 1946 as first class seaman. His pre-med training was at the University of Minnesota from 1946-1947 and he received his M.D. degree from the University of Pennsylvania in Philadelphia, June 1951.

His internship and two years of residency in internal medicine was at Geisinger Hospital and Foss Clinic from July 1951-July 1954; residency was at Cleveland Clinic, Cleveland, OH from July 1954 to 1955; and he joined the Missouri Valley Clinic, Bismarck, ND as an internist in July 1955. The clinic merged in 1971 to become Mid-Dakota Clinic, P.C. He is still practicing at this clinic.

A lieutenant junior grade in the USNR Med. Corps (inactive) June 19, 1951, he later retired as lieutenant senior grade on Aug. 1, 1967. He was president of the North Dakota Medical Assoc. from 1981-1982; Governor North Dakota Chapter, American College of Physicians, 1987-1991.

Married Betty Jean Bell on Sept. 11, 1950, and they have four children and 10 grandchildren.

ROBERT M. LINDGREN (LTJG) was born

in Medford, MA on Nov. 13, 1925. He entered the V-12 Program at Tufts College, Medford, MA on July 1, 1943, and was commissioned ensign on March 6, 1945, at Ft. Schuyler Midshipman School, Bronx, NY. He continued his training at Advanced Line Officer School at Miami, FL; Gunnery School at Norfolk, VA; and Advanced Gunnery School at Treasure Island, CA.

Assigned to the USS *Putnam* (DD-757) as assistant gunnery officer, he served in the Occupational Forces of the Japanese Empire. He returned first to the West Coast then to the East Coast to train gunnery crews out of Newport, RI until he was released to inactive duty in Boston, MA on June 15, 1946.

Lindgren completed three semesters at Tufts College and received a BS in civil engineering in January 1948. After working one year as a CE, he became a manufacturer's representative in 1949 and formed Lindgren Corp. in Camp Hill, PA until its sale in 1993.

Currently, he resides with his wife, Joyce, in this town during the winter and in East Dennis,

Cape Cod, MA in the summer. They have three sons.

EDWARD M. LINKER (LT) was born in

Chapel Hill, NC on May 29, 1926. He enlisted on April 13, 1944, and entered the V-12 Program at Duke University.

He left active duty June 25, 1946. His awards include the American Defense and WWII Victory Medal.

Linker and Jean Stone were married on June 20, 1953, and have four children. He retired from Tultex Corp., Martinsville, VA.

JEROME RICHARD LISANKIE (LTJG)

was born in Brooklyn, NY on Nov. 18, 1924. He entered the Navy V-12 Program Rensselaer Polytechnic Institute in July 1943 and was commissioned ensign on Nov. 19, 1945. He served on the USS *Lauderdale* (APA-179) from Dec. 1, 1945 March 27, 1946, in the Asiatic-Pacific Theater. He was released from active duty in June 1946.

Returned to RPI in September 1946 and graduated in February 1947 with a mechanical engineering degree. He was appointed lieutenant (jg) on Nov. 19, 1948, and was honorably discharged from the USN on Oct. 22, 1958.

Lisankie worked 37 years for Foster Wheeler Corp. and retired in May 1987 as director of contract operations, Fired Heater Div., Livingston, NJ. He died Oct. 28, 1991, while playing golf, his favorite sport. He is survived by his wife, Kathryn, five children and six grandchildren. Mrs. Lisankie resides at their home in Long Valley, NJ. *Submitted by Kathryn Lisankie.*

W. GARLAND LOFTIS (COTTON) (MAJ)

enlisted in the USMC in April 1942 and entered the V-12 Program at Duke University in July 1943. He left with the first group to Parris Island, SC in October 1943, attended Officer's Training School at Quantico, VA and promoted to 2nd lieutenant. He was assigned to the 3rd Mar. Div. on Guam in October 1944 and reassigned to the 6th Mar. Div., C Co., 1st Bn., 22nd Mar. Regt., 6th Mar. Div. in December 1944.

Landed on Okinawa April 1, 1945, as a rifle platoon leader and received a Silver Star and Purple Heart. In June 1945 he served with C Co., 1st Bn., 22nd Mar. and as rifle company commander was awarded a Bronze Star. Loftis returned to Guam as battalion supply officer and finally to China in October 1945. He was discharged in April 1946; returned to active duty during the Korean conflict in 1952; discharged as a captain in 1955 and retired as major in 1961.

Became territorial vice-president for Allstate Ins. Co. and was awarded the Most Outstanding Manager in 1972, 74 and 75. He retired in 1983. He was captain of high school basketball team which won the national championship in 1940, after winning 72 games without a defeat. He played at Duke University with undefeated Freshman team and Southern Conference Championships in 1941-42 and 1942-43 seasons.

Loftis married Marion Wilkins in 1943 and they have two children, Gary and Delane, and three grandchildren. His hobbies are tennis and gardening.

ROBERT B. LOWTON

ROBERT B. LOWTON entered the V-12 Program at Columbia from 1943-1944 and Yale from 1944-46. He was commissioned ensign in February 1946; followed by training cruise in USS *Cleveland* (CL-55); was CIC officer in USS *F.T. Berry* (DD-858) based at Tsingtao, China.

Discharged in June 1947, he returned to Yale and received BE in June 1948 and BS for NROTC work. He joined the Kellex Corp. (name changed to Vitro Corp. in 1950) as job engineer, project engineer, constructions engineer, project manager and assistant director of International Dept.

Joined General Electric in 1964 and worked on nuclear power as site resident manager, project manager, manager of System Design, manager of advanced engineering and manager ARWR Project until retiring in 1989. He resides at San Jose, CA with his wife Geraldine. They have three sons: Philip, Andrew and Gordon, who also live in California.

FRANCIS X. LOYER

FRANCIS X. LOYER (FIREMAN 1/c) was born in Mingo Junction, OH on May 8, 1926. He started active duty on July 1, 1944, at Dennison University, Granville, OH and served one year there. Most servicemen were in the battle in the South Pacific, while Loyer stayed in the States and served in the battle of Buckeye Lake (the Playground of Ohio).

Served his second year at Great Lakes, then went to Seattle, WA where he was assigned to the destroyer, USS *Porterfield* (DD-682), which was assigned to the South Pacific Theater. Japan surrendered at that time, so he was sent to San Diego, CA Navy Yard where the ship was put in moth balls. He stayed there nearly a year until finishing his two-year enlistment, then was honorably discharged in June of 1946.

Returned to his home in Mingo Junction, and in 1948, he married Aleda Moats. He has been in the appraisal of damaged motor vehicles and has operated his own business, the Ohio Valley Auto

Damage Appraisal Co., for the past 25 years. He is now an associate with Irvine and Associates Adjusting Co. out of Wheeling, WV.

Loyer and his wife, Aleda, reside in St. Clairsville, OH. They have four married children and five grandchildren.

WILLIAM P. LUCE

WILLIAM P. LUCE (QM3/c) was born in Philadelphia, PA on Oct. 18, 1924. In February 1943 he entered the V-12 Program at Colgate University. Other duty stations included Bainbridge, MD; Terminal Island, CA; and the South Pacific.

His memorable experience was sailing in the spring of 1944 from Guadalcanal, where he was on a crash boat, to Colgate University where he joined the V-12. The war ended while he was in his fourth semester and he was discharged in May 1946.

Retired as supervising news editor from the *New York Times* and now lives in Englewood, NJ with his wife, Carolyn Darrow.

RICHARD W. LUEKING

RICHARD W. LUEKING entered the V-12 Program on July 1, 1943, at Berea College, KY. He was commissioned ensign at Northwestern, Chicago, IL on March 29, 1945; attended Tactical Radar School, Hollywood, FL and Fighter Director School at Barber Point. Lueking reported to the USS *Guam* (CB-2) in Inchon, Korea in August 1945; returned with ship to Bayonne, NJ in December 1945 and was discharged in March 1946. He had no battle action.

Returned to Berea College and graduated in January 1948. Procter & Gamble Marketing, 1948-1963, brand management Tide, Ivory, et al and he introduced Cascade, Ivory liquid. He was vice-president of advertising of briefly rejuvenated Eastern Airlines, 1964-1965; advertising agencies, 1966-1967; Lever Bros., varied upper mid-management positions, 1968-1978; president, Westport Marketing Group, 1979-1983; and helped to initiate and plan the Statue of Liberty Centennial celebration.

Lueking retired to San Diego County in 1984. He lives with his wife, Joyce, in Leucadia, CA. They have five daughters: Barbara, Beverly, Amy, Mary Beth, and Laura, living in Connecticut, Utah and Maine. His first wife, Winifred, is the mother of his daughters and lives in Connecticut.

REEVES A. LUKENS

REEVES A. LUKENS (CAPT) born in Philadelphia, PA on Aug. 7, 1926. He graduated from Roxborough HS in Philadelphia in June 1943; attended the University of Pennsylvania; enlisted in the V-12 Program on Dec. 16, 1943, and was assigned to Trinity College, Hartford, CT on March

1, 1944. He transferred to NROTC Brown University, Providence, RI on July 1, 1945.

Commissioned in June 1946 and assigned to the USS *Wheatear* (AM-390) Charleston, SC. He transferred to the regular Navy in April 1947 and was released on Feb. 23, 1949.

Lukens was employed by Travelers Ins. Co. in April 1949 and served in Hartford, CT, NYC and Columbia, SC. He was employed by Pilot Life Ins. Co., Greensboro, NC in December 1955. Pilot Life merged with Jefferson Standard Life in 1987 to form Jefferson-Pilot Life. He retired Aug. 31, 1991, as 2nd vice-president Group Marketing Services.

Rejoined the Reserve after active duty; affiliated with Intelligence Program August in 1957 and retired in August 1984 with 40 plus years of service. He and his wife, Peg, have a son and daughter, Jan and Kathi. He retired from Jefferson-Pilot Life Ins. Co., Greensboro, NC.

FRANK ARNE LUNDBLAD

FRANK ARNE LUNDBLAD (LTJG) was born in Fredrikstad, Norway, Europe. He joined the service on Nov. 1, 1943; entered the V-12 Program at Dartmouth College/Harvard Business School. Other stations/schools include the Naval Supply Depot, Spokane, WA; VR 7 NAS, Honolulu, Supply & Disbursing Officer; AKS13 Hesperia, Bikini Atom Bomb Test; and the University of Oslo in Oslo, Norway.

After his discharge on June 9, 1947, he served in various marketing and sales positions with Carter's Ink Co.; Gillette Safety Razor Co., Ratheon Mfg. Co., Overseas Chem. Div. & Domestic Div. of W.R. Grace & Co., was senior management consultant for Arthur D. Little, Inc. and president of Van Melle Inc., Cornerstones Energy Group and LCA Mgmt. Consulting Associates.

He currently resides in Brunswick, ME with his wife, Deborah. They have two daughters and a son.

CHARLES DONALD LUNDERGAN

CHARLES DONALD LUNDERGAN was born in Washington, IN on Sept. 24, 1923. He enlisted Dec. 1, 1942; was assigned to V-12 on July 1, 1943 at the University of Notre Dame; commissioned at Camp Endicott, RI on Feb. 6, 1945; and served at Adm. Nimitz Adv. HQ and 109th Constr. Bn. both on Guam.

Returned to Notre Dame to receive a Master's in physics and taught physics and mathematics at St. Louis and Texas A&M Universities. He joined the staff of Sandia National Laboratories in 1956 and was the supervisor of the Shock Wave Physics Div. and served on the management staff. He completed his professional career in the science and technical activities of foreign intelligence; currently, he is a consultant to Sandia National Laboratories.

Now living in Albuquerque, NM with his wife, Elaine. Of his four sons: Michael is a teacher, Timothy a computer scientist, Donal is with Albertsons Stores and Dan is a hospital administrator.

MARSHALL H. LUTGEN

MARSHALL H. LUTGEN was born and raised in Auburn, NE. He entered Doane College V-12 Program after serving as a pharmacist mate for two years. He had Pacific sea duty aboard the USS *Hamblin* (APA-114) and received an honorable discharge in December 1945.

Lutgen attended the University of Nebraska and the University of Minnesota where he received his degree. Then returning to Lincoln, NE where he was a Cessna aircraft dealer and an auto dealer for many years. During 1961 he changed his career to that of a stock broker. The next 30 years he spent in management, served as an officer in a major mutual fund, and owned his own broker/dealer organization.

Currently, semi-retired, he and Marjorie, his wife of 50 years, live at Hiwan County Club, Evergreen, CO. They have two daughters and three grandsons. The daughters both married career naval flight officers and two of the grandsons are currently in college. One on a military scholarship and the other seeking a career in flying.

GEORGE C. MACGILLIYRAY

GEORGE C. MACGILLIYRAY (LT COL) was born in Boston in 1922. He enlisted in the USMC in 1941, joined the 7th Regt., 1st Mar. Div. and went to Samoa (April 1942) and to Guadalcanal. After four months of combat with Lt. Col. Puller's battalion, 7th Mar., he left Guadalcanal in January 1943 for Melbourne, Australia. He received the Navy Commendation Medal from Adm. Halsey for valor and was promoted to sergeant.

Participated in New Guinea and Camp Gloucester campaigns (1943-44); was sent back to Officer Candidate Bn. (Camp Lejeune), then to Dartmouth College V-12 in November 1944. WWII ended in 1945 and the V-12 was disbanded, so he transferred to Boston Navy Yards and was discharged in 1947.

Returned to Dartmouth on the GI Bill and graduated in 1948; he spent two years at Clark University for a MA in geography/cartography; joined the Central Intelligence Agency in 1950 and retired in 1982.

Married Gloria (deceased) and they had three children: Jeffrey, James and Kathleen, and five grandchildren. He now lives in Bethesda, MD and is a life member of the 1st Mar. Div. and Guadalcanal Veterans Associations. For the past 12 years he has been a volunteer at the Marine Corps Historical Center, Washington, DC.

ROBERT N. MADDOX

ROBERT N. MADDOX was born in Winslow, AR on Sept. 29, 1925. He graduated from Winslow HS in April 1941; entered V-5 Program in July 1944 at Iowa State College; and was commissioned ensign on Oct. 24, 1945. He served as engineering air maintenance officer with Sqdn. 1A, primary training from January-May 1946 and as material officer

from May-July 1946 at Cabaniss Field, TX. He was released to inactive duty in July 1946 and discharged in December 1958.

Maddox received B.Sc. from Univ. of Arkansas in 1948; M. Chem. E., Univ. of Oklahoma in 1950; Ph.D. from Oklahoma State Univ. in 1955 (all in chem. engr.); and Sc.D. (honoris causa) from Univ. of Arkansas in 1991. He was faculty member at Oklahoma Univ. from 1950-86, served as dept. head from 1958-77; did research in natural gas sweetening and sulfur production, computer applications, process design, transport properties; is the author of more than 175 tech. papers and seven books and is widely and well-known as a chemical engineer educator and consultant. He retired in 1986, but is still active as an author and consultant. He has received numerous awards and recognitions.

He and Paula Robinson were married in 1951; she passed away and he married Pauline Razook in 1987. He has two children, Deirdre O'Maddox Dexter (attorney) and Robert Dozier (CPA), and four grandchildren: Daniel and David Dexter and Caitlin and Robert Logan Maddox.

HAROLD I. MAGOUN JR.

HAROLD I. MAGOUN JR. was born June 11, 1927. He entered the V-12 Program as a pre-med on July 1, 1944, three weeks after his 17th birthday; the youngest one in his unit. They drilled for two hours that first afternoon in Dallas in 114° weather, an experience he will never forget.

Returned to inactive duty on Oct. 26, 1945, and spent one more year at South Methodist University, then completed his education at the Kirksville College of Osteopathic Medicine. He has practiced in Denver, CO and is board certified in manipulative medicine.

Magoun has two daughters and five sons, none of whom served in the military or became a doctor. He is very grateful for the education he received in V-12.

ROBERT T. MANLOVE

ROBERT T. MANLOVE (BOB) was born in Newark, NJ on March 28, 1923. He moved in 1925 to the Jersey Shore, lived there until WWII and graduated from Asbury Park HS in 1941. In the spring of 1940 he joined the USNR unit at Perth Amboy, NJ and was called to active duty on July 1, 1942.

Late November 1942, he was assigned for sea duty to PC-488. In May 1944, he was ordered to report to

pre-midshipman facility at Asbury Park, NJ (his hometown). Two memories of the time spent ther was awaking one morning to the PA telling ther of D-day and a visit from a high ranking office from Washington telling them about the V-1 Program.

On July 1, 1944, he was assigned to the V-1 Program at Princeton University and stayed ther until the war's end and completed the equivalent o two college years. VE and VJ occurred during thi period. In October 1945 he requested return to rank and reassignment and returned to regular dut on Nov. 1, 1945.

Manlove received his honorable discharge a the separation center at Lido Beach, Long Islan on Dec. 18, 1945. He then went to the Midwest i September 1946; finished school at the Univ. o Minnesota; lived in Illinois for 10 years and i Wisconsin since 1959.

Married in 1948 and has three children an six grandchildren. He had a wonderful 35 yea career with a leading business forms manufac turer, doing office work simplification. He is nov a retired widower and spends winters in Florid and summers in Wisconsin.

L. BROAS MANN JR.

L. BROAS MANN JR. (BUD) (ENS) wa born in Detroit, MI on Sept. 24, 1925. He joine the V-12 Program at Illinois Institute of Technol ogy in July 1943; graduated with BSME and wa commissioned ensign in February 1946. He spen four months on a training cruise in the USS *Cleve land* going to Atlantic coast ports and Bermuda He spent five harrowing days in the hurricane o 1946, during which the ship's doctor performed a appendectomy. He was discharged on July 1, 1946

He was an instructor in mechanical engineer ing at Lawrence Institute of Technology, Detroi MI, while acquiring MSME at Northwestern Univ Evanston, IL. He joined the Chrysler Corp. in 195 as a research and development engineer on th automotive gas turbine program. He was manage of the Research Design Dept. in 1964 and o Advance Powertrain Engineering in 1980.

Mann married Marion DeSutter in 1949 an they have two sons and two daughters: Cathy pre-school teacher; Tim a programmer/analys Lisa a nurse practitioner; and Chris a golf stor manager. There are eight grandchildren. He re tired in 1987, but was retained as powertrai engineer consultant and is still active in that capac ity.

ROBERT MANN

ROBERT MANN (LTJG) was born in Mem phis, TN on Jan. 28, 1924. He joined the USN i July 1943 and entered the V-12 Program at th University of the South (Sewanee) and USN Midshipman School at Columbia University.

Mann participated in the Okinawa invasio for 20 days and was discharged in July 1946.

Single, he never married. He is a retired rea estate appraiser.

WILLIAM R. MAPLE

WILLIAM R. MAPLE (CAPT) was born i Princeton, NJ on Dec. 2, 1924. He enlisted in Jun

1943; entered the V-12 Program at Princeton Univ. in November 1943; attended Pre-Midshipman School at Asbury Park in November 1944 and Midshipman School at Northwestern Univ. in January 1945; commissioned ENS, USNR, July 30, 1945, then served aboard the USS *Bogue* (CVE-9) until Aug. 16, 1946.

Served in naval reserve units from 1947-84 in Trenton, NJ; Durham and New Bern, NC; Eureka Springs, AR; Pensacola and Norfolk. He commanded units in the last four until retirement in 1984.

Maple earned an AA from Princeton, BS from Rutgers and Master of Forestry from Duke. He was employed by the U.S. Forest Service from 1951-75 and Martin Community College, Williamston, NC from 1976-89.

Retired in 1989 and lives with his wife, Dottie, in New Bern, NC. They have two daughters, Mary and Betty; a son, Bobby; and a granddaughter, Betsy.

GEORGE JOSEPH MAROTTA (ENS) was born in Boston, MA on March 24, 1925. He enlisted in September 1942 in the NROTC and in July 1943 for the V-12 Program. He received a BS in mechanical engineering in June 1945 from Tufts College in Medford, MA.

His memorable experiences include placing a naval reserve ring on his wife's finger at their ring dance 50 years ago, and seeing the captain's gig of AKA-54 in Tsingtao Harbor after the radio did not find the AKA-54 in the harbor.

Marotta was discharged in April 1946 and retired in June 1983 from AT&T as Mfg. Engineering Dept. supervisor. He owns and operates a part-time tour business. He and his wife, Blanche June, have a daughter and a son.

PETER S. MARRA (LCDR) was born in New York City on June 15, 1918. He enlisted Sept. 10, 1943, and entered into active service Nov. 1, 1943, Medford, MA. Service schools completed include: V-12, Fire Fighting, Midshipman and CECOS. He has a BCE degree and a MCE degree (sanitary engineering).

Vessels and stations served on include: USNTU, Tufts College, Camp Peary Midshipman School, Camp Endicott Officers School, 133rd NCB, CECOS and LantDivBuYards&Docks, New York. With active duty from 1943-49, he served overseas at Oahu, Maui, Iwo Jima and Guam. At present he is on inactive duty in the CEC, USNR with the rank of LCDR (RET). His awards and medals include the Asiatic-Pacific, American Theater, Navy Unit Commendation, WWII Victory and Navy Occupation Medals.

From 1959 to 1985 he owned and operated a consulting civil engineering and surveying business. He was partially disabled by a stroke in 1984, and from 1985 to present he has been semi-retired but performs pro-bono for former clients.

Marra married Gloria M. Chistoni on Sept. 13, 1947, and they have two children, Kenneth (CDR, USN) and Janice. Kenneth has three daughters and Janice has two daughters and a son.

HOWARD C. MARSH JR. (ENS) was born in Detroit, MI on Nov. 27, 1924. He entered the V-12 Program on July 1, 1943, at Central Michigan College and was commissioned ensign in January 1945 at Northwestern Midshipman School. He entered the Naval Gunfire Liaison Officer Training Program and served aboard the USS *Birmingham* (CL-62).

Participated in action at Iwo Jima and Okinawa, where their ship was hit by a kamikaze. Later, he was assigned to the USS *Vicksburg* (CL-86) as a communication officer. He was released in September 1946 and graduated from Michigan State University in June 1947.

Marsh retired from Cigna Corp. as a marketing vice president in 1985. He is a widower and lives in Vero Beach, FL.

ROBERT L. MARSHALL (CAPT) was born on a farm near Haverhill, KS on June 2, 1925. He enlisted in the USN in March 1943 as a V-5 candidate; attended CMSU from November 1943-1945; completed Midshipman School at Columbia Univ., New York City and in July 1945 was assigned to the USS *Barton* (DD-722).

Discharged in July 1946, but he was recalled during the Korean War. He served aboard the USS *The Sullivans* (DD-S37) and later taught at the Midshipman School in Newport, RI. He obtained the rank of captain in 1968.

Marshall graduated from Kansas University in 1948 with a BA in mathematics and taught at Hill City, KS for three years. Following administrative assignments in El Dorado, KS (1953-56); Kansas City, MO (1956-61); he received his doctorate from Kansas University in 1961; Washington, DC (1961-67); he joined the faculty of CMSU, Warrensburg, MO in July 1967 and served as Dean, School of Public Services for 20 years. He has lectured or served as a consultant in many counties on traffic safety. In 1987 he became president, Safety, Health and Environmental Resource Center International which is located on the campus of Central Missouri State University.

He and Mary Anne were married in 1946 and live in Warrensburg, MO. Their daughter, Suzanne Broussard, is a systems safety engineer in New Orleans where she and her husband reside with daughter, Adriane, and son, Guy. The Marshall's son, Robert C., is an artist in New Orleans.

GEORGE L. MARTIN enlisted in the Navy V-5 Program in 1944 and joined the V-12 unit at Middlebury College. In 1945 he returned to V-5 status, served "tarmac duty," and was at Pre-Flight School when released from active duty. He entered Lehigh University in 1946, graduating with a mechanical engineering degree in 1948.

One memory of his V-12 days happened during the winter of 1944-45. There was two feet of snow on the ground and the temperature was about minus 20°F at 0100 a.m. when a fire drill was called. Out of their warm bunks, dressed in skivvies and pea coats, they tumbled out of the dorms to stand muster. The next morning there was a very large reporting to "sick-bay." Needless to say, the Navy MD was upset and never again was such a drill called under those conditions.

For 36 years he was employed in Alabama and Maryland in the aerospace industry. He was a pioneer in solid propellant rocketry, involved in the design, development and manufacture of many propulsion systems for military and space applications. For many years he was also division procurement manager.

Martin retired as senior staff rocket engineer and lives with his wife, Kathleen, in Middlebury, VT.

JAMES A. MARTIN JR. (LTJG) entered the V-12 University of North Carolina, Chapel Hill on July 1, 1943. He was assigned to Asbury Park, NJ Pre-Midshipman School from November 1944-February 1945 and commissioned ensign at Ft. Schuyler, NY Midshipman School on July 3, 1945. He was assigned as fire control officer in the USS *Springfield* (CL-66) in the Pacific, returned with the ship in January 1946 to San Pedro, CA, and was released to inactive duty in March 1946 at Memphis, TN.

Graduated from Vanderbilt Univ. School of Engineering with a BE degree, electrical in June 1948. He was employed by Dupont Co. of Wilmington, DE, at the Old Hickory, Tennessee Plant in June 1948. Martin worked as engineer, manufacturing supervisor and personnel administrator at Old Hickory, Kinston and Cape Fear in North Carolina and Wilmington, DE. He retired as personnel manager at Old Hickory at the age of 60, in December 1984. He never participated in the naval Reserve; he resigned his commission as LTJG by letter to Secretary of Navy in 1954.

Presently resides in Madison, TN with his wife of 50 years, "Kibba." They have two sons (engineer and attorney), one daughter (horticulturist) and three grandchildren.

EDWARD A. MASON

EDWARD A. MASON was born in Rochester, NY on Aug. 9, 1924. He entered the V-12 Program at the University of Rochester, July 1, 1943, Columbia Midshipman School, June 1945, served on the USS *Calvert* (APA-32) and was discharged in May 1946.

Graduated from the U of R in June 1945; received SM in 1948; ScD in 1950 in chemical engineering from MIT; was assistant professor, Chemical Engineering MIT, 1950-53; and director, Research, Ionics Inc. to 1957. He returned to MIT in 1957 as an associate professor, chemical engineering; professor in 1963; department head of nuclear engineering in 1971; commissioner, U.S. Nuclear Regulatory Commission, 1975-77; and then vice president Research, Amoco Corp. until he retired in 1989.

He is a member of the National Academy of Engineering and American Academy of Arts and Sciences. Currently he is director of Unicom Corporation and Symbollon Corp. and a consultant.

Mason resides in Osterville, MA (Cape Cod) with his wife, Barbara. They have six children and 10 grandchildren.

JOSEPH D. MATARAZZO

JOSEPH D. MATARAZZO (CAPT) was born on Nov. 12, 1925. He enlisted in the USN and from November 1943-January 1944 he completed 12 weeks of boot camp at the US Naval Training Station in Sampson, NY. While completing boot training he was chosen by his commanding officer as that naval base's representative to the V-12 Program and was ordered to report to Columbia University. During February-October 1944 he completed one year of college courses in Columbia's V-12 unit; following which he was ordered to the NROTC Midshipman Program at Brown University where, during October 1944-June 1946, he both earned his AB degree and commission as an ensign. He remained in the Navy an additional year while serving on the fleet tanker, USS *Manatee* (AO-), until his transfer to the USNR-Ready Reserve in July 1947. He remained a Ready Reservist carrying out ad hoc assignments out of the offices of the Surgeons General of the USN, Army and Air Force for 30 years until his retirement to the Retired Reserve as a Navy Captain in 1986.

After his 1947 transfer to the Ready Reserve, he studied psychology at Brown University (1947-48); and Northwestern University (1948-52) from which he received his Ph.D degree in 1952. During

his entire professional career he has been a medical school faculty member: 1952-55 (assistant professor, Washington University); 1955-57 (Research Associate, Harvard Medical School); and 1957-present (professor and chairman, Dept. of Medical Psychology, School of Medicine, Oregon Health Sciences University, Portland, OR.

He is the author of three books, and over 200 scientific publications, and served as the elected president of the American Psychological Assoc. during 1989-90. He and his wife, Ruth, who also is a professor of medical psychology, have a son, two daughters and two grandchildren, all of whom live in Portland, OR

ROBERT A. MATTHEWS

ROBERT A. MATTHEWS (CAPT) was born Aug. 8, 1925, in Holden, MA. He entered the USNR on March 15, 1943, and went on active duty in the V-12 Program at the University of Pennsylvania on July 1, 1943. During this tour of duty he was a member of the varsity football, wrestling and lacrosse teams. He was commissioned in November 1944 at Northwestern University Midshipman School and proceeded on orders to the Solomon Islands, South Pacific for his first assignment afloat as XO LCT-1320 Flotilla 6, in support of 1st Mar. Div. ops.

In April 1945 he left the Solomons in a convoy of 18 LCTs, three PCs and three LCIs stopping at Tarawa, Kwajalein and Eniwetok and eventually arriving at their destination in the Marianas after a 36 day 3,600 mile trip. His most memorable experience was the transporting of 8" shells in his LCT to the USS *Indianapolis* in late July 1945, at Tinian, coincident with the off loading of the atomic bombs onto another landing craft. They didn't realize at the time what had taken place until the *Indianapolis* was sunk near the Philippines and the full story of their mission was revealed.

Final four months of LCT operations was in the Aleutian Islands after which he flew home to be separated from active duty in July 1946. He joined an active Reserve unit and retired in September 1975. His medals include the American Campaign, Asiatic-Pacific with one Battle Star (for consolidation of the Northern Solomons), WWII Victory, Meritorious Service, Armed Forces Reserve and Naval Reserve Medals.

Sara P. Matthews, his wife, served as a USN Hospital Corpsman during the Korean War and they have a daughter, three sons and 11 grandchildren. He is currently employed as manager contracts, Boeing Helicopters Div., Philadelphia, PA.

CHARLES H. MAYFIELD (CHUCK)

CHARLES H. MAYFIELD (CHUCK) (RADM) was born in Oil City, OK on Jan. 14, 1924. He entered the V-12 Program at Southwestern Louisiana and Tulane University as a first class radioman. Other stations include Great Lakes NTC, IL; West Coast, Pearl Harbor, Pago Pago, New Caledonia, Samoa, Fiji Islands, Australia and Tokoyo.

His memorable experiences include the beautiful sight of the Golden Gate Bridge upon return

to the U.S. after a tour of duty in the South Pacific. Over three decades of service, including duties as commanding officer, Navy Reserve Supply Corps Units.

Personal awards include: VP of Hibernia Bank, New Orleans; VP of 1st Nat. Bank of Commerce, New Orleans; president of Acadian Chap of Naval Reserve Assoc. and the Legion of Merit Medal. He retired in 1981.

Charles Mayfield passed away in May 1983. He is survived by his wife, Mary Ellin; son, David; daughter, Martha; and four grandchildren.

JAMES PATRICK McALLISTER

JAMES PATRICK McALLISTER was born in Brooklyn, NY in August 1925. He entered the V-12 Program at Union College, Schenectady, NY in May 1943; attended Stevens Tech. in New Jersey and graduated at Ft. Schuyler, NY in March 1945. He joined LST-1041 in Pearl Harbor; LST-1041 operated in the northern Marianas and Okinawa and after the war in North China. He was released from duty in June 1946.

McAllister graduated from Union College in 1949 with a BA. He joined the family towing company in New York as a deckhand, captain and docking pilot. President of McAllister Towing and Salvage, Ltd., Montreal, Quebec, he retired in 1982 as vice president. The LST-1041 had a 50th reunion at Atlantic City with 20 men attending.

Presently lives between Stamford, CT and Abaco, Bahamas with his wife, Kathy. They have five children: James (wife Cathy), Marianne (husband Bill), Jeffrey (wife Stacy), Mark (deceased) and Helen (husband Jonathan) and three grandchildren: Jessica, Corey and Daniel.

BARRY McCABE

BARRY McCABE (LTJG) was born in Detroit in 1924. He enlisted in the V-12 Program in July 1943 at Central, MI and was commissioned ensign from Columbia in October 1944. He volunteered for UDT, trained at Fort Pierce and Maui, joined UDT #21 in July 1945 and was among the first to land on Japan before the surrender was signed on Sept. 2, 1945. This mission, destroying suicide boats and making sure it was safe for Marines to land, was most memorable.

McCabe transferred in November 1945 to UDT #12, returned to the U.S. and was assigned to LSM-107 in Astoria, OR as a gunnery/communications officer. He was reassigned in February 1946 as a negotiating officer for the USN in Washington and New York and discharged as lieutenant (jg) in August 1946.

Graduated from the University of Michigan in 1948. His career was mostly in ad agencies and he is now an ad/marketing consultant. Married 41 years to Kathy, they have three children and four grandchildren and have lived in Westport, CT for the past 28 years.

WILLIAM E. McCANN (CDR) was born in Conception Jet, MO on Jan. 14, 1922. He joined the service on Aug. 29, 1942, and entered the V-12 Program at SE Missouri State College. Other stations and Schools include: Great Lakes, IL; Univ. of Oklahoma Medical School; Ready Naval Reserve, Oklahoma City and Tulsa, OK; NAS Memphis; and the Navy Recruiting Station at St. Louis, MO.

Served aboard the USS *Rochester* (CA-124) from 1951-52; Ready Reserve Naval Medical Co. in Kansas City, MO and he flew with the Naval Air Transport Sqdn. VR-881 NAS Olathe, KS. On July 1, 1971, he went to the Inactive Reserve until he retired in January 1982. He received the C.D. Henry Memorial Award CAMA Aviator Flight Surgeon in 1969.

McCann and wife, Betty, have six daughters.

JOHN J. McCARTHY (DUTCH) (CDR) enlisted in August 1942 in Philadelphia. He attended the Villanova College V-12 Program and Northwestern Univ. Midshipman School. Commissioned ensign in March 1944, he served aboard the USS LCI(L) and (G)46 as XO and CO for 26 months.

His service included American, European Theater (Italian Campaign) and picket duty in the Pacific. He was assistant personnel officer, Bainbridge Trng. Station, MD and was released from active duty in July 1946 as LT(jg).

McCarthy worked for the Mobil Oil Co. for 37 years in various managerial positions. He retired from Mobil in December 1983 as manager of Light and Heavy Products, Valley Forge, PA. During the same period he continued association with the NR (in spite of 10 Co. movers, three out of state).

Retired as commander with 26 years service in September 1981. He served as training and as CO of surface divisions in Eddystone, Folsom, Pittsburgh, PA areas. Married 48 years to Mary, they have six daughters and seven grandchildren. Presently he runs a tree farm (Pennsylvania Stewardship Program). He participated as a world class master runner in four world veteran games and three senior olympics. The events consist of 200 meters up to marathon.

His memorable experience was traveling "space available" to run in the Berlin marathon (thru the Wall) three days prior to reunification of East/West Berlin on Oct. 3, 1990. Also memorable was the *Stars in Stripes* article on the six McCarthy brothers.

DONALD P. McCLAIN (LTJG) was born in Aberdeen, WA on March 29, 1923. He graduated from Stadium HS, Tacoma, WA on June 9, 1941. He enlisted in the USNR V-1 in May 21, 1942; reported to active duty V-12, Willamette, Univ., July 1, 1943; and transferred to pre-Midshipman

School, Asbury, NJ, March 1, 1944. He was commissioned ENS, D-V(G), Columbia Univ., Aug. 10, 1944.

Reported aboard the LSM-72 at Houston, TX on Sept. 17, 1944, and he toured many forward combat areas of the Pacific Theater. He retired as lieutenant jg with a permanent physical disability on June 17, 1946. He received his BA degree in business administration at the U of W and MBA degree at Univ. of Puget Sound.

After 28 years with Merrill Lynch, he retired as vice-president. Married 46 years he has one son, one daughter and three grandchildren. He enjoys aerobics, lap swimming, wilderness backpacking, winter sports, local political and civic offices, photography and extensive global traveling.

JOHN P. McCULLOUGH was born in Dallas, TX on May 10, 1925. He joined the Corps in July 1943; graduated from Notre Dame Midshipman School in November 1945; served on the USS *Herald of the Morning* (AP-173) in the Pacific area and was discharged in July 1946. McCullough graduated from OU in 1945 with a BS in chemical engineering and in 1949, he received a Ph.D. in physical chemistry from Oregon State University.

Dr. McCullough first worked at the U.S. Bureau of Mines, progressing to Chief of its Thermodynamics Branch. He joined Mobil Oil Corp. in 1963, holding a series of technical management positions before retiring in 1989 as general manager of Environmental Health and Safety, Mobil Oil Corp. and vice-president, Mobil Research and Development Corp. He served as chairman or member of boards of the Chemical Industry Institute of Toxicology; International Petroleum Industry Environmental Conservation Association; World Environment Center; Institute of Health Policy Analysis, Georgetown Univ.; Gordon Research Conferences; and American Chemical Society Committee on Chemistry and Public Affairs. He was a director of Mobil Foundation and of Heico Corporation. He is a trustee of the Stony Brook-Millstone Watershed Assoc. and a Woodrow Wilson Foundation Visiting Fellow.

He has contributed 90 articles to scientific journals and books. His honors include the Distinguished Service Award of the U.S. Dept. of the Interior, the Huffman Award of the International Calorimetry Conference and the American Chemical Society's Award in Petroleum Chemistry and its Leo Friend Award.

John McCullough met his wife, Ann Calvert, at OU and they now live in Princeton, NJ. They have three daughters and three grandchildren living in Korea, Andover, MA and New York City.

KEITH D. McGOWAN JR. (ENS) was born in Waycross, GA on Dec. 24, 1921. He enlisted in the USN in 1942 and entered the V-12 Program at Duke University in 1944. Other duty stations included Jacksonville; Pensacola and Atlantic City Naval Air Stations; Newport, RI NTC and USS *Columbia* (CL-56). After receiving his commission and being discharged in 1946, he returned to Duke and received his BS degree in mechanical engineering in 1947.

Employed by Hercules Inc. in 1947, he worked in corporate engineering and at plant locations; and provided design and supervision for instrumentation and automatic controls for chemical processes. He retired in 1986 as principal instrumentation consultant. He then worked in a similar position for Ciba-Geigy Corp. for three years and does volunteer work at the Veterans Affairs Hospital.

He and Udell were married in 1949 and live in Wilmington, DE. They have two daughters, Deborah and Patricia, and a son, Paul.

CHARLES McKENZIE washed out of the V-12 Program at Valley City, ND due to a physical condition. He left there and went to Great Lakes for boot camp then to Norman, OK and worked in the galley. He was therefore a great help in the kitchen during his 40 years of married life which ended Sept. 14, 1988, due to a heart attack in Laguna Hills, CA.

After his discharge he attended the U of Minnesota, graduated with a BA in mechanical engineering (1949) and worked at Armour and Co., Honeywell, Toni Co. and Hollister in Chicago. He bought his own company, McKenzie Repair Inc. (a gas welding equipment repair company), in St. Paul in 1967. At the time of his death he was attending a seminar, on repairing equipment for ambulances, in California.

He married Bea, an ensign in the USNR, on May 22, 1948, and they had two children, Cheryl and Kent.

THOMAS P. McKEOWN was born in New York City on Feb. 21, 1925. After enlisting in the USN on Jan. 17, 1944, he was assigned to the V-12 Program on Feb. 28, 1944. He was commissioned ensign, Supply Corps and trained at the Naval Supply Corps School in Bayonne, NJ.

Served in the Atlantic and Caribbean areas in the USS *Macomb* (DMS-23), USS *Macon* (CA-132) and with the 6th Fleet in the Mediterranean aboard the USS *K.D. Bailey* (DD-713). He completed his naval service at the Newport Training Station and was discharged Jan. 14, 1949.

McKeown received a BA in business administration and an MA in personnel administration from George Washington Univ. After 10 years in industrial relations, he was appointed in 1961 to the U.S. Dept. of Labor, Boston, where he held several administrative positions until his retirement in 1986. He and his wife, Ruth, married in 1949 and have three children and four grandchildren. They have lived in Wrentham, MA since 1961.

WALTER McLAREN (LTJG) was born in Dayton, OH on Jan. 5, 1922. He entered the V-12 Program at Hampden - Sydney, July 1, 1943. Other stations and schools include the Solomons, Maryland, San Diego, Pearl Harbor, Okinawa and Columbia University.

His memorable experience was going ashore at Naha, Okinawa and viewing all the devastation shortly after it was secured. He was discharged in May 1946 and received one Battle Star.

McLaren is a realtor; he is married and has two daughters and one son.

EDWIN DONALD McNEES (LTJG) was born in Dallas, TX on Aug. 24, 1925. He joined the USN Supply Corps on Feb. 12, 1943; entered the V-12 Program at Mercer Univ., Macon, GA and the Univ. of North Carolina at Chapel Hill. Other stations and schools included Transient Officers Center, Liason Island Command, Guam; USS *Hermitage* in Pacific area and Harvard University at Cambridge, MA.

Memorable experiences include quartering the USO (show biz) personnel and troops in transit in South Pacific and performing on Guam and "small stores" officer USS *Hermitage*, a converted Italian luxury liner for troop transport. He was honorable discharged and received the American Theater, WWII Victory and the Asiatic-Pacific Theater Medals.

McNees and his wife, Mary, have a daughter and a son, Donna and Brian. He is the assistant district attorney of Dallas County, TX.

JACK A. MELLENTHIN (LT) was born in Riverside, CA on June 29, 1924. He enlisted Oct. 28, 1942; was called to active duty July 1, 1943; attended the V-12 Program at the Univ. of Texas and Midshipman School at the Univ. of Notre Dame. Other military assignments included Catapult School, Philadelphia Navy Yard, November-December 1945; USS *Lexington* (CV-16), January-May 1946; and USS *Point Cruz* (CVE-119) May-June 1946.

His memorable experiences include: transferring from *Lexington* to *Point Cruz* at Pearl Harbor; when *Point Cruz* developed trouble with reduction gear and ordered to proceed to Bikini, where preparations were being made for the first atom bomb test; and visiting Japan and seeing some of the fire bomb destruction in the Tokyo area.

Released to inactive duty July 15, 1946, he remained in the Reserve and completed 20 years

before retiring. He returned to the Univ. of Texas and received his degree in aerospace engineering in August 1947. Employed by the National Advisory Committee for Aeronautics from 1948-77; at Moffett Field, CA for Ames Research Center, his last seven years were spent as project engineer for various wind tunnel tests of space shuttle models.

Mellenthin retired from NASA in January 1977 and moved to Nampa, ID where he still resides with his wife, Rosalyn. They have a son, a daughter and five grandchildren.

HARRY A. MERLO (2/LT) was born in Stirling City, CA on March 5, 1925. He entered the Southwestern Louisiana Institute in January 1944 and was commissioned a 2nd lieutenant in 1945 at Quantico, VA. He was injured at Camp Pendleton, CA while preparing for Pacific Theater action. He was discharged in March 1946.

After receiving a BS from the Univ. of California, Berkeley in 1949, Merlo joined the Rounds Lumber Co. in Cloverdale, CA and has been in the lumber business ever since. He served as chairman and president of Louisiana-Pacific Corp., a Fortune 200 company recognized for its leadership in innovative building materials, since its spin-off from Georgia-Pacific in 1973. He has been an executive at G/P since 1967.

Today, Merlo resides in Portland, OR and owns a ranch in northern California, managed by his son, Harry Jr. He has received numerous awards for leadership in his industry and in many philanthropic programs dedicated to the youth of America.

EARL ROBERT MEZOFF (LTJG) was born in McKeesport, PA on Feb. 10, 1925. He entered Baldwin-Wallace College, V-12 Program on July 1, 1943, and was commissioned ensign in November 1944 at Columbia Midshipman School, NYC, NY.

Served in the South Pacific on the USS *Nevada* and as XO of PC-1136. He was naval attache at formation of the United Nations, San Francisco, CA in 1945. He was promoted to lieutenant jg in 1949.

Mezoff received his AB degree in English from Thiel College in 1947; MA in psychology from Michigan State University in 1948 and Ed.D. in higher education from the Pennsylvania State Univ. in 1965. He was president of Dana College, Blair, NE from 1971-78 and president of Tusculum College, Greeneville, TN, 1978-88.

Retired in 1988, he currently resides in Greeneville, TN with his wife, Joan M. They are the parents of Margaret Holman and Patricia Magoon.

WILLIAM H. MIDDENDORF entered the V-12 Program at the University of Louisville, transferred to Virginia and graduated with BS in 1946 and MS from Cincinnati in 1948, where his teaching career began. In 1960 he earned a Doctorate from Ohio State. He holds 27 patents for circuit breakers and other devices, and has published numerous articles on design methodology, engineering management technique and electrical insulation. Two of his articles on design won best paper of the year prizes.

He has written six books including *What Every Engineer Should Know About Product Liability* (with J.F. Thorpe) and has participated in many liability cases. He is editor of the above mentioned series (presently 31 volumes) and co-editor of the *Electrical Engineering and Electronics* series (85 volumes) As a Fellow of the Institute of Electrical and Electronic Engineers since 1968 he has been especially helpful to the engineers of Underwriter Laboratories and to the National Electrical Manufacturers Association.

DONALD T. MIESBAUER (CDR) was born in Milwaukee, WI on May 28, 1926. He entered Lawrence College V-12 Program on July 1, 1944, transferred to V-5 Program on July 1, 1945, and was released to inactive duty from flight training in 1946. He resumed flight training in September 1951, was commissioned ensign and designated aviator in April 1953. He served with the Atlantic Fleet in Air Antisubmarine Sqdn. 27 until September 1955. He joined the Reserve Air Antisubmarine Sqdn. 22 based at NAS Glenview, IL in 1956.

Miesbauer attended the Univ. of Wisconsin, received a BBA in 1949 and was admitted to practice as a CPA, state of Wisconsin in September 1967. He served in various administrative positions of Wisconsin Savings & Loan Assoc. for 20 years, the last 10 years as secretary, then president and then was senior vice-president of 1st Savings Assoc. for several years. He was treasurer and director of RS&A, LTD, an investment management and brokerage firm for five years prior to retirement in 1991.

Presently resides in Milwaukee, WI with his wife. He has three daughters and one granddaughter. After 22 years of active federal service, he retired from the USNR as a commander.

BEN F. MILES II (CAPT) was born in Chicago, IL on Aug. 14, 1927. Miles entered the V-12/V-7 Program at Hampden Sydney; attended

Flight School in 1948 and participated in Korea.

His memorable experiences include the 1952 bootleg missions, 4605 TAC and 1962 Air America. His awards include the Distinguished Flying Cross, Air Medal, Unit Citation and others.

Married, he has four children (two of whom are lieutenant colonels in the USMC). He is a farmer.

ARTHUR MILLER JR. (LCDR) was born in Washington, DC on April 6, 1924. He enlisted in the USNR as A/S in Navy V-12 Program at Penn State in April 1943; commissioned ensign at Midshipman School, Columbia University, October 1944 and qualified as Combat Information Ctr. officer at Tactical Radar School, Hollywood, FL, January 1945.

Served aboard troop transport USS *Gasconade* (APA-85), carrying troops and supplies from the West Coast to the Philippines and was anchored at Tokyo Bay during the formal Japanese surrender ending WWII. He completed active service writing training manuals at the Bureau of Naval Personnel, Washington, DC. He remained active in Naval Reserve, serving surface warfare and public information units in Washington, DC.

After the war, Miller followed a career in journalism as a newspaper reporter, writer and editor for Navy Dept. *All Hands Magazine*, U.S. Information Agency and National Geographic Society. He was regional public affairs officer with the National Park Service in Philadelphia and author and co-author of five books and numerous magazine articles.

Married to the former Marjorie Lyman, daughter of a Navy family, they have four daughters: Nancy Miller (lieutenant in the USN), Sue Smith, Kathi Thomas and Jan Martin.

RALPH WAYNE MILLER (LT) was born in Carmi, IL on May 31, 1921. He entered the Navy V-12 Program at Columbia University in 1943; served aboard the USS *Doneff* (DE-49) in the Aleutian Islands and Central Pacific, 1944-1945, as communications officer; and attended Combat Information Center School at Pearl Harbor, HI in January 1945.

Returned aboard the *Doneff* at Mare Island, CA in March 1945 and served in the Guam, Saipan and Okinawa areas until the end of WWII. In 1945 he went to Truk and Marcus Islands, still aboard *Doneff*, to accept the surrender of Japanese garrisons on those islands and was a guest of the Canadian navy at commissioning of the HMCS *Ville de Quebec* (FFH-332) at Quebec City on July 14, 1994.

Currently, he broadcasts sports on Radio Station WSLM, Canadian Broadcasting Corp. radio network and Australian Radio. He resides in Indianapolis with his wife, Judy.

RICHARD H. MILLER (LT) enlisted Nov. 25, 1942 and entered the V-12 Program on July 1, 1943 at Trinity College, Hartford, CT. He transferred to Portsmouth, VA in October 1943, awaiting assignment to Midshipman School; transferred to Columbia University, USNR Midshipman School in December 1943; was commissioned in April 1944; assigned to the Amphibious Training Base, Ft. Pierce, FL in October 1944 and shipped out to Waimanalo ATB, Oahu.

He became a communications officer, and later was assigned to ComPhibGrp-14 with duty in Manila Bay in September 1945. Later he was assigned to the Communications Staff of Commander, 4th Naval District and was released to inactive duty on June 30, 1947. He volunteered for active duty in August 1950 and reported to Little Creek ATB, VA. He was assigned to the Communications Staff of ComPhibGrp-4 and later assigned to NavBeachGrp-2, HQ Unit as XO.

Released from active duty in February 1954, he returned to the Univ. of Miami to complete a BA degree. He was employed at the Letterkenny Army Ord. Depot, Chambersburg, PA as a clerk-typist, then as management analyst. He transferred to the Army Mgmt. Engr. Trng. Agency, Rock Island, IL in 1957 and taught courses in procedures analysis, organization analysis, computer programming, ADP systems analysis, and data processing mgmt. He transferred to FWPCA in 1966 and to the National Park Service in 1970 as data processing division chief.

Retired in 1978 and became a tree farmer in Virginia where he now lives.

WYMAN F. MILLS was born in Table Rock, NE on April 12, 1924. He enlisted in the USN in 1942 in Portland, OR and entered the V-12 Program at Willamette Univ., Salem, OR on July 1, 1943. He completed boot camp at Farragut, ID and boarded the USS *Intrepid* aircraft carrier at Treasure Island, bound for Pearl Harbor.

Assigned to the USS *Devastator* (318) minesweeper in June 1944 as quartermaster 3/c petty officer, he swept minefields at Iwo Jima and Okinawa before D-day. He went through the complete battles. His ship was hit by a Jap suicide plane and they secured behind the battle lines for repairs and was soon back in combat. He did convoy duty and swept minefields in the China Sea.

Discharged in January 1946, he attended Oregon State Univ. School of Engineering, graduating with a BS degree in 1949. He retired in 1982 from Simpson Timber Co., manufacturer of forest products, after 33 years in purchasing and fleet management.

He owned and operated wholesale manufacturer Shadowbox Picture Frame Co., 1983-91. Mills lives with his wife in Seattle, WA; they raised five sons: Michael, Robert, Steven, John and David. Mike served in the Army during the Vietnam War.

LAURENCE W. MILNES (LCDR) was born in Riverdale, CA on Aug. 30, 1924. He enlisted in the Navy V-1 Program in August 1942; was ordered to V-12 Program at the University of Texas on July 1, 1943; attended Camp Endicott Midshipman School at Davisville, RI in July 1945; and commissioned ensign, Civil Engr. Corps in August 1945. He served as a construction officer, 125th Naval Construction Bn. in Okinawa and was released to inactive duty in June 1946.

Milnes received a BS in civil engineering from the University of Texas, Austin in June 1945 and MS in civil engineering from the University of California, Berkeley, in 1948. A registered California CE; engineer and watermaster of Fresno Irrigation District, CA for five years; design and traffic engineer, Fresno County, CA for two years; Public Works Director/City Engineer, city of Walnut Creek, CA for four years; Public Works Director/City Engineer for 15 years and assistant city manager for 14 years for the city of Fremont, CA. Since retirement in 1988, he has been a capital project management consultant.

Presently resides in Fremont, CA with his wife, Elinor. They have two daughters, two sons and eight grandchildren. He was honorably discharged from the USNR as LCDR in 1960.

RONALD E. MINTZ (BARON) (COL) is a native of Rocky Mount, NC. He enlisted in the USMC on Feb. 23, 1943, and entered Duke Univ. Marine V-12 Program on July 1, 1943. He was directly commissioned as a regular officer in the USAF. He served on active duty until he retired in the grade of colonel on Sept. 1, 1974.

Military service included various operational and staff assignments in engineering, technical intelligence, research & development, missile, space and reconnaissance operations. Post-retirement career was as conservation scientist in Research Triangle, NC and Seattle, WA.

Retired since Oct. 26, 1990, he resides at Air Force Village, San Antonio, TX. He is a member of Ordre Pour le Merite and listed in *Who's Who is South and Southwest*.

JOHN W. MITCHELL (CAPT) was born in Clearfield, PA on July 23, 1926. He enlisted in the USNR in January 1944 as a naval aviation cadet and was called to active duty as V-12(a) and reported to V-12 unit, Bethany College, WV on July 1, 1944. After June 1945, he served at NAS Bunker Hill in Indiana, North Carolina pre-flight, St. Mary's pre-flight and NAS Corpus Christi, TX until May 1946. He graduated in business administration, Penn State University in January 1949.

Mitchell served in the U.S. Army with duty at the Pentagon (Ft. Meyers) and elsewhere for two years during the Korean War and is still active as mission pilot in the USAF Aux. with the rank of captain.

In civilian life, he spent 30 years in the publishing business and retired from American Broadcasting Co. Publishing Co. as VP, Sales in 1984. He is still a self-employed realtor in Bermuda Dunes, CA. He married Joan Betterly, Bloomfield Hills, MI in 1955 and they have two children, John (an engineer with UNYSIS) and Janet (a nurse).

RONALD B. MITCHELL (CDR) was born in Findlay, OH on July 17, 1926. He entered the Navy V-5 Program at Oberlin College on July 1, 1944; transferred to Navy V-12 Program Southwestern Louisiana Institute in November 1944; and transferred to Tulane University Navy V-12 NROTC in November 1945.

Discharged in June 1946, he graduated from Tulane Univ. Medical School in 1951; interned at USNH, Philadelphia in 1951; MSTS Pacific, 1952-1954; and OB/GYN in Akron, OH from 1954 to present. Currently clinical professor of OB/GYN Northeastern Ohio University College of Medicine. He retired as CDR, MC, USNR.

He married Patricia Joan Thompson in 1951 and they have four children.

CHARLES G. MONNETT (1/LT) was born in Greensboro, NC on July 3, 1923. He enlisted in the USMC on Nov. 12, 1942, as a PFC while attending Guilford College. On July 1, 1943, he started the V-12 training program at Duke Univ. In January 1943 he went to boot camp at Parris Island, SC. In March 1944, he attended Officer Candidate Applicant School at Camp Lejeune, NC and in July 1944, he attended Special Officers Candidate School.

Promoted to platoon sergeant on Sept. 21, 1944; appointed 2nd lieutenant on Sept. 30, 1944; and assigned to the 17th Plt. of the 33rd Repl. Draft on Nov. 11, 1944. He departed from San Diego, CA on Dec. 27, 1944, aboard the *Sea Boss* and

arrived on Guadalcanal on Jan. 11, 1945. He was assigned to the 29th Marines on March 14, 1945, and took APA-25 to Okinawa on April 1, 1945, and worked in shore party units.

Monnet became platoon leader of 3rd Plt., G Co., 3rd Bn., 29th Marines on the northern end of Okinawa and served until the island was secured. He traveled on LST 229 to Guam and served with the 6th Motor Transport Bn. In September 1945, he was reassigned to the 22nd Marines on Guam. In October 1945 he traveled on the *Montreal* to Tsingtao, China.

In China he graduated from Special Services School on Nov. 28, 1945, and served as regimental education officer of the 22nd Marines. He organized school at Shantung Univ. where high school, college and GI Bill courses were taught. He left China on April 1, 1946 on the USS *Gen. H.W. Butner* for San Francisco, CA and was discharged in May 1945 at Camp Lejeune. He served until 1950 as a 1st lieutenant in the USMCR.

A graduate of Duke University, a member of Lions International, Marine Corps League, Odd Fellows, Shriner, Methodist, author and amateur archaeologist, he has received two first time world-wide awards. He was president and owner of Monnett Carpet and Draperies. He is married to the former Diamando Katsinas and they have four children and five grandchildren. He lives in Greensboro, NC and in North Myrtle Beach, SC.

EUGENE ROBERT MONTANY was born in Malone, NY on June 14, 1926. He was raised and schooled in Chester, VT and entered Rensselaer Polytechnic Institute, Troy, NY in January 1943. He entered the V-12 (a) Program at RPI on July 1, 1944. He received B Aero Eng and was commissioned ensign in February 1946.

Engineering officer, Naval Air Transport Svc. Sqdn. VR-1, NAS Patuxent River, MD from February 1946 to August 1946; USNR Active Reserve NAS, Port Columbus, OH from 1946-1949; he was medically retired in 1951. A research engineer for Curtiss-Wright Corp., Columbus, OH from 1946-1949; preliminary design engineer, North American Avn., Downey, CA from 1949-1950; engineering and management at United Aircraft (Technologies) Corp. Research Laboratory and Pratt & Whitney Div., East Hartford, CT from 1951-1987; retired as VP technology and strategic planning in September 1987.

Presently resides in Palm City, FL; East Orleans, MA; and Manchester, CT. He married the former Martha Jane Pfanz of Columbus, OH in March 1947 and they have five children and 14

grandchildren. He remains active as a jazz/swing musician.

RICHARD D. MOOG (LTJG) was born in Eveleth, MN on June 5, 1925. He entered the V-1 Program on July 1, 1943, and received a degree in aero engineering. He served in Panama at Coco Solo NAS in flying boat overhaul and repair until discharged in July 1946. From 1947-1948, he was an instructor at the Univ. of Minnesota, Aero Engineering Dept.

After five years with Braniff Airways, he joined North American Aviation and worked in flight test of the X-10 research test vehicle and became S-10 flight test project engineer at Edwards AFB in 1957.

Joined Martin Marietta in 1960 in Denver, CO and worked there as a systems analyst and unit head until retiring in 1990. His most notable achievements were the analyses of the terminal descent of the Viking Mars lander and the recovery of the Space Shuttle solid rocket boosters. He was on the flight support team of the first Space Shuttle, STS-1.

He and his wife, Nancy, are both retired in Littleton, CO.

RODNEY T. MOONEY (CDR) enlisted in the USN in 1943 from Quincy, MA and entered the V-12 Program at Harvard, transferred to the NROTC Program at Tufts in 1944, graduating in 1946. Coincidentally, this reversed the path of his Dad who many years earlier, started at Tufts and graduated from Harvard.

On active duty through 1968, Mooney served in seven ships (skipper of two) and on many shore and staff assignments including Postgraduate School, two bureau tours and duty on the staffs of both CINCPACFLT and CINCLANTFLT and an advisory committee to SECDEF. He was twice awarded the SECNAV Commendation Medal.

After retirement from the Navy, he was employed by United Technologies Corp. as a marine gas turbine engineer and later as an aerospace engineer. Since the untimely death of his wife Irene, in 1986, he has devoted his life to a "third career" of volunteer medical work; to his five children; and to travel. He makes his home in Andover, CT.

C. NEWELL MOORE was born and raised in Danvers, MA. He joined the Navy in 1943; started the V-12 Program in July 1943; received his degree in mechanical engineering and was commissioned in the USN in 1945. His last Navy assignment was as harbor patrol officer of Subic Bay, Philippine Islands in 1946.

Went into plumbing, heating and A/C and started his own business in 1959 which he sold to his employees in 1988 when he retired. Married in 1947 to the former Mary Ringrose of New Britain, CT, he has two children, Paul and Carolann, both college graduates and doing well. They have two grandsons in high school in Ashland, OH and one granddaughter who is a sophomore at Bryn Mawr in Pennsylvania. He stays busy with his hobbies of golf, cards and travel. He and Mary have traveled in 46 countries to date.

LEWIS KINGSLEY MOORE was born in
Glen Cove, NY on June 24, 1924. In September 1942, he entered Tufts College Naval ROTC Program which was incorporated into the V-12 Program on July 1, 1943. He graduated from Tufts College and was commissioned ensign in February 1945.

Served on LST-466 and was based in Morotai, in the Dutch East Indies carrying Australian troops in the invasion of Borneo. At war's end he was assigned to AM-102 sweeping mines in the Formosa straits. He returned with ship to San Diego for de-commissioning in January 1946 and was released to inactive duty in Boston, MA in June 1946. He resigned his commission and was discharged in February 1955.

After five years in business, he attended Boston University, graduated with a MA in history in 1952 and entered teaching at Culver Military Academy in 1952. He retired in 1992 and presently resides in Culver, IN with his wife, Eleanor. They have two daughters, one living in Culver and one in Columbus, OH; three grandchildren; and one great-grandson.

STEPHEN G. MOORE entered the USN
from Burlington, VT on July 1, 1943. He joined Dartmouth College V-12 unit, the largest of the initial V-12 Programs. He was commissioned ensign at Northwestern Univ. (Abbott Hall), Midshipman School in November 1944. He served as deck officer aboard USS LST-1048 (later USS Morgan County) in the Western Pacific and East China Sea. His memorable experience of V-12

days was using gas ration coupons to pay off dorm chiefs for special privileges.

In March 1946 he returned aboard the USS LST-913, served briefly with harbor unit in Boston, MA until going on inactive duty in July 1946. He graduated from Dartmouth in February 1947 and began a career in banking in New York City. He was a bank executive in Vermont from 1955 to retirement in 1979. He has served on several corporate and social agency boards.

With his wife of 50 years, he travels extensively from residence in tranquil Charlotte, VT bordering Lake Champlain. They have three successful children and four grandchildren.

DANIEL MORGAN (LCDR) was born in
Norwood, PA. He enlisted in the USNR in September 1943 as S2/c, V-5, TARMAC, Olathe NAS; transferred to V-12 in February 1944 at Missouri Valley College; transferred in 1945 to NROTC, Notre Dame and was commissioned in June 1946.

Served aboard the USS Gherardi until 1947, returned to inactive duty and received BSC degree from Notre Dame in 1950. During the Korean conflict, he served in the USS Gallup and fleet training groups in Norfolk and Guantanamo.

Morgan returned to inactive duty in 1954 and married Navy nurse LTJG Nancy A. Hamlen. He selected Regular Navy from inactive duty in 1955 and served in the USS Scanner and USS Isherwood; was head of Naval Science Dept., California Maritime Academy from 1957-1959; was XO of USS Ashtabula until 1962; Surface Force Team Leader, Navy SubSection, Taiwan and on staff of Commander Western Sea Frontier where he retired as LCDR in 1967.

He held several Human Resource managerial positions for Magnavox Co. before retiring in 1988. Morgan and his wife, Nancy, reside in Ft. Wayne, IN.

BEN L. MORTON (LT) was born in Oklahoma
on Sept. 28, 1925. He entered the V-12 Program at Oklahoma Univ. in July 1943 and completed the program at Colorado Univ. in February 1946 with BS in civil engineering. He served in the USS Columbia as ensign until released from active duty in July 1946. He returned to active duty in August 1950 as lieutenant jg on the USS Cavalier which participated in the Inchon Landing on Sept. 15, 1950, and later with 1st Anglico of Marine Corps as liaison officer until October 1951. He was discharged from the USNR as lieutenant in May 1959.

As a civilian he worked for Cities Service Oil Co. and after receiving a MBA from Oklahoma

Univ., he had a 30-year career with Exxon in Houston, New York and London. He retired as financial executive in December 1985. He now lives in Houston with his wife, Joan. They have one son and two daughters: Bruce, Janet and Kimberly.

CHARLES R. MUIR enlisted in the Navy and
entered the V-12 Program in November 1943. He was commissioned as a naval reserve officer in November 1945, served as communications and ships services officer on the USS PCE-868 in the South Pacific, and was released from active duty in February 1947.

The most memorable experiences from his Navy service and the life long friendships developed during his Navy V-12 days.

He graduated from the University of Louisville with a BS and MBA degree and was employed in sales, production and administration with divisions of the Brown-Forman Company for 41 years, retiring as a senior vice-president in 1989.

He and his wife, Dorothy, live in Louisville, KY. They have three children and four grandchildren.

RAPHAEL MUR (RAY) (CAPT, JAGC) was
born in Brooklyn, NY; was commissioned and received a BS degree at Tufts College in June 1946 and served in the USS Massey (DD-778) from June 1946-June 1947. He received an LL.B. degree from Harvard in June 1950 and a LL.M. from NYU in June 1955.

He was employed as a civilian attorney in the Office of the General Counsel of the Navy, 1950-1961, in Bureau of Ships, Military Sea Transportation Service and New York Branch. In 1961, he joined the Legal Department of Grumman, retiring in 1990 as Vice President, General Counsel and Secretary of Grumman Aerospace Corp. He completed the Advanced Management Program at Harvard Business School in 1977.

Mur commanded three Naval Reserve law units and retired as CAPT, JAGC, USNR. He is practicing law as a specialist in alternative dispute resolution. He lives with his wife, Sonia, in Wantagh, Long Island, NY.

JOHN WILLIAM MURPHY (LTJG) was
born in Lawrence, MA on Sept. 18, 1923. He joined the service July 1, 1943, and entered the V-12 Program at Brown University. Other duty stations include Coronado, Oceanside, San Bruno, Oahu, Okinawa, Guam and Plattsburg.

121

He was discharged July 3, 1946 and received four awards.

Murphy is married and has two children and two grandchildren. A research chemist, he retired in 1985 and is now in security.

TIMOTHY J. MURPHY JR. attended Fordham University from January 1943 to June 1943; the University of Rochester V-12 from July 1943 to June 1944; Plattsburg July 1944 Northwestern Midshipman School and was commissioned Nov. 22, 1944.

Other duty stations include San Francisco to Hollandia New Guinea on USAT *Sea Cat* - January 1944; LCT-84, Leyte Gulf, February 1944-January 1945; commanding officer, LCI-1095, February 1945-June 1946; Hawaii to Charleston, SC. An engineering and supply officer, he was discharged in June 1946.

Murphy attended Fordham from September 1946 until graduation in June 1948; received MBA from New York University; worked for IBM HQ, New York City and Armonk from 1948 until retirement in 1982. He was adjunct professor at Fordham Graduate School of Business from 1968-1973.

He has been married to Doris Williams since 1949 and they have five children and nine grandchildren. Currently, he is a certified financial planner and golfer.

HARRISON C. MURRAY (CAPT) was born at Harwood, MD on Feb. 26, 1925. He enlisted in the V-5 Program in January 1944 and then transitioned to the V-12 Program.

From March 1944-June 1946, he attended the V-12 Programs at Emory and Henry College, the University of the South and the University of Oklahoma. He was commissioned ensign USNR from the NROTC Unit at Oklahoma in June 1946.

Murray transferred to the Regular Navy and served on active duty in the USN for 30 years, retiring in 1974 as a captain. He served in the Pacific during both the Korean and Vietnam conflicts. He was commanding officer of a destroyer escort, a guided missile destroyer, a destroyer division and a USN Station in the Republic of the Philippines.

Awards include the Legion of Merit and two Bronze Stars with Combat V.

In retirement he worked for 10 years as a senior systems analyst with the Navy Dept.

WILLIAM A. MYERS (LTJG, ELT9) was born in Akron, OH in 1924 and raised by CWO3 USN Don Myers and Marion (Vigh) Myers. He was admitted to Caltech in 1942 via Dr. Carl Anderson-Nobel Loreate. Retired in 1990. Configured weapons from 1951-1984. Taught visibility analysis from 1946-1993.

Enlisted in the USNR-V1 August 1941, V-12 Caltech 1943; Midshipman, Cornell, 1945; line officer, USS *Iowa*, 1945; honorable discharge USNR ENS, Aug. 15, 1946. Resigned LTJG, 1955; GS-13 in 801 NWCCL 1969; Caltech BS-Eng. 1942-1945; USC MS-Ed 1951. Engineered building and 70-90mm production-line systems. Worked 1946-1984 in engineering, teaching, think-tank modeling; taught 1984-1993 in sec. schools, colleges and trade school.

Worked in industrial-military complex for USN, USAF, Army, DOD. Ident./Eval. new weapons config. by logical analyses. Developed weapons by management, design, test, evaluation. Systems were bombers, fighters, missiles, decoys, drones, frag-explosives, air defense, ECM/ECCM, insurg.-opns, resp., B-70, F5A, SHRAM + Std-ARM. Quail, AN/USD2, Blue-7, Vietnam AAA/SAM, EA6B, Abt Corp.

He has two children, Melodie Marie Kersten and John Anderl Myers.

RICHARD R. NEILL was born in New York City on June 20, 1925. He enlisted February 1943 and entered the V-12 Program at Princeton University on July 1, 1943; attended Asbury Park (NJ) Pre-Midshipman School, July-August 1944; commissioned ensign in December 1944 at Columbia Midshipman School, NY; and attended Communications School from January-April 1945.

Neill joined Communications Unit 460 in Hawaii in August 1945. The unit flew to Tokyo Bay on V-J Day (an awesome sight from air) and established Navy communications shore operations there. He served as CWO at Yokosuka Naval Base until April 1946, then completed active duty at 3ND Communications HQ in NY until June 1946. He never heard a shot fired in anger.

Graduated from Princeton with AB in English in 1948; MA from NYU in 1953; joined Prentice-Hall Publishers as editor in 1948; and retired in 1985 as president, Executive Reports (P-H business publishing subsidiary). He has continued as fund-raising consultant for various non-profit organizations, and he resides in Tarrytown, NY with his wife, Patricia. One son, Robert, is in broadcasting and data processing in Florida.

FRANK NEISH entered the V-12 Program on July 1, 1943, after three semesters at Penn State. He was sent back to the same school for another three semesters in the program and received his commission at Columbia Midshipman School in October 1944. He returned to Penn State in the fall of 1946 and received his bachelor's degree in journalism the following June.

The next 12 years were spent working in the advertising department of *The Daily News*, McKeesport, PA. In 1959 he opened his own advertising agency, Frank Neish Advertising, Inc., handling local, national and international accounts until 1990, when he retired from the active business. Unmarried, he is still active with many local institutions as board members and committee head.

Service time was spent with Amphibious Forces (LCTs and LSMs) in the Pacific from 1944-1946.

ROY C. NEWMAN (LTJG) was born in Bakersfield, CA on Dec. 5, 1921. He enlisted in the USN as an apprentice seaman on April 18, 1942, and served at the Naval Ammunition Depot, Fallbrook, CA. He entered the V-12 Program July 1, 1943, at University of the South, Sewanee, TN and was commissioned ensign on June 20, 1945, out of the ROTC Program at University of Texas in Austin.

Newman served on the USS *Dyess* (DD-880) as assistant engineering officer and assistant gunnery officer in both the Atlantic and Pacific. The *Dyess* was outfitted for picket duty and served as part of the occupational fleet in Japan after V-J Day.

Released to inactive duty May 19, 1946, he resigned as lieutenant junior grade on Nov. 10, 1954. In 1946 he returned to his career with Pacific Telephone Co. in California and retired as district manager in 1982 after 42 years service.

Newman currently lives with his wife, Jayne, in Tustin, CA. They have a son, daughter-in-law and grandson living in Seattle, WA.

DONALD L. NICHOLS enlisted in the USN in 1943. He entered the V-12 Program at Worcester Polytechnic Institute and received the BS degree there. He was discharged in July 1946 following commissioning and indoctrination. He joined the Submarine Reserve Div. in New London, CT and was employed as an electronics engineer at the USN Underwater Sound Laboratory.

Later he was head of the Marine Physics

Section at General Dynamics/Electronics and subsequently became chief engineer for Antisubmarine and Undersea Warfare. After that, he was manager of Raytheon's Antisubmarine Warfare Product Line.

At the USN Underwater Systems Center, he became associate technical director and retired from there in 1979. He then worked as a consultant in management and engineering until 1990. He also lectured in value engineering seminars nation-wide.

He and his wife live in Oxford, ME and have seven children. His oldest son, Alan, served in the Gulf of Tonkin during the Vietnam War.

ROBERT G. NICKELS (LTJG) was born in Chicago, IL on May 29, 1925. He enlisted in the Navy on May 19, 1943, and eventually became a Supply Corps Disbursing Officer for ComSubPac. Military stations and schools include: V-12 Unit, Notre Dame; NSOTC, ABATU, Lido Beach, Long Island, NY and Bayonne, NJ; NCSC, Harvard; Submarine Bases at Midway, Pearl Harbor and Guam. He was commissioned June 27, 1945, at the Navy Supply Corps School, Harvard, Boston, MA.

His memorable experiences include surviving a typhoon and tidal wave on Guam; losing power in one engine of an RSD Skymaster on flight from Hawaii to the U.S. and returning before half-way point to Honolulu for repairs.

Honorably discharged on July 17, 1946, his decorations include the American Theater Campaign, Asiatic-Pacific Theater Campaign and the WWII Victory Medal.

After 38 years with AT&T, Nickels retired as accounting supervisor in May 1984. He is a travel consultant with Sun Travel Service, Park Ridge, IL and a part-time income tax consultant for John Panagakis, CPA, Chicago, IL. He married Mary Louise Earll on Aug. 13, 1953 and they have two children, Leslie Lynn Phillippi and Scott Earll. Leslie Lynn and Martin Phillippi have two children, Daniel and Amber Nicole. Scott and his wife, Kimberly, have three children: Kelli Anne, Gregg and Jamie Ryan.

WILLIAM HARRISON NIMS JR. (CAPT) was born in Fort Mill, SC on Jan. 5, 1924. He enlisted in the USMC in 1941 and served Marine Barracks MCAS, St. Thomas, V.I, attaining the rank of platoon sergeant. He was selected for V-12 Program in July 1943; attended University North Carolina V-12 from October 1943-April 1945; OCS at Quantico, VA; commissioned in October 1945; and assigned MB Camp Pendleton disbanding divisions returning from Pacific. He was com-

missioned USMCR in 1946 and released from active duty.

Discharged July 1, 1958, his awards and medals include the Marine Corps Good Conduct Medal, American Theater Medal, WWII Victory Medal, China Service Medal II, National Service Defense Medal, Korean Service Medal and the United Nations Medal.

The best thing that happened to him in V-12 was meeting and marrying Theresa Speaks. Nims attended University Georgia and graduated in 1949 with Bachelor degree in forestry. He worked as assistant district forester fire control (21 counties).

Recalled to active duty in February 1951, he served tour of duty with 2nd Mar. Div. and attended Basic School at Quantico. He served tours of duty with 3rd Mar. Bde., combat cargo officer, USS *Winston* (AKA-94) for 26 months; 16 months in Western Pacific, MB Camp Pendleton, 3rd Mar. Div., Japan. He attended MTC Infantry School, Ft. Benning, was #1 in the class and received a Letter of Commendation from the commanding general. Served tour of duty with 1st Mar. Div. (Reinf.) and was discharged as captain, USMC in July 1958.

Nims was employed by Eagle-Picher Industries, Inc. as plant manager of lead oxide plant, a zinc oxide plant and agricultural chemical plant. He retired in 1982, a division manager as president of wholly owned subsidiary. Was employed in Joplin, MO, Galena, KS and Picher, OK.

Currently resides in Joplin with his wife, Theresa. They are parents of two sons and one daughter: Bill III (shop foreman of large metal fabricating shop); Kenneth (independent contractor building custom cabinetry and furniture); and Cecilia (office manager of Mid-America Truck Maintenance, Inc.). They have two granddaughters, one grandson and two great-grandsons.

PAUL F. NOBLE was born in Vienna, MD on Oct. 28, 1921. He enlisted in the USNR on March 9, 1942; was assigned to District Security Office in Baltimore; entered the V-12 Program on July 1, 1943, at Mt. St. Mary's College; transferred to Franklin & Marshall College, then to Ft. Schuyler Midshipman School. After one month he was separated and yeoman first class rating was restored.

He was assigned to the Military Government at Ft. Ord, CA and shipped to Okinawa. After one year he left Okinawa and in 1946 was discharged in Maryland. Noble graduated in 1949 from Franklin & Marshall College and joined Equitable Life of Iowa as a field life underwriter. In 1968 he received CLU designation from the American College of Life Underwriters and in 1984 received Chartered Financial Consultant designation. In 1983 he retired from Equitable Life of Iowa.

Presently resides in Baltimore, MD with his wife, Mary. They had three daughters (one deceased), one son and four grandchildren.

ROBERT C. NOE (LTJG) was born in San Diego, CA on July 6, 1925. He enlisted April 7, 1943; started active duty July 1943, V-12 NROTC,

UNM, Albuquerque, BSME & commissioned June 22, 1945. Other military stations and schools include: North Carolina State College, Naval Training School (Diesel and Electric); Philippine Islands (1946), USS YMS, gunnery, Ships Service, Supply, Commissary. Task Force was assigned to mine sweeping in small bays and inlets, and he participated in Rio Grande Fleet (1943-1945).

Honorable separation to inactive duty on Aug. 11, 1946; promoted to lieutenant jg Dec. 31, 1948; transferred to active Reserve May 23, 1952, to Ready Reserve Jan. 1, 1953, to the inactive Reserve Jan. 1, 1955, and was honorably discharged on Oct. 5, 1959. His awards include the American Area, Asiatic-Pacific Area and the WWII Victory.

From 1946-1988 he worked as mechanical engineer, Power Plant, Phelps Dodge Corp., Morenci, AZ; T.C. Noe, Contractor, Gallup, NM; electric utility director, city of Gallup; retired 1988. Pr. chm. of the board, Noe Enterprises; member AIPE and AIME and in *Who's Who in the West* and *Who's Who in Technology*.

Married on Aug. 14, 1945 (V-J Day), he lives with his wife, Sally, in Gallup, NM; they have three children: Kathe, Bill and Tom, and three grandchildren.

JAMES M. NOLAN (CAPT) of Loudonville, NY entered the USN on Nov. 21, 1942, and was sent to the USNTC at Sampson, NY. He entered the V-12 Program at Trinity College, Hartford, CT in September 1943; graduated from Midshipman School at Columbia University in June 1945; commissioned on June 5, 1945; assigned to the USS *Dortch* and served until discharged on June 6, 1946.

Joined the Naval Reserve in 1947, he was recalled to active duty in September 1950 and served in Japan and Korea on the staff of COMNAVFE until November 1952. He remained active in the Naval Reserve until his retirement in 1981. During his naval service he served in various command billets and rose through the ranks until he was appointed a captain in 1974.

Nolan was employed as a senior investigator with the office of the New York State Attorney General until his retirement in 1985. He is married to the former Jean Deuel and they have one daughter and two grandchildren.

PAUL R. NORDSKOG completed six semesters in the V-12 Program at St. Ambrose College

and in pre-chaplain training at Oberlin College. He sang in the First Church Choir and played in the Oberlin College Symphonic Band. At the Fleet Home Town News Center, he was a copy editor at the world's busiest copy desk.

Completed an MA degree at the University of Iowa and taught in Spencer and Britt, IA. His 32 years in teaching included four years of administrative work, but mostly he taught mathematics and science in Park Ridge, IL.

Nordskog continued church work and music, recently singing in the 200-voice Apollo Chorus of Chicago, usually in Orchestra Hall. He originated the first Illinois smoking ban law; his part-time jobs included selling garages; and he is a do-it-yourselfer.

Married since 1948, his wife, Shirley, taught music for 32 years. They have three children who all completed college.

CARL J. NORDSTROM III (LCDR) was
born in Seattle on Aug. 3, 1924. He was inducted into the USN in 1943; entered the V-12 Program in 1944 at University of Washington; received BSEE in February 1946; commissioned upon graduation; attended indoctrination training at Newport, RI and served onboard USS *Montpelier* (CL-57). He was assigned to the USS *Henry W. Tucker* (DDR-875) as communications officer until released in November 1946.

Recalled in 1950, Nordstrom served on USS *Agerholm* (DD-826) and on USS *Harry E. Hubbard* (DD-748) until October 1952. He worked for the Boeing Company for 32 years, mostly in the missile and space sector. He received his PE license in 1970 and for several years performed commercial and industrial electrical design until rejoining Boeing in 1977.

Retired from the Naval Reserve in 1971 and from Boeing in 1988. He and his wife, Virginia, also a Seattle native, have three daughters and seven grandchildren.

SAMUEL STRUDWICK NORVELL
(TUT) was born in Florence, AL on Dec. 11, 1924. He joined the V-1 Program in September 1942 to attend Florence State Teachers College. Then he was called to active duty in July 1943 for the V-12 Program at the University of the South at Sewanee, TN.

Attended Midshipman School at Columbia University, New York City; received his commission in October 1944; reported to USS LCT Group 40, Flotilla 14 in the South Pacific; took command of LCT-876 and helped secure the islands of Peleliu and Angaur.
124

Memorable experiences include helping to pick up survivors of the torpedoed and sinking USS *Indianapolis* in late July 1946 and the sudden shock of the atomic bomb that ended the war in the Pacific within a week or two. He returned to the States and was discharged from active duty in May 1946. He remained in the USNR.

Norvell graduated from the University of Alabama in Tuscaloosa; went to the University of Alabama in Birmingham; and received a medical degree in June 1953. After internship at St. Vincents Hospital, Birmingham, AL, he returned to Florence, AL to enter into the family practice of medicine.

Married to the former Martha Neal Key of Russellville, AL on Aug. 30, 1948; they have four sons: Samuel Jr., William Key, Carter Clarke and Neal Key. Norvell is still in the family practice of medicine.

ROBERT CHARLES NUCCIO (QM3/c)
was born in Los Angeles, CA on June 18, 1925. He entered the service on Oct. 31, 1943, USNR V-6, attended V-12 NTU, Arizona State Teachers College

Participated in action in the Asiatic-Pacific, the Philippine Liberation and the American Area. Nuccio was discharged March 30, 1946. He passed away July 20, 1987.

ROBERT T. O'BRIEN was born in Chicago,
IL on Aug. 4, 1925. He graduated from Loyola Academy in June 1943 and left two weeks later for the Marine V-12 Program at Western Michigan College, Kalamazoo, MI. He stayed there the full four terms, then left for Parris Island, SC on Nov. 1, 1944, and then on to Camp Lejeune, NC early in 1945.

He was in pre-OCS at Lejeune when the war ended and selected the option for discharge on Oct. 12, 1945. O'Brien finished college at Loyola University, Chicago and graduated in June 1947 with BSC degree. He entered the stock and commodity brokerage business, bought full membership on the Chicago Board of Trade and is a life-long member to this date; he also managed trading office for Unilever at the Board of Trade.

O'Brien is married and has three sons and one daughter and three grandchildren. He took his two oldest sons to Parris Island in 1981 for a nostalgic visit. They could do the obstacle courses easily (Dad just watched). He attended the 50th V-12 Wartime Reunion at Western Michigan University in September 1993 and had a great time. It seemed like only yesterday.

FRANCIS A. OLSEN (CPL) was born in
Manchester, NH on May 15, 1924. He joined the USMC on Jan. 31, 1943, and entered the V-12 Program at the University of North Carolina, Chapel Hill, NC on July 1, 1943. Other military locations included Parris Island, SC; Long Island, NY; Camp Lejeune, NC (for pre-OCS); MB MCS, Quantico, VA (for PCS); Camp Pendleton, CA; Pearl Harbor, HI; Guam, North China occupation (6th, 1st

Mar. Div.) Oct. 16, 1945-June 2, 1946 (including surrender of Japanese Forces.

Olsen received an honorable discharge on July 4, 1946 from NTC, Great Lakes, IL. His awards include Marskman and the Good Conduct Medal. His memorable experiences include his college education, UNC and duty in North China.

As a civilian, he was a mechanical senior designer for 38 years at Electric Boat Div./General Dyn. Corp. in Groton, CT and helped in design of first nuclear submarine, SSN-571 *Nautilus* and first FBM submarine, SSBN-598 *George Washington* and others. He retired May 31, 1989, and presently resides in Ledyard, CT with his wife, June. They have four children and five grandchildren.

MAX EUGENE OLSEN (LTJG) was born in
Pottawattamie County, Hancock, IA. He graduated from Treynor High School in 1942, attended the Navy V-12 Program and was discharged from the service Aug. 15, 1956. After graduating from Iowa State University, the University of Southern Idaho and Creighton University School of Medicine,  he rose in the medical profession to chief of staff at both Jennie Edmundson Memorial Hospital and Mercy Hospital in Council Bluffs.

Besides being chairman of the Pottawattamie-Mills County Medical Society, a district counselor for the Iowa Medical Society Rural Health Committee, he was also active in farming and livestock breeding.

Dr. Olsen passed away in December 1992. He is survived by his wife, Elaine; two daughters, Maureen and Teresa; two sons, Van and Craig; two grandchildren; and two step-grandchildren.

RICHARD MARTIN OLSON was born in
Chicago, IL on Dec. 26, 1925. He enlisted in the USNR on July 1, 1943; attended V-12, Institute of Technology, BS Ch.E., October 1945; and was commissioned ensign in November 1945. He attended the Naval Gunnery School, Washington, DC and was deck officer in USS *Tuscon*. Olson was discharged on July 23, 1946.

After his discharge he was employed by CPC International on Oct. 1, 1946. Involved with research, development and process engineering, he has seven patents, two publications and did process and project engineering in 10 countries. He retired Jan. 1, 1986 as senior research engineer.

Community service included work with the

Village Plan Commission, High School Board of Education, Village Trustee and water commissioner. He and his wife, Doris, have three sons: Kenneth, Donald and Kurt, and four granddaughters. They reside in North Riverside, IL

WILLIAM J. ORLEY (CAPT) was born in Detroit, MI on Dec. 10, 1920. He enlisted in the USN in September 1942 and entered the V-12 Program at the University of Michigan on July 1, 1943. He graduated in June 1944 with BSE (Aero); attended pre-Midshipman School at Asbury Park, NJ; Midshipman School at Notre Dame, South Bend, IN and was commissioned Oct. 9, 1944. He attended Advanced Base Aviation Training Schools at Norfolk, VA and Lambert Field, St. Louis, MO. He was assigned to Fleet Air Wing 4 (FAW4) in December 1944 and served active duty in Attu, Kodiak, and Whidby Island, WA as aircraft maintenance officer and O&R officer.

Discharged on July 14, 1946 as lieutenant jg, he joined the Naval Reserve in July 1946 and remained active until Dec. 11, 1980. Drilled at NAS Grosse Ile, MI and NAF Selfridge, Mt. Clemens, MI, his billet assignments were aircraft maintenance officer, avionics officer, leadership training officer, administration officer and commanding officer of NARDIV 732, NARS Y1, RRU5114 and GVTU 5114.

After being discharged in 1946, he returned to the University of Michigan; worked for the Packard Motor Co. in jet engine research and completed his Masters Degree in business administration in 1949. Subsequently worked for Ex Cell O Corp., Ford Motor Co., Huck Manufacturing Co. and started his own companies in 1953. He joined Hydralink Corp as executive VP (1963); joined Condamatic Co. (1984) and retired as president in March 1990. Currently he is president of Applied Hydro-Dynamics Inc. and Terra-Seal Corp.

He is married to F. Susan and has three adult step-children.

ROBERT D. OSBORN (LCDR) was born in Pana, IL on Aug. 10, 1926. He joined the USN March 4, 1944; attended the V-12 Program at Howard College, Birmingham, AL and Ottumwa Iowa pre-Flight School. Other stations include Pensacola, FL; Kaneohe, T.H.; Sangley Point, P.I.; Westover AFB, MA and Blackbush, England.

Designated a naval aviator on Aug. 15, 1947, he participated in the sinking of PBY Buno 64028 615 NM west of Johnson Island and evacuation of U.S. Embassy, Canton, China. He retired Oct. 4, 1966, and received the WWII Victory Medal.

He married Gloria M. Broadhurst and they have four children and 10 grandchildren. He is retired in San Diego, CA.

JOHN A. OSTROM enlisted in the USN in 1943 and was assigned to the V-12 Program from February 1944 to June 1946 at Washburn, Swarthmore and Princeton (NROTC) units. He received BS degree in civil engineering from U.C. Berkeley in 1947, commissioned USNR, on inactive duty, CEC Reserve to 1958.

Employed by Bechtel for 40 years in engineering and construction of utility power, mining and civil infrastructure works. Assistant project manager SF Bay Area, Rapid Transit Project, Manager of Services Mining Div., Project manager Skagit Nuclear Power Plant, manager of corporate training and development manager of several renewable energy projects.

He has been retired since 1993 with his wife, Betty, in Belvedere, CA, managing real estate investments there and in La Jolla, CA. They have a daughter, Mary Anne, and sons John and Tom.

SAMUEL H. PACKER (RADM) was born on Sept. 8, 1925. He enlisted in the USN on May 24, 1943; attended Dartmouth College in the V-12 Program and graduated from Tufts University NROTC in 1945 at which time he was commissioned. He served in the USS *Dennis J. Buckley* (DDR-808) and USS *E-PCE* (R) 852 prior to attending Submarine School. He then served on submarines USS *Sea Owl*, USS *Cero*, USS *Remora*, USS *Sirago* and commanded USS *Harder*.

Commanded Submarine Division 53, Submarine Development Group One (deep submergence operations), and the Middle East Force and was an instructor at the Submarine School (operations). He served on the Staffs of Commander Submarine Force, U.S. Atlantic Fleet (intelligence); Commander Submarine Force, U.S. Pacific Fleet (operations); Commander in Chief, USN Fleet Forces Europe (plans); Commander U.S. Military Assistance Command Vietnam (security assistance); the Chief of Naval Operations (programs); and for two tours on the Joint Staff, Organization of the Joint Chiefs of Staff (plans and policy). His most recent active duty assignments were director of operations, U.S. European Cmd., and assistant and acting Deputy Chief of Naval operations (plans, policy and operations).

Retired in 1983 and has subsequently been working as an independent consultant in national security affairs in the Washington, DC area, principally for the Institute for Defense Analyses. His awards include the Distinguished Service Medal (three awards), Defense Superior Service Medal, Legion of Merit (three awards), Meritorious Service Medal (two awards), Joint Service Commendation Medal, and Navy Commendation Medal.

He is a graduate of the Naval Postgraduate School, and the Naval War College and has MA degree in international affairs from George Washington University.

Married the former Peggy Rebhun of Cincin-

nati, OH and they have five children and six grandchildren.

WILLIAM O. PARKER JR. (LTJG) entered the USNR, on Sept. 16, 1943, at the Great Lakes Naval Training Station. He transferred to the Illinois Institute of Technology, V-12 Program and was commissioned an ensign on Feb. 21, 1946. He attended Steam Engineering School, Newport, RI and was assigned to USS *Henry R. Kenyon* (DE-683), San Diego until the *Kenyon* was decommissioned.

Released to inactive duty in June 1947 and resigned his commission as LTJG, USNR in 1960.

A BSME degree and experience as assistant engineering officer in the USN led to his career with Duke Power Company. He retired in 1990 as vice president of fossil production in Charlotte, NC after 42 years of service. He still resides in Charlotte, NC, with his wife Anita E. Parker. He has a daughter employed by Harmon Commonwealth, Baltimore, MD and a son employed by Crown Controls, Charlotte, NC.

ROBERT JAMES PASCOE (ENS) was born on Sept. 4, 1925, in Butte, MT. He entered Montana School of Mines V-12 unit on July 1, 1943; transferred to University of Washington V-12 unit in March 1944. He was commissioned an ensign at the University of Washington in February 1946 and was assigned to the USS *Montpelier* in March 1946, for "Salt on the Stripe" training cruise, Cruiser Div. 14.

He was discharged to the Naval Reserve in July 1946.

In October 1946, he joined Westinghouse to begin his 38 year career in sales and marketing management. At retirement in 1984 he was national OEM sales manager, located in Chicago. Moved to Reno in November 1984 and taught managerial sciences at the School of Business, University of Nevada until 1992, when his career was interrupted by quintuple coronary artery bypass surgery.

Currently resides in Reno with his wife of 47 years, Virginia. They have two daughters who have given them four granddaughters and three grandsons.

WILLIAM BRADSHAW PATRICK (LTJG) was born on Nov. 29, 1923, Indianapolis, IN. V-12 IL St. Normal U. July 1, 1943-June 30, 1944. Commissioned ensign SC Nov. 1, 1944, as member Navy's Mid Off. Candidate Class 5, in 12 mos. Supply Corps Program at Harvard Graduate School of Business July 6, 1944-July 2, 1945.

Shipped out USS *Fallon* (APA-81) S/F CA, Aug. 3, 1945, for Japan invasion.

Arrived Pearl Harbor Aug. 9, 1945. En route atom bombs dropped on Hiroshima and Nagasaki. Sept. 8, 1945-June 11, 1946, General Stores and later "frozen" (for want of "replacement") disbursing officer, serving Bomb Disposal, Carrier Aircraft Service #43, Construction Bn., Marine Air Group #21, Naval Air Transport Service & Photography.

Reconnaissance units and base personnel stationed at NAB #943 Agana, Guam. With surplus "points" earned and arrival of "replacement," reassigned to USA and released to inactive duty on Aug. 3, 1946, after Aug. 1, 1946, Alnav promotion to LTJG. The Principia A.B. 1947 Harvard Law LLB 1950. General practice of law, Indianapolis, IN, emphasis probate, from 1950 admission to Indiana Bar to present.

Resides in Noblesville, IN with his wife, Ursula. They have three children, all IU graduates: William (MBA in finance, employed CNA Ins. Co., Chicago); Ursula B. Moul (Major, JAG USAF, Pentagon, Washington, DC); Nancy B. Patrick (material planner, Thomson Electronics, Marion, IN); and two grandchildren, Audra and Spencer Moul.

FRANK H. PATTERSON (LCDR) entered the Navy V-1 Program ere draft, through a helpful CPO recruiter. On July 1, 1943, he was ordered to enroll with the Navy V-12 unit at Middlebury College, VT. He was assisted over minor undergraduate rough spots by a benevolent XO, and his career possibilities became brighter. In September 1944 he transferred to Northwestern Midshipmen's School and met a young woman from Iowa at Chicago Ave. and Rush Street. He was commissioned an ensign in January 1945, followed by Communications Officers School at Harvard. He served on a decommissioned CVE, then transferred to the USN Supply Corps. He enjoyed sea duty on *Tapa*.

In 1950 he married the girl from Iowa. He filled aviation supply billets and in 1963 retired as LCDR (SC), with thanks to a lot of COs, sailors and colleagues.

Earned his BA and MBA degrees and was employed in responsible positions in industrial, academic, service and consulting organizations. He is designated a certified professional logistician (CPL) and currently performing as volunteer accountant, community foundation trustee and is treasurer of two non-profit organizations. He resides in historic community of New Harmony, IN, with his grandmother from Dubuque.

EARLE WHITAKER PAYLOR JR. (ENS) was born on July 24, 1925, in Scottsburg, VA. He entered the Navy V-12 Program at Emory and Henry College on July 1, 1943. Attended Naval Reserve Officers Training Corps at the University of South Carolina in March 1944. Commissioned ensign, USNR in September 1945 and assigned to patrol duty in Sasebo Harbor, Japan, from Novem-

ber 1945-March 1946. He was discharged in July 1946.

Attended Duke Divinity School from 1946-1949, and served as pastor of United Methodist Churches in Virginia at Richmond, Hampton, Fishersville, Arlington, Hopewell, Mechanicsville and Glenn Allen. Following retirement, he serves as associate pastor of Welborne United Methodist Church in Richmond.

Currently resides in Richmond, VA with wife, Edna K. Paylor. They are the parents of sons, David, employed by state of Virginia; Robert, LCDR in the USN stationed in Norfolk, VA; and daughter, Mary Rebecca Hepler, of Louisa, VA. They have four grandchildren.

ROBERT LOWELL PAYNE (ENS) was born Dec. 16, 1924, in Kankakee, IL. Entered the USN on Dec. 31, 1942, at the Great Lakes Recruit Training Center, then went to Electrician Mate School at Purdue University. He spent two semesters in the V-12 Program at Oberlin College followed by three semesters in NROTC. He was commissioned Feb. 25, 1945, at Notre Dame; Pre-Radar School, Brunswick, ME; Radar School, MIT; flight operations, Camp May, NJ and Norfolk, VA.

Released from duty on March 2, 1946, received his BSEE degree in 1948 and MSEE degree in 1950, from Purdue University. He taught engineering for four years; spent one year at North American Aviation heading NATO Radar School in Germany; and 27 years at Hughes Aircraft Co. in system engineering, test engineering and program management. Retired in 1981 and now lives in Leisure World, Laguna Hills, CA. He is a member of the Leisure World Board of Directors and does some part-time accounting and plays with computers.

WILLIAM SPENCER PAYNE, M.D. was born on March 22, 1926, in St. Louis, MO. He entered DePauw University V-12 (pre-med) Program from July 1944-October 1945. Graduated from Washington University Medical School in 1950 and was commissioned MC USNR. Served in the Korean War, 1952-1954, as medical officer of the flagship for the amphibious force of the Pacific (CTF-90), USS *Eldorado* (AGC-11) and subsequently as assistant pathologist, USN Hospital, Great Lakes, IL. He was discharged USNR in 1954.

Completed seven years postgraduate training for MS (surgery) and certification in general sur-

gery and thoracic and cardiovascular surgery an became a member of the surgical staff of the May Clinic, advancing to professor, Mayo Medica School and head, general thoracic surgery, May Clinic, Rochester, MN. He has participated in th training of many young men and women in surger and has authored over 275 publications and ha been the recipient of many honors including th Howard Gray Award, the James C. Masson Name Professorship, and a directorship of the America Board of Thoracic Surgery, as well as listing i *Who's Who in America* and Pekkanen's *Be Doctors* and holding offices in national and inte national professional associations. After 40 year he retired on June 30, 1990.

CLARK W. PECK, (LT) was born on April 1 1925, in Lakewood, OH. He entered Baldwi Wallace College V-12 Program in July 1, 194 Berea, OH. He attended Columbia Universit Midshipman School; commissioned an ensign o April 26, 1945; and assigned to USN Trainir Center, Miami, FL, May-July 1945. Assigne temporary duty, USS *Audubon* (APA-149), Sa Francisco, CA to Leyte Gulf, P.I. Reported fo duty, USS *Lark* (ATO-168), Manus Island, Sou Pacific, September 1945-February 1946. Deconmissioned AK-70, USS *Crater*, Pearl Harbor, Jun 1946.

Released from active duty in August 194 and discharged from the USNR in 1953 with th rank of lieutenant.

Graduated from Baldwin-Wallace Colleg with BA degree in 1948 and the Ohio State Unive sity School of Dentistry in 1953. Worked as clin cal professor at Case Western Reserve Universit School of Dentistry, 1965-1993. Currently, has private practice that he began in 1954 in Nort Olmsted, OH.

Received the Citizen of the Year Award i 1965 in North Olmsted, OH; International Colleg of Dentists, 1976; American College of Dentist 1988; board of directors, American Academy Fixed Prosthodontics, 1989-1992; and presiden Greater Cleveland Dental Society, 1991.

Resides in North Olmsted, OH with wif Beryl. They have two children, Kathy Hildenbran Dallas, TX; and CDR Scott R. Peck, USN Denta Corps; and three grandchildren: Jeffrey and Bre Hildenbrand and Heather Peck.

LOUIS NATHAN PENER (PO) was bor Aug. 3, 1922, in Kansas City, MO. After gradua tion from Manual High School, he enrolled in th Navy V-12 Program at Park College (Nickel Ba racks) in Kansas City and graduated in 194 During WWII he served as the radioman on hi ship and was stationed in Hawaii and Japan durin his service. He was training for the invasion Japan aboard an LCT when the atomic bombs wer dropped on Japan.

His brother, David Paul Pener, also served i the USN during WWII, and was aboard the US *Cowpens* when Japan surrendered.

After the war, he went on to the University o

Wisconsin and the University of Missouri-Kansas City where he graduated from Law School. He practiced law in Kansas City until his death with the law firm of Pener, Eveloff and Geller.

Married Bertha Boresow, also from Kansas City, MO, and they were married over 40 years when Louis Pener passed away on June 4, 1993. He is survived by three children: James, Debra and Gary; and nine grandchildren, all residing in the Kansas City area. Among many other organizations, he was a member of the Jewish War Veterans.

STANLEY LAWRENCE PERKINS (LTSG) was born on Sept. 8, 1925, in San Francisco, CA. He joined the Navy in June 1943 and entered the V-12 Program. Attended Washburn, UCLA and Navy Supply Corps School and was stationed in Japan. He served in the Korean War from 1951-1952.

Participated in several battles off the East Coast of Korea, Wonson, Hungnam, etc. and served in LSD-5, *Gunston Hall*, LSD-26, *Tautuga* as disbursing and supply officer as well as coding officer.

Discharged on Sept. 29, 1956, with the rank of lieutenant, senior grade.

Married Harriet and has two sons and three grandchildren. He retired from magazine publishing business.

LOUIS N. PERNOKAS (CAPT) was born on Dec. 21, 1926, in South Boston, MA. He joined the Navy V-12 Program at Dartmouth College in July 1944 and was commissioned an ensign in 1945. He attended Harvard Medical School in 1950 and trained in general and vascular surgery at Boston City Hospital.

Recalled to active duty in 1951 and was stationed in Korea, 1951-1952, with the 1st Bn, 1st Mar., 1st Mar. Div. He was at USNH, Newport, RI, 1953-1954, at the time of the explosion aboard the USS *Benning* in 1954. After discharge in 1954 he served many years in the Reserve force and multiple Marine Reserve Units. Retired from the Reserves as a captain in 1989. He was discharged in July 1944 and retired as captain in 1989. His awards include the Commendation Medal and Combat V.

Practiced general and vascular surgery in Boston, MA until 1968 and also taught at Harvard Medical School during that period. Moved to Rhode Island in 1968 to continue his practice and in 1994 was selected by his peers as a leading doctor in his field. Retired from active practice in May 1994.

Lives with his wife Charlotte in Saunderstown, RI; they have a son, Nicholas (living in Texas).

LINDSEY J. PERRY JR. (ENS) was born Jan. 2, 1926, Mt. Olive, NC and grew up in Reidsville, NC. He enlisted in the Navy V-12 Program on May 20, 1943. On July 1 he entered the University of Virginia for two terms then transferred to the University of Michigan for six terms. He was discharged as an ensign in August 1946.

Graduated University of Michigan in February 1946, BSE (AERO) and Elon College, May 1948 with BS in business. Worked at NACA Langley, Air Proving Ground Command at Eglin AFB, North American Avn., Bell Aircraft and Northrop Corp.

Participated in the design and/or flight testing of the B-70, F-108, F-109, F-5, F/A-18 and F-123 stealth fighter. He retired from Northrop Corp. as director of the Competitive Aircraft Assessment Group, ATDC on Dec. 31, 1991.

He and his wife Dale live in Rancho Palos Verdes, CA and have three children: Patricia, Becky and Lindsey III; three granddaughters and three grandsons.

RALPH MARTIN PETERS was born May 9, 1926, Knoxville, TN. He enlisted in the Navy V-5 (V-12a) in February 1944, and was assigned to Milligan College in July 1944. He was reassigned to the Central Michigan University in July 1945 then to the University of Michigan in November 1945. He was honorably discharged in June 1946 when the V-12 Program was terminated.

Enrolled at Lincoln Memorial University in September 1946; BS degree conferred in 1949; graduate study at the University of Tennessee; and MS degree conferred in 1953; and Ph.D. in 1960.

Taught and coached at Clinton High School in Tennessee, 1949-1956; joined faculty and staff of Lincoln Memorial University in 1956, serving successively as director of admissions and alumni affairs, chairman of Department of Education, and executive vice president. He joined faculty and staff of Tennessee Technological University in 1963, serving as dean of students, 1963-1968; dean of the graduate school, 1968-1989; full professorship, 1963-1989; and is presently a Graduate Dean Emeritus and teaching graduate courses part-time.

Resides in Cookeville, TN with his college sweetheart and wife Lorraine D. Peters, who is TTU Emeritus Professor of home economics. They have two daughters and three grandchildren. He has numerous professional and civic memberships and awards.

SIDNEY K. PEVETO (LT) was born on Oct. 22, 1922, in Orange, TX. He entered the Navy V-12 Program at Central Missouri State University in July 1943 and was commissioned an ensign at Columbia University in February 1944. He served as a communication officer, LST-669 and transported invasion troops to Eniwetok, Admiralty Islands, Iwo Jima, Philippines and Okinawa. They were underway to invade Japan when the bombs were dropped.

Graduated from North Texas University in 1949; received his Ph.D. in psychology and history from East Texas University in 1969; assistant professor, Cameron University, OK, 1969-1973; and instructor at Grayson College, 1973-1986.

His memorable experiences include being anchored next to a ship on beach at Leyte Gulf when MacArthur waded ashore and when Gov. Winthrop Rockerfeller traveled with them.

Retired from the USNR in 1969. His awards include the WWII Victory Medal, Japan Occupation Medal, American Theater, Asiatic-Pacific (three stars) and Philippine Liberation (two stars).

Resides in Sherman, TX with wife, Patsy Scott. They have two children, Major Ronald L. Peveto, USAF flight surgeon; Linda Peveto Warmann, petroleum engineering business manager; and five grandchildren. He retired from teaching in 1986.

AUGUST C. PFITZER (LCDR) was born on April 5, 1923, in Chattanooga, TN. He joined the Navy on Dec. 1, 1942; entered the Navy V-12 Program at the University of South Carolina. Stationed at USC, Asbury Park, Cornell, Bowdoin, Mare Island and served in the USS *Mullany* (DD-528). Graduated on Oct. 15, 1944, with a BS degree in electronic engineering.

Released from active duty in July 1946, stayed in the Reserves and was recalled on Feb. 20, 1951, to the USS *Mullany*. He was released on Oct. 20, 1952.

Employed by TVA in 1946 and retired on April 13, 1984, as chief of the civil engineer and design branch.

Married Maxine Frink and they have three children. Retired with pay from the USNR and presently lives in Chattanooga. He loves to travel.

JOHN T. PHILLIPS (CAPT) was born on Jan. 28, 1927, in St. Paul, MN. He entered the Navy V-12 Program at the University of Wisconsin on July 1, 1944, after high school graduation. Transferred to the V-12 unit at Yale University on July 1, 1945, and was released to inactive duty on June 23, 1946, as apprentice seaman. Returned to Yale in September 1946 as a Midshipman in the NROTC Program. On June 6, 1947, he was commissioned a 2nd lieutenant USMCR, received BA degree in economics and was released to inactive duty.

Started a 38 year career in trust banking in the twin cities with First Trust Co. of St. Paul. He attended night law school, received an LLB degree in 1951 from St. Paul College of Law, and was admitted to the Minnesota Bar.

Recalled to active duty in 1951 and 1952,

serving stateside as a squadron and battalion adjutant and legal officer. He was discharged from the USMC as captain in 1956.

Retired from First National Bank of Minneapolis as senior vice president in 1985 after serving as head of the Personal Trust Services Department. He and his wife then moved to Prescott, AZ. They have three daughters and two sons.

RAY C. PHILLIPS (2/LT) attended Millsaps College, then entered Cumberland University, Lebanon, TN in 1941. He joined the USMC in 1942; entered the Navy V-12 Program at Millsaps College, Jackson, MS in 1943; transferred to Miami University, Oxford, OH for one semester and was commissioned 2nd lieutenant, Quantico, VA, 1944. After serving in the occupation forces in Japan, he was discharged in December 1946.

Received his baccalaureate degree from Middle Tennessee State University in 1947; MS degree from Vanderbilt/Peabody in 1950; attended the University of Fribourg, Switzerland, 1951; received doctorate from Auburn University, 1961, and accepted a permanent position at Auburn University, along with his wife, Dr. Phyllis Phillips, where they remained until their retirement as associate and full professors. He received numerous awards and recognition's, i.e., "Outstanding Educator in Alabama," "Outstanding Graduate Faculty Member at AU."

After returning to his ancestral home in Tennessee, he was persuaded to accept a position as vice president for academic affairs at Cumberland University, 1983-1986. He retired a second time, but remained on the University Board of Trust. In 1991 he was again persuaded to relinquish retirement and became president of the University which he had entered as a freshman 50 years before. He and Phyllis have three children: Jim, Gina Murray and Lisa Ham.

HANK PISANKO (0-8) was born on March 14, 1925, in Trenton, NJ. He entered the Navy V-12 Program at the University of Notre Dame in July 1943; was stationed in the Pacific Rim and attended ROTC at University of Notre Dame in 1945.

His most memorable experience was working with Moe Berg (Princeton 1923) in Intelligence.

Discharged in September 1946. His awards include the Michai Award in Singapore and the Distinguished Leadership Award from the American Biographical Institute.

Married Sophia E. Zudnak. He is chairman of the board (Emeritus), P.K. Co., LTD, Hong King in 1995.

A.L. PLEMONS enlisted in 1943. He went to Cnst. C, Warrensburg, MD for the Navy V-12 Program from 1944-1945. He was discharged in May 1946 and served again from 1951-1952. He had some impressive experiences while aboard the USS *Juneau* with Task Force 77 in Korean waters.

He married Blanca and they have a son, Blair, who has three children; daughter, Alana, who has three children; adopted son, Michael, who has five children; and grandchildren. He is now retired and lives in West Corina, CA.

ERLING B. PODOLL (ENS [D] L) was born on Feb. 24, 1925, in Westport, SD. He joined the Navy Oct. 10, 1942, and entered the V-12 Program, USNRMS, MSTC, Minot, ND. His military stations included Tinian, Mariana Islands; USS LCT Group 39; LCT Flotilla 13 USNRMS, Northwestern University, Chicago, IL.

His memorable experience was being a passenger on board the USS LCT-991, which on July 26, 1945, made two trips out to the anchored cruiser *Indianapolis* at Tinian, Mariana Islands. The first load was the canister of U-235 and the second load was a crate of bomb parts. He was released to inactive duty on July 1, 1946, and discharged on June 9, 1959.

He is married and has three children and six grandchildren. Retired in South Dakota after working as a wildlife biologist for 36 years in South Dakota, Wisconsin and North Dakota.

CHARLES V. POMATTO enlisted in May 1943 and completed boot camp at Great Lakes Naval Training Station. He entered the Navy V-12 Program at the University of Virginia in November 1943. Discharged in June 1946, but remained in the Naval Reserve until April 1961.

Upon his return home, he married Irene and began work at Westclox, Division of General Time until 1951. He then joined General Electric Co. at Syracuse, NY, transferring to Oklahoma City and Philadelphia, before leaving in 1970 for Dallas with Docutel Corp. He finished the last 12 years of work with Northern Telecom, retiring in 1990.

He and Irene reside in Dallas; they have three children, two live in Dallas and one lives in Tampa.

FRANCIS P. POWERS (LTJG) was born on Jan. 13, 1926, in Brooklyn, NY. He entered the Navy V-12a Program on March 1, 1944, at Middleburg College, VT; transferred to Tufts University, Medford, MA for V-12 and ROTC Navy. Commissioned an ensign and awarded a BA degree in naval science from Tufts University in May 1946.

He was assigned to the USS *Amphion* (AR 13) at Norfolk, VA; transferred to the USS *Portsmouth* (CL102), 6th Fleet (Mediterranean). He was released to inactive duty (USNR) as lieutenant (jg) in May 1947 and awarded a BA degree history from Tufts University in February 1948.

He was an executive for Equitable Life Assurance Society of the U.S. based in New York City and retired in 1983 after 35 years of service. He was a career consultant and project director for Right Associates (NYC) and retired in 1993 after 10 years of service. Currently resides in Rossmoor (Jamesburg), NJ with his wife, Gertrude, and has four daughters and eight grandchildren.

WILLIAM SHELTON PRITCHARD JR. (LT) was born on Dec. 24, 1924, in Birmingham, Jefferson County, AL. Enlisted on Dec. 23, 1942, entered the Navy V-12 Program at Howard College (Samford University) then went to Northwestern Midshipman School. He spent one year of sea duty on the USS *Willard Key* 775, two years on the USS *Brinkley Bass* 885 and participated in the Korean Campaign.

Discharged on July 10, 1946, and again on Oct. 6, 1952. His awards include the Korean Service Ribbon, UN Ribbon, Japanese Occupation Ribbon, American Theater of War Medal, Asiatic Pacific Area Medal and WWII Victory Medal.

Married Ann Adams and they have three sons: William Shelton III, Franklin Adams and Thomas McCoy. He is currently practicing law in Birmingham, AL and was a past president of the Birmingham Bar Association.

JULES RICHARD "DICK" PRIMM (CAPT) was born on May 4, 1923, in Bogota, NJ. He enlisted in the USN as an apprentice seaman on Dec. 12, 1942, and completed boot camp at Great Lakes, IL. This was followed by Diesel School, Navy Pier Chicago; on completion he was assigned to the USS *Atherton* (DE-169) doing convoy duty in the Atlantic; promoted to motor machinists 2nd class and selected for officer training in the Navy V-12 Program at Stevens University, Institute of Technology, March-October 1944. He transferred to Denison University from October 1944-February 1945; then transferred to NROTC University of New Mexico, Albuquerque, NM, March 1945-June 1946.

Commissioned in June 1946 and ordered to the USS *Henley* (DD-762) serving as supply officer in various billets ashore and afloat. His memorable experiences were his entire Navy career, some experiences were more memorable than others.

Retired on Aug. 1, 1973, with the rank of captain and commanding officer, Naval Supply Center, Charleston, SC. His awards include the Navy Good Conduct Medal, EAME Campaign Medal, WWII Victory Medal, Navy Occupation Service Medal, Expert Rifleman Medal, Armed Forces Expeditionary Medal, Joint Service Commendation Medal, American Campaign Medal, National Defense Service Medal with Bronze Star and Meritorious Service Medal.

He moved to Charlotte, NC and became manager of a warehousing distribution company and first director of Charlotte Foreign Trade Zone 57. Married Helen Catherine Primm and they have four children: Julie, Camille, Mark and Richard. He retired in January 1990 and became the president of his own international business consulting firm, Prima Enterprises.

ROBERT HODGES PROUTY (SUP SGT [QM]) was born on March 31, 1923, in Rolla, MO. He enlisted in the USMC on Oct. 31, 1941, and served one year in the clothing room in the recruit depot and one year in QM office at Camp Elliott, both in San Diego. He then spent 14 months in HQ, FMF, Pacific G-4 unit at Pearl Harbor and Camp Catlin. He entered the Marine Unit of V-12 at Purdue University from fleet in November 1944. He was discharged on Nov. 10, 1945.

His most memorable experiences occurred while serving as squad leader for the nine black V-12'ers in the USMC. He will never forget observing John E. Rudder consistently saying grace before each meal and Fred C. Branch's remarks following finger nail inspection "They couldn't tell if my nails were clean, because I am black."

Retired from the U.S. Dept. of Education, Regional Office, Kansas City, MO, and lives in Eldon, MO, about 15 miles from the Lake of the Ozarks, a popular vacation area in the Midwest.

EUGENE R. PUCKETT (LT COL) was born on Jan. 8, 1927, at Parris Island, SC. He entered the Navy V-12 Program at Duke University on Jan. 8, 1944. As an NROTC student, he graduated from the University of Louisville and was commissioned to active duty in the Marine Corps in June 1947.

Served on shipboard duty for the first time in later years, as a Marine, and served in Korea and Vietnam with HQ USMC and joint service duty along the way.

Retired on June 20, 1974, and has since established himself in data processing as a second career.

He and his wife, Wanda, live in San Diego, CA and have three grown and successful children scattered throughout the southwestern U.S.

WILLIAM MUNDEN PUTMAN (SK2/C) was born on March 12, 1925, in Midlothian, TX. He graduated from Miles High School, Miles, TX in 1942; enlisted in the USN in February 1943, taking boot camp at San Diego; and Storekeeper School at Toledo, OH. He served with the Supply Depot, Navy 140, Epiritu Santos, New Hebrides, and attained the rank of storekeeper 2/c.

After being accepted in the V-12 Program, he returned to the States, and attended Mississippi College, Clinton, MS, three semesters, and NRTOC Unit at Duke University, Durham, NC. He was discharged from the Navy in January 1946.

Received AB degree in accounting from Duke in 1947, employed by Ernst & Ernst, Ft Worth, TX, and became a CPA in 1949. He enrolled in

Lutheran Seminary, Capital University, Columbus, OH, 1950; BD, 1954, and after ordination, served Lutheran parishes in Victoria, Weslaco, Texarkana and Houston, TX. He retired in 1990.

Married Sarah Haynes and they reside in Houston, TX. They have three children: Marianne Gould, Rockport, TX, band teacher; Mark Putman, Albuquerque, NM, technical director, Drama Department, UNM; William David Putman, Ramstein AFB, Germany, electronics tech; three grandchildren and two step-grandchildren.

JAMES L. QUALLS (LT COL) was born on April 6, 1923, in Putnam, TX. He enlisted in the USMC in November 1942 and was called to active duty on July 1, 1943. He entered the Navy V-12 Program at North Texas Agricultural College; attended SOCS at Camp Lejeune in 1944; and Junior School at Quantico in 1951.

He served with I Co., 9th Regt., 3rd Mar. Div. in 1945 and with 1st TK BN, 1st Mar. Div. in Korea, 1951-1952. Wounded at Iwo Jima on March 20, 1945.

His memorable experience was receiving notice of graduation from Abilene Christian University with a BS degree in business administration while in training at Camp Lejeune in June 1944.

Retired from the Reserves on April 6, 1983, with the rank of lieutenant colonel. His awards include four Battle Stars for Korea, Bronze Star and Purple Heart.

Married Ruth and they have three daughters: Vickie Norris, Donna Stone and Janis DiPaolo; and eight grandchildren. Retired, he enjoys playing golf. He spent 34 years in accounting and is now a preacher at the Church of Christ in Beaufort and to recruits at Parris Island, SC on Sunday mornings.

EDWARD C. RAFFERTY joined the USN in June 1944 and entered the Navy V-12 Program in June 1945. He was discharged in June 1946 and received his BEE degree in electrical engineering from Cornell in 1950; MS degree in mechanical engineering from Lehigh in 1959 and is a registered New Jersey professional engineer.

Joined Ingersoll Rand Co. in 1953 and held several management, engineering and manufacturing positions. He was general manager of the Axi Div. and vice president, operations, worldwide, of the Turbo Div. before retiring in May 1985. His work involved the application and op-

eration of high pressure centrifugal compressors driven by aircraft derivative gas turbines. This work involved traveling to remote areas of the world.

He has been president of the Pohatcong, NJ School Board, vice-president of the Pohatcong Planning Board, president of the Pohatcong Playground Assoc., Board of Trustees of Easton (PA) Hospital, past grand knight and past faithful navigator of Warren Council, Phillipsburg, NJ, Knights of Columbus.

Married Marion over 42 years ago and they have five sons, a daughter and eight grandchildren. They spend their time in volunteer work and traveling. They have visited over 75 countries.

CARL V. RAGSDALE (CAPT) was born on May 16, 1925, in Illmo, MO. He attended Washington University, St. Louis, MO, June 1942-December 1943. He enlisted in the Navy V-5 Program in December 1943 at Denison University, OH and transferred to the V-12a Program. Commissioned an ensign at USNMS Ft. Schuyler, NY, July 1945; served in CINCPAC Staff, XO, Pacific Fleet Camera Party, O-in-C Western Pacific Fleet Camera party, 1945-1947. Was photo officer, Atlantic Fleet Camera Party, 1948-49. Returned to the Naval Reserve and graduated from Denison University in June 1950.

Recalled to active duty for Korea, 1950-1953 (seven Battle Stars on two tours) onboard the USS *Eldorado* (AGC-11) and Pacific Fleet Combat Camera Group and as director and producer at the Naval Photographic Center. Returned to the Reserves and worked as a director and producer in the motion picture industry in New York City, 1954-1976.

He was a pioneer in the production of television commercials, and was awarded every major motion picture award for excellence, including the Academy Award Oscar in 1966. He was nominated for another Oscar in 1967. He founded Carl Ragsdale Associates, and owned Sun Dial Films, Inc., a division.

Moved to Houston, TX in 1976 and retired from the Navy as a captain in May 1985 with 41 1/2 years of active duty and Naval Reserve service. He was awarded the Navy Meritorious Service Medal, Navy Commendation Medal (two awards), and the Silver Anvil Award from the Public Relations Society of America for the best community relations program of 1983.

Captain Ragsdale is still active in Naval affairs and video productions in Houston, where he currently lives with his wife, Dr. Diane Ragsdale. His son, John, lives in Andover, NJ and his daughter, Susan Nevers, lives in Glen Allen, VA.

DARELL S. RANK (ENS) was born on March 11, 1926, in Reading, PA. He enlisted in the USN on May 3, 1944, entered the Navy V-12 Program at Swarthmore University and NROTC at the University of Pennsylvania.

He was discharged on June 27, 1946. Married

Gloria E. Dercole and they have a son Kevin. He is retired.

CHARLES E. RATLIFF JR. (CDR) was

born on Oct. 13, 1926, in Morven, NC. He entered the Navy V-12 Program at the University of North Carolina, Chapel Hill, from Davidson College in July 1944. He was commissioned an ensign at Navy Supply Corps School, Harvard University, November 1945.

Attended Aviation Supply Officer School, NAS, Jacksonville, and served until August 1946, as aviation supply officer, accounting officer and briefly as acting Protestant Chaplain, NAS, Midway Island. He retired from the USNR as Commander.

Graduated from Davidson College in June 1947; received his Ph.D. in economics from Duke University, 1955; has been on faculty of Davidson College since 1947, became Kenan Professor Emeritus of Economics in 1992. He was Gold Medalist, CASE 1985 Professor of the Year; received Hunter-Hamilton Love of Teaching Award in 1992; was author of books and articles on public finance and development; professor of economics, Forman Christian College, Lahore, Pakistan, 1963-1966 and 1969-1970.

He and wife, Mary Virginia Heilig, have three children: Alice Ratliff, faculty, UNC School of Law; Katherine Ratliff, faculty, Belmont College; and John Ratliff, director of athletics, University of North Florida.

FRED J. RAVENS JR. entered the Navy V-12

Program on June 28, 1944, at the University of Washington and the Navy V-5 Program on June 30, 1946. He returned to school in 1947 and graduated with a BS degree in electrical engineering in December 1947. Later obtained MS degree in engineering management from Northeastern University.

His career started with Kellex Corp. and Jackson & Moreland. Was later employed by the Corps of Engineers for 35 years, retiring as chief of the design branch.

In 1985 he joined The Maguire Group, an international engineering organization and traveled world-wide for the company. He retired in 1989 as vice president and director of design. He is now a facility manager and resides on Cape Cod at Falmouth, MA, with his wife, Catherine. They have six children, including a dentist, a doctor and two Ph.D. candidates, one in engineering and the other in history.

EUGENE T. RAYMOND (LT) entered the

Navy V-12 Program at the University of Washington on July 1, 1943, and received his BSME degree in October 1944. He completed studies required for shipboard engineering officers at the USS Prairie State Engineering School at Columbia University Midshipman School and was commissioned an ensign on March 5, 1945.

130

Served as assistant air ordnance officer on the carrier *Lexington*, leading 150 men arming 96 airplanes. He was awarded three service medals and the Presidential Unit Citation for the *Lexington's* role in the South Pacific leading to the defeat of the Japanese Empire. Between 1949-1952, he served as ordnance and gunnery officer in a Weekend Warriors Air Patrol Sqdn. and was promoted to lieutenant.

In July 1946, he joined the Boeing Co., beginning a 42 year career in the development of aircraft hydraulic, mechanical and actuation systems. He was a senior group engineer and had a leadership role in several related technical societies. Since retirement, he has done some consulting and authored the book *Aircraft Flight Control Actuation System Design* published in 1993 by SAE International.

FRANK M. REAVES (ENS) entered the Navy

V-12 Program at the University of Louisville on July 1, 1943, receiving a bachelor of mechanical engineering degree. He was discharged on Aug. 1, 1946, as an ensign.

Returned to school in 1946 and received an MA degree from the American University in 1965. He joined R.M. Thornton, Inc., contractors in 1948, where he remains in 1994, in various positions from estimator to chairman of the board. Thornton specializes in commercial industrial institutional construction and remodeling and particularly complex mechanical systems.

Resides in Potomac, MD with his wife Jane. They have two sons, one is president of Thornton and the other is a playwright in New York City. Their daughter is an assistant project manager with Thornton.

EBERHARDT RECHTIN (ENS) joined the

Navy V-12 Program at the California Institute of Technology in 1943, graduating with a BS (EE) in March 1946. He was commissioned as an ensign (USNR) in June 1946, served briefly on the *Columbia*, left active duty for the volunteer Reserve shortly thereafter, and resigned in 1958 when the reserve was disbanded.

In 1946, he returned to CalTech for a Ph.D. (EE) cum laude in 1950, joined CalTech's jet propulsion laboratory and became an assistant director, directing NASA's deep space network.

In 1967 he became director of the defense advance research projects agency, then in succes-

sion, chief engineer of Hewlett-Packard, presiden of the Aerospace Corp. and professor of engineering at the University of Southern California, h present position.

He received the Navy's Distinguished Publi Service Award in 1983 for technical contribution as chairman of the Chief of Naval Operation CIACT and the National Research Council's Na val Studies Board. He is a member of the Nationa Academy of Engineers and a Fellow of the AIAA IEEE and IES.

PAUL EDWARD REEVES, MD was bor

on June 25, 1925, in Wheeling, WV. He entere the Navy V-12 Program at Colgate University March 1943. He entered the University of Roche ter Medical School in 1945 and was discharge from the Navy during his first year of medic school in 1946.

Started active duty in the U.S. Army in 195 never saw combat duty, but served in the occupa tion forces in Germany, 1951-1953. He was dis charged from the USAR in 1955.

Did his internal medicine residency in A bany VAH from 1953-1955, and started his pr vate medical practice in 1955. Memorable exper ence was being summoned to the 7th Army Inspec tor General in 1952, at Stuttgart, regarding prob lems in the 1st Med. Bn.

He has one son, three daughters and fou grandsons. Retired from his medical practice an now works part-time in a 19th Century Museum

EDWARD E. RENFRO III (RADM) entere

the Navy V-12 Program on July 1, 1943, at Carro College, Helena, MT. He was commissioned midshipman and ensign at Harvard Graduat School of Business Administration, followed b service in South Pacific on the USS *Rendov* (CVE-114). In early 1946 he converted to USI upon assignment to complete his master's degre in the first post war class at Harvard Graduat School of Business Administration.

Served 32 years and retired in 1975 with th rank of rear admiral. Principal proficiencies, nuclea power (early member of Adm. H.G. Rickove staff), shipbuilding and logistics, with time out fo combat duty in Korea and Vietnam.

Following Navy retirement, he spent six year as a member of the faculty and director of researc at Georgia Institute of Technology and 10 years a director of nuclear operations, Florida Power Corp He fully retired in 1992 and currently resides i Crystal River, FL with Shirley Ann, his wife of 3 plus years.

J. ALLEN REYNOLDS JR. was born o

May 1, 1925. After his freshman year at Aubur University and his acceptance for the V-5 Pro gram, he was initially assigned to Millsaps Co lege V-12 Bn. July 1, 1943-1944. Served as battal ion commander and tapped by the ODK and Et Sigma Scholarship Society. Completed USNRMS at Columbia University and commissioned Oc 26, 1944. Subsequent service was largely min warfare afloat aboard YMSs in the Atlantic, Pacific and Far East. Released from active duty on Aug. 12, 1946.

Attended University of Alabama, 1946-1950, and received his AB degree in mathematics and LLB degree. He had a 30 year career with First Alabama Bank of Montgomery, with the last 20 years as EVP and director. He was selected as Alabama's Outstanding Young Banker in 1963.

Holds memberships and recognitions in numerous civic and professional affiliations, including Silver Beaver Awd. and listings in various *Who's Who* publications. Retired in Montgomery, AL, and enjoys travel, roses, fishing, hunting, University of Alabama football and volunteering.

Married Jane St. Clair Ball of Montgomery, AL on Sept. 13, 1952, and they have three children and five grandchildren.

JOHN MARVIN REYNOLDS (LTJG) was
born on June 25, 1924, in Washington County, VA. He entered the Navy V-12 Program at Emory & Henry College and Midshipman School at Columbia University. He served in the USS *Little Rock*.

Memorable experience was touring South America in the USS *Little Rock* as good will ambassador for President Truman. He was discharged in June 1946 and again in June 1953.

Married and has two children and two grandchildren. He is a retired university dental professor.

ROBERT M. RHOADES (LTJG) was born
on Dec. 5, 1921, in Cleveland, OH. He joined the USNR on Nov. 27, 1942, and entered the V-12 Program at Purdue University.

Served on the USS *Pillsbury* (DE-133) and participated in the sinking of the U-546 at 1844 hours on April 24, 1945, after the sub had sunk their sister ship the USS *F.C. Davis* (DE-136) at 0855 hours. They took surrender of U-858 on V-E Day.

He was discharged in February 1946.

Married and has five children and 12 grandchildren. He is retired.

CHARLES C. RICH (LTJG) was born on Dec.
8, 1922, in Cincinnati, OH. He was raised in Springfield, OH and graduated from Wittenberg University with an AB degree in 1945. On July 1, 1943, he entered the Naval Reserve and attended the V-2 unit at Denison University, Northwestern University Midshipman School, Small Craft Training Center, Miami and Antisubmarine Warfare School, Key West.

Served aboard the mine sweeper, USS *Clamour* (AM-160) in the Western Pacific, was released to inactive duty in August 1946 and retired from the Reserve as lieutenant (jg) in 1966.

Following active service he completed graduate study in geology at Harvard, receiving MA degree in 1950 and Ph.D. degree in 1960. During 1951-1956 he did field work and taught in New Zealand. He also did field work in Antarctica, Alaska and the western U.S. From 1958-1992 he taught geology at Bowling Green State University and retired as professor emeritus. He is married and has one son and one daughter.

PAUL A. RIEPMA (LT) was born on Feb. 22,
1924, in Kalamazoo, MI. He entered the Navy V-12 Program in July 1943 at Western Michigan University, Kalamazoo, MO. Transferred in February 1944 to the University of Illinois V-12 unit, Champaign, IL, graduating in June 1945 with a degree in electrical engineering. He was commissioned an ensign at Notre Dame Midshipman School in November 1945, followed by sea duty aboard the USS *La Salle* (AP-102), providing military transport to and from Japan, China, the Philippines, etc. in the aftermath of WWII. He was released to inactive duty in June 1946.

Recalled for the Korean War and served two years as lieutenant on the communications staff of the Chief of Naval Operations at the Pentagon, Washington, DC.

His civilian career consists of nearly 40 years with Consumers Power, Michigan's largest utility company. He has held a variety of middle management positions at several Michigan cities including the general office at Jackson, MI. He retired in 1985.

Married Gloria over 46 years ago and they reside in Vero Beach, FL in the winter and Lake Michigan in Muskegon, MI in the summer. They have four children, nine grandchildren and one great-grandchild.

IRA W. RIMEL (WES) (LTJG) was born on
Jan. 10, 1921, in Wibaux, MT. He enlisted as a seaman in September 1942, trained at Faragut, ID (storekeeper). Entered V-12 at Montana School Mines in Butte, November, 1943; transferred to the University of Washington, Seattle, Supply, July 1944; transferred to Harvard Graduate School Business, November 1945; commissioned February 1946; (paymaster), Naval Prison, Norfolk; and released to inactive duty in August 1946.

Recalled during the Korean War in September 1950 and put three frigates in commission at Japan, two to South Korean Navy and served in

third, the USS *Burlington* as supply officer. Served at the Navy Supply Depot, Clearfield, UT, industrial relations, and wrote a civilian personnel manual. He was released to inactive duty as lieutenant (jg) in September 1951.

Retired as a real estate negotiator and appraiser in 1986 and spent 30 months as technical writer of military equipment. He has sold writings since the age of 20; is currently writing a column *Fishing and Hunting News* for magazines such as the *Montana Journal*; and he expects his biography and history book to be on the market in 1996-1997.

DAVID D. RING (LTJG) was born in February
1925, at Cedar Rapids, IA. He entered the Navy V-12 Program in July 1943 at Park College, MO and completed his BSE degree in math at the University of Michigan. He was commissioned at Columbia Midshipman School and served in the Aleutian Islands until August 1946 when he was released to inactive duty and discharged as lieutenant junior grade.

Graduated from the University of Michigan Law School in June 1949; moved to San Francisco and continued in law practice. He is now senior partner in Fabris, Burgess & Ring, a general practice law firm, with a second office in San Diego.

He and his wife live in Tiburon. In 1980 they flew their own airplane back to Kodiak and found that the town and Navy facilities of WWII now exist only in photos in the museum. In 1990 they flew their airplane around the world for their 40th wedding anniversary. They have three children, all graduates of the University of California, and five grandchildren.

HAROLD M. RISENER (CDR) was born on
April 3, 1924, in Greenville, TX. He enlisted in the USN on Nov. 27, 1942; entered the Navy V-12 Program at North Texas Agricultural College; followed by Midshipman School at Northwestern University.

Stationed in the USS LST-605; USS LST-459; USS *Electron* (AKS-27); and USN Station, New Orleans, LA. Participated in at Leyte, Mindoro, Luzon and Okinawa operations. During WWII, his ship, LST-605, destroyed 10 enemy aircraft. He became CO of his own ship, LST-459, at the age of 21.

Retired April 3, 1984, with the rank of commander. Awards include the Navy Unit Commendation, Philippine Liberation Medal with three stars, Asiatic-Pacific Area Campaign Medal with four stars, American Defense Service Medal, WWII Victory Medal, Navy Occupation Service Medal,

Korean Service Medal, UN Service Medal and Naval Reserve Medal.

After his active duty in the Navy, he was in the fire sprinkler business in San Antonio for over 30 years. His firm, Texas Automatic Sprinkler Co., was one of the largest independent sprinkler companies in the U.S. He retired from business in 1989 and is now enjoying golf and the retired life.

He married Bennie L. Risener over 47 years ago. Is or has been member or officer in numerous business, professional and social organizations.

ALBERT GEORGE RISTAN (CDR) was born on June 12, 1924, in Passaic, NJ. He entered the Navy V-12 Program at Princeton University in July 1943; transferred to Wharton School at the University of Pennsylvania in 1944; and was commissioned an ensign in December 1945 at Harvard University. Transferred to the Regular Navy while stationed at Guantanamo Bay in 1946.

Served at shore activities in Guantanamo, Argentina, Newfoundland, Paris, France, Oakland, Mechanicsburg, Norfolk, New York and aboard the USS *Antares* (AKS-33) and *E.G. Small* (DD-838) during the Korean War.

Retired in 1967 as a commander in the Supply Corps and began a second career in university planning at the University of Maryland at College Park. Retired a second time as assistant athletic director at Maryland in 1989. Presently resides in Adelphi, MD and has a son, two daughters and five grandchildren.

DAVID W. ROBERTS (LTJG) was born on Dec. 4, 1924, in Amherst, MA. He entered the Navy V-12 Program at Tufts University and transferred to Ft. Schuyler Midshipman School where he was commissioned on July 3, 1945. Military stations included flight training at NAS, Dallas, TX; Corpus Christi; Pensacola; and Quonset Point.

His memorable experience was carrier qualification aboard the USS *Saipan*. Discharged on Oct. 1, 1956, his awards include the WWII Victory Medal and American Area Campaign Medal.

Married, he has three daughters and six grandchildren. He graduated from the University of Massachusetts in 1950 with an ornamental horticultural major. He is now retired.

JAMES W. ROBERTS (CAPT) enlisted in the USN and entered the V-12a Program in 1944. He flew combat missions in Korea from the USS *Bon Homme Richard* as squadron commander of VA-164; in Vietnam from the USS *Oriskany*.

Received BS degree from the University of Minnesota and a MS degree in political science from Auburn University. He

graduated from the Armed Forces Staff College in 1963 and the Air War College, Maxwell AFB in 1970.

Awarded the Distinguished Flying Cross, numerous Air Medals, Purple Heart and various additional awards over 31 years of active duty.

After retirement in 1975, he was a financial advisor in Orlando, FL. Presently active as an advisor and area coordinator for the Florida State Golf Association.

He and his wife Jean live in Winter Springs, FL. They have two daughters, Susan Lee Handley of Rochester Hills, MI and Karen Ann Smith of Clearwater, FL.

WAYNE A. ROBERTS (ENS) was born on Feb. 17, 1924, in Los Angeles, CA. He enlisted in Naval Air Corp in November 1942. He could not pass the eye test when called to active duty in March 1943 and was assigned to company 43-104, San Diego Naval Training Station for boot camp. Selected for the V-12 Program near the end of boot camp and initially stationed at the University of the South, Sewanee, TN, July 1, 1943.

Transferred to California Institute of Technology, Pasadena, CA on Nov. 1, 1943; received BS degree in engineering in June 1945 and midshipman training at Notre Dame. He was commissioned an ensign in October 1945, with additional training at Harvard School of Business. Served temporary duty at Treasure Island; sailed to Japan on the *General Butner* (AP) as a troop officer; assigned to fleet activities, Yokosuka, Japan; served as flag secretary, (Commodore Briggs); administrative aide to chief of staff (Capt. Milton H. Anderson); and subsequently to chief of staff (Capt. Anthony L. Rorshach). Returned to CIT in October 1946 in geophysics and received MS degree in earth sciences in 1948.

He has many memories of the V-12 Program and other assignments. Of those related to the V-12 Program, probably the most significant is related to his initial assignment and subsequent transfer. He was scheduled to be sent back to CIT, which he had attended prior to enlisting in the USNR, but two typographical errors in his records indicated that his major was American history (instead of biochemistry) and that he was to be sent to the University of the South instead of Milligan College, near Knoxville, TN. The correspondence with BuPers concerning his apparent mis-assignment and the subsequent transfer indicated the goodwill and fairness inherent in the USN.

On his trip to Japan and his assignment with fleet activities, the *General Butner* passed within 50 miles of a typhoon center. He stood on the upper deck, about seven or eight feet above the main deck which had 33 or 34 feet of free board, and estimated the wave height to be at least 80 feet crest-to-crest. This typhoon sunk destroyers to the west that had lost steerage way. He still remembers the look of the waves and the flexure of the *Butner* as they drove diagonally through the waves.

Married LaVerne Matthews in 1948 and they have five children: Bruce, Wayne, William, Candy

and Valerie. He worked for the U.S. Geological Survey, Douglas Aircraft, Climax Uranium, Boeing Aircraft, Dennison Mines, Uravan Minerals, Strauss Exploration, Moore & Tabor, Atomic Energy Commission (later Energy Research and Development Agency, Dept. of Energy) and retired to Seattle in 1985.

LLOYD EMERY ROBIDEAUX JR. (CDR) was born Nov. 2, 1923, in Bend, OR. He entered the University of Idaho Southern Branch V-1 Program on July 1, 1943, and was commissioned an ensign on Dec. 14, 1944, at Columbia Midshipman School in New York, NY.

Commissioned LSM(R)-529 at Houston, TX in September 1945 and served as communication executive and commanding officer. He was released from inactive duty at Seattle, WA in July 1946.

Attended Oregon State University, Corvallis and received his BS degree in science in June 1948. Recalled to active duty in July 1951 and assigned to USS *Greer County* (LST-799) as commanding officer. In July 1954 he transferred to the USNR Center in Las Vegas, NV and served as its commanding officer. He was released to inactive duty in March 1958 and retired from the USNR in November 1983 as CDR.

Was mathematics teacher and department chairman at Shafter High School, Shafter, CA until his retirement in June 1986. Currently resides in Bakersfield, CA with wife Christy.

H. CARTER ROBINSON (CDR) was born on May 17, 1926, in Augusta, GA. Entered the Navy V-12 Program at Tulane University and was commissioned on May 17, 1943.

Stationed in the USS *Sabine* and the USS *Currituck*. Discharged in 1947 and again in 1952.

Married Nancy and they have a son Clay.

REMBRANDT C. ROBINSON (RADM) was born on Oct. 2, 1924, Clearfield, PA. He attended Pennsylvania State University before enlisting in the USNR on June 5, 1943. Appointed midshipman and attended the Naval Reserve Midshipman School on board the USS *Prairie State*. Commissioned ensign in the USNR on Oct. 26, 1944, served in LST-485 and saw action in the Pacific Theater during WWII, including the invasion of Okinawa. Transferred to regular Navy in 1946 and served successively in LST-1032, LST-601, and LST-912.

In June 1948, he joined the USS *English* (DD-896) as engineer officer. Received the Bronze Star with Combat V for meritorious service.

Assigned at the Bureau of Naval Personnel from 1952-1954; served as XO of the USS *Walke* (DDE-517) until July 1956; attended the Armed Forces Staff College and became assistant head of the Command Policy Section, Strategic Plans Division, Office of the Chief of Naval Operations in January 1957. Assigned for fitting out of the USS *Charles Berry* (DE-1035), taking command in November 1959. He commanded the USS *Bradford*

(DD-545) until January 1962 when he returned to the Office of the Chief of Naval Operations, in the Strategic Plans Division.

Received a BA degree in social sciences from George Washington University in 1964 and was executive assistant and aide to the CINCPAC. Took command of Destroyer Squadron 31 in September 1968 and awarded his second Bronze Star for meritorious achievement on Yankee Station. Assigned March 1969 to the chairman's staff group, Office of the Joint Chiefs of Staff where he was promoted to flag rank.

In July 1971 RADM Robinson assumed command of Cruiser Destroyer Flotilla 11. He was killed on May 8, 1972, in a helicopter crash in the Gulf of Tonkin during a landing approach to the USS *Providence* (CGL-6) while coordinating the first surface bombardment of the port of Haiphong, North Vietnam during the Vietnam War.

RALPH EUGENE ROBISON (LTJG) was
born on Aug. 23, 1925, in Carbondale, IL. He attended the University of Illinois V-12 Program, June 1944-June 1945, and was named the university's outstanding civil engineering graduate. He went to Midshipman School at Camp Endicott, RI; served in Okinawa, October 1945-August 1946, and supervised hospital construction in the Pacific during WWII.

Received his master's degree in engineering and joined the Kansas City office of Howard Neddles Tammen and Bergendoff Inc. Transferred to the company's Cleveland office in 1956 and eventually became vice-president of the corporation.

He won many awards such as the Greater Cleveland engineer of the year by professional societies. He is an avid golfer; collector of buffalo nickels; and a reader of historical figures. He was discharged Sept. 6, 1956.

Robison passed away on Aug. 28, 1993. He is survived by his wife Evelyn; children: Dr. Richard E. of Sierra Madre, CA; Kevin S. of North Olmsted; Rebecca Field of Stratford, NJ; Dr. C. Jeanne Robison of Siloam Springs, AR; Judith DeSanits of Pittsburgh; Catherine Ranney of Canton; Karen Hitt of the Marshall Islands; and 15 grandchildren.

WESLEY B. ROOT JR. (LTJG) was born on
July 16, 1925, in Baton Rouge, LA. He enlisted in the USNR on July 1, 1943; attended Tulane University Navy V-12 Program and Midshipman School at Ft. Schuyler, NY.

Discharged on Aug. 12, 1946, with the rank of lieutenant (jg). His awards include the American Theater of War Campaign Medal, Asiatic-Pacific Area Campaign Medal and WWII Victory Medal.

He and wife Marilyn have four children and six grandchildren. He has an engineering degree from Tulane University and his master's degree from LSU. Employed by Exxon Corp. from 1947-1986, he is now retired and enjoys his family, camping in Tunica Hills, St. Francisville, LA and traveling.

R.D. RORABAUGH (ROCKY) (2/LT) was
born on May 21, 1923, in Bellaire, KS. He enlisted in the USMC on Dec. 4, 1942, and entered Bowling Green State University, OH, July 1, 1943. He played in the National Invitational Basketball Tournament in Madison Square Garden with Bowling Green State University and was a hurdler on an undefeated BGSU track team. He was commis-

sioned a 2nd lieutenant on March 28, 1945, at Quantico, VA.

Served as platoon commander at Camp Lejeune, NC and in a security detachment in Trinidad, British West Indies for nine months as recreation officer, post exchange officer and training officer. Graduated from Ft. Hays State University, Hays, KS, in June 1947 and received his master's degree from the University of Wyoming in August 1952. He was discharged at Great Lakes Naval Station on June 26, 1947.

He was named to the Ft. Hays State University Athletic Hall of Fame; served as athletic coach and teacher for nine years; as school administrator (superintendent of schools) for 31 years and retired in June 1986.

Married to Mija for 45 years when she passed away. He is now married to Joyce and they reside in Hays, KS. He has four sons: Steve a computer programmer in Ft. Collins, CO; Chris a athletic coach and teacher in Ellis, KS; Kevin a city mail carrier in Russell, KS; and David a police officer in Hays, KS.

RICHARD G. SALOOM, MD (CAPT) was
born on March 7, 1924, in Lafayette, LA. He entered the Navy V-12 Program on July 1, 1943, at Southwestern, LA and transferred to Tulane University Medical School in February 1944. He was released from inactive duty on Jan. 17, 1946.

Graduated with a MD degree in 1947 and served as flight surgeon SAC USAF 1952-1954 with rank of captain. He is still active in family practice in Lafayette, LA and serves as senior FAA medical examiner.

He has three sons, two of whom are physicians, and two daughters. He was fortunate to have been a recent guest on the USS *George Washington* (CVN-73) and experienced the great thrill of tailhook landing and catapult takeoff.

CHARLES S. SANDERS (TM2/c) was born
on Aug. 10, 1922, in Washita, CO. He enlisted in the USN on Dec. 28, 1941. It was his intention to become an naval aviator, but instead he was sent to boot camp at Great Lakes Naval Training Center. Following three weeks of immunization procedures, he was shipped to Pearl Harbor and assigned to the USS *Aylwin* (DD-355) on March 8, 1942.

The USS *Aylwin* was soon involved in the Coral Sea Battle, subsequently seeing action at Midway, The Aleutian Island Operation and nu-

merous conflicts in other Pacific Island Operations. During this time he advanced to torpedo man 2/c. In August 1944 he was transferred to pre-midshipman school at Asbury Park, NJ. Upon completing this duty, he was sent to V-12 at Arkansas A&M College for three semesters. He was then transferred to NROTC at the University of Louisville. By 1945 he had accumulated more than enough points and was discharged on Nov. 18, 1945.

His Navy experiences were memorable, at times frightening, often monotonous. However, being selected to become a part of the V-12 Program was a privilege and greatly appreciated. Also memorable was being seasick on the USS *Aylwin* for one week and being scared during the Japanese air attack at the Coral Sea Battle.

Graduated in 1948 from the University of Oklahoma with a degree in petroleum geology, worked for 37 years as an exploration geologist and retired in 1986. He now lives in Santa Fe, NM.

Married and has three children and five grandchildren, he is presently retired.

KENNETH M. SASSEEN (XO) entered the
V-12 Program at the University of Redlands, CA on July 1, 1943; graduated from Cornell University Midshipman School on Jan. 24, 1945 and was assigned to the amphibious training base, Coronado, CA.

Received orders to the USS *Andromeda* (AKA-15) in May 1945, but spent approximately three months between islands and temporary duty aboard various ships (including a destroyer, cruiser, LST and aircraft carrier) before catching up with the *Andromeda*. He became executive officer of the USS *LCS*(L)-21 in 1946 and was discharged in July 1946.

Returned to the University of Redlands, where he received a BS degree in mechanical engineering and physics in 1949. He was project engineer with the Fluor Corp; and later took the opportunity to become a partner in a small engineering/construction company, involved with petroleum process and production systems. After the company was sold, he formed his own engineering/construction company, until his retirement in 1990.

FRANCIS FIELD SAUCIER (FRANK)
(DECK OFFICER) was born on May 28, 1926, on a farm near Leslie, MO. He graduated from Washington High School, Washington, MO at the age of 16. A week later he was sworn into the USN on May 29, 1943, at the age of 17. He went on active duty in the V-12 Program on July 1, 1943, at Westminister College and entered the USNR Mid-

shipman School at Notre Dame on Nov. 6, 1944. Capt. J. Richard Barry, commanding officer, waived the age 19 requirement and allowed Saucier to graduate with his class on March 8, 1945, at the age of 18, making him one of the youngest deck officers commissioned in the modern USN.

His first assignment was in amphibious forces as skipper of Beach Party #76, and he served in the Philippines and Japan. His second assignment was in China aboard the USS *Mt. Olympus* (AGC-8), as deck watch officer and second division officer. He was released to inactive duty in August 1946.

Graduated from Westminster in 1948, with a BA degree in mathematics. He was in the oil and gas business and was an outfielder with the St. Louis Browns when recalled to active duty, via a collect telegram, in April 1952.

Served as battalion commander, HQ Cadet Bn, NAS Pensacola, until released to inactive duty a second time, April 1954, at which time he voluntarily retired from the Baltimore Orioles Baseball Club.

Former CEO, president and owner of Chemical Industries, Inc. and various privately held companies, he is currently a financial consultant.

He and his wife of over 43 years, the former Virginia Lee Pullen of Wichita Falls, TX, live in Amarillo, TX. He has a daughter, Sara, and a son, John, who live in Amarillo; and two granddaughters, Mary Claiborne and Alexandra. He is grateful to the USN for its positive, practical and productive effect on his life.

ROBERT L. SAUNDERS (2/LT) was born on April 12, 1923, in Opp, AL. He entered the V-12 Program at Millsaps College on July 1, 1943. After two semesters he transferred to Franklin and Marshall College and was commissioned a 2nd lieutenant, USMCR, Quantico, VA, April 11, 1945. After a brief assignment in Guam, he was assigned to the 8th Service Regt. in Sasebo, Kyushu, Japan. He was discharged in September 1946.

Recalled to active duty, 1951-1952, and again served in the 8th Service Regt., Camp Lejeune, NC. He was discharged in 1952.

Worked as a high school teacher of physics and chemistry in Tallahassee, AL, 1947-1951 and 1952-1954; principal, 1954-1956; administrative assistant, Tennessee Study of Higher Education, 1956-1957; professor and associate dean of education, Auburn University, AL, 1957-1970; and dean of education, Memphis State University, 1970-1988.

He is president of several state and national professional organizations, published in professional journals, co-authored three college textbooks and helped design the Tennessee Career Ladder Program. He retired in 1988 and is currently residing in Somerville, AL, where he writes, consults and gardens.

RICHARD R. SAVAGE (CO) was born on Dec. 12, 1925, in Brimfield, IL. Entered the V-12 Program at the University of Illinois on July 1, 1943, and transferred to DePauw University in

134

March 1944. He was commissioned an ensign on March 6, 1945, from Ft. Schuyler, Bronx, NY. He was assigned to the USS *Swenning* (DE-394) operating out of Guam. He became the last commanding officer decommissioning *Swenning* at Green Cove Springs, FL, June 1946.

Returned to the States in early 1946 going through the Panama Canal on their way to Boston for overhaul prior to decommissioning.

Received a BA degree in mathematics from DePauw in May 1948 and also had graduate studies at Northwestern and College of Insurance. After 38 years of service, he retired from Insurance Services Officer in New York as executive vice president. Subsequently he retired from The Home Insurance Co. after seven years as senior vice president. He is the author and/or co-author of most standard policy forms in the property and casualty fields of insurance. He is now offering services as expert witness (Consolidated Asbestosis, Shell Oil, Safeway Stores, etc.) and consultant (lectures at Lloyds of London, CPCU, RIMS, NAIC).

Married Patricia Dungan over 44 years ago and they have two daughters, one son and 10 grandchildren. He and his wife reside in Florham Park, NJ.

ROBERT Y. SCAPPLE (ENS) entered the California Institute of Technology V-12 Program in July 1943. He received BS degree in electrical engineering in June 1945 and commissioned an ensign at Notre Dame Midshipman School in November 1945. Served overseas at Peleliu Island (harbor master), and Guam (maintenance, yard craft) and released from inactive duty in 1946.

Attended Stanford University and received MS degree in electrical engineering in 1947 and a degree of engineer, metallurgy in 1950. He worked for Westinghouse Electric Corp. for four years doing metallurgical studies of steam turbines, ship propulsion gears, Navy rapid fire gun, subsonic and transonic wind tunnel compressors and nuclear submarine development.

Spent 34 years with Hughes Aircraft Company. He had numerous patents, awards, technical papers in materials research and development for airborne radar systems and space programs. He was key engineer in establishing Hughes' first R&D Hybrid Microcircuit Facility where they developed all hybrid materials technology for Navy F-14 Program, and subsequent programs using hybrids; as manager of Hybrid Microcircuit Facility; manager, materials technology laboratory, responsible for most materials science and materials engineering for Hughes. He retired in 1988.

Resides in the mountain village of Wrightwood, CA with his wife Iona and Gamay Beaujolais VIII, a lovely standard poodle. They have a daughter, Karrin who is a professor of political science, Southwest Missouri State University.

ALBERT D. SCHMIDT (LTJG) was born on Nov. 16, 1925, in Alpena, SD. He entered the V-12 Program in October 1943, at Miami University, Oxford, OH. He was winning company commander and commissioned in August 1945 at Columbia University. He served on the USS *Hollandia* (CVE-97) and various LSTs in the China area. He was released to inactive service in August 1946 with the rank of lieutenant junior grade.

Graduated in June 1949 from the South Dakota School of Mines and Technology, with a BS degree in electrical engineering, with honors. His entire career was with Northwestern Public Service Company, gas and electric utility company, in South Dakota and Nebraska. He served 41 years with 25 1/2 as chief XO. Currently chairman of the board of NWPS and retired in 1990 to a new home in Sioux Falls, SD. Was former director of 12 national and regional organizations. He is listed in 11 Who's Who type directories.

Married Joyce Anderson in November 1946 and they have two children, Roxanne Rae Eiser and Janet Jay Foss and five grandchildren. His hobbies include woodworking, hunting, golf and travel.

RUSSELL CALVIN SCHMIDT (ENS) was born on May 18, 1925, in Martin, ND. Entered Minot State University and the Navy V-12 Program on Nov. 1, 1943. Commissioned an ensign in February 1946 at Harvard University, Cambridge, MA.

Served as the supply officer aboard the USS *Samuel Chase* (APA-26) transporting personnel to Pearl Harbor, Guam and Saipan. Decommissioned the USS *Samuel Chase* in Norfolk, VA and released to inactive duty in August 1946.

Earned his bachelor's, master's and doctoral degrees and taught at two Liberal Arts Colleges and one Theological Seminary. The balance of his career was spent as a congregational (UCC) clergy and as educational consultant in the Michigan Dept. of Education.

Presently resides in Grand Rapids, MI and St. Petersburg, FL since retiring in 1986. He has two daughters: Gloria Seabold and Rebecca Dobler and five grandchildren: Danielle, Chad, Derk, J.B. and Lee. An encounter with poliomyelitis in 1948 required a wheelchair for mobility. The V-12 training provided a foundation on which to build a career.

PHIL H. SCHOGGEN (ENS) was born on Aug. 28, 1923, in Tulsa, OK. He entered Park College, MO V-12 Program on July 1, 1943; commissioned an ensign in June 1944 at Navy Midshipman School, Plattsburg, NY and served as OinC of LCT-1202. He participated in the Okinawa operation from D-day until January 1946. He received a Navy Commendation Ribbon and was released to inactive duty in February 1946.

Completed his BA degree in psychology at Park College in June 1946. Returned to active duty in June 1950 for the Korean conflict. He served as OinC of a group of three LCTs, renamed Landing Ship Utilities (LSU). He participated in Inchon landings on D-day and continued until again released to inactive duty in January 1951.

Completed his Ph.D. in psychology at the University of Kansas in 1954 and held faculty appointments at the University of Oregon, George Peabody College, York University and Cornell University, Ithaca, NY. He published numerous scientific papers and two books on research in ecological psychology.

Retired as professor emeritus from Cornell University in 1990. He now lives with wife Maxine in Nashville, TN. They have four children: Leida, in San Francisco; Christopher, in Wisconsin; Ann, in Berkeley, CA; and Susan, in Nashville.

LEON J. SCHWEDA (ENS) was born on September 10, 1924, in Milwaukee, WI. He entered Marquette University, Milwaukee, WI, in the V-12 Program on July 1, 1943. Received his BSME degree and was commissioned an ensign in March 1946. He is a member of Tau Beta Pi and Pi Tau Sigma.

Served in the Atlantic on the USS *Denver* and in the Pacific on the USS *Moale* (DD-693). The *Moale* was at Bikini Atoll for the atomic bomb tests in July 1946. He was discharged in May 1947.

Spent 38 years with Bell System; 34 years with Wisconsin Bell; and four years with AT&T. His work assignments included cost engineer, accounting manager, annual charges, depreciation engineer, rate case statistics, appraisal engineer, construction budget, departmental budgets, total company budget and computer system development and/or implementation.

He took five years of accounting courses at Marquette University night school. Member of Beta Alpha Psi. He retired in April 1986 and moved to Aurora, CO in June 1986.

Married June Rose Schmitt, RN, in May 1949. His wife graduated from the Seton School of Nursing at Penrose Hospital in Colorado Springs, CO in 1948. They have three children: Nancy, Susan and James.

RUSSELL L. SEARS (LTJG) enlisted in the Navy in May 1943, and entered the V-12 Program at Brown in July 1943. He attended Midshipman School at Columbia University and was commissioned in October 1944. He served in the Pacific aboard the USS *Alderamin* (troop transport and cargo) until May 1946; in Iwo Jima and Okinawa;

and transferred to active Reserve in 1950 as a lieutenant junior grade.

Received his chemistry degree from Brown and worked in Automotive Coatings Research at Inmont Corp. He worked as technical director; international research director; vice president and general manager, R-M division until retirement in 1987.

Married Frances Straub in 1947. They have five children: Carol, Joan, Ann, Russell Jr., and Ellen; and four grandchildren. He is a licensed pilot, but has retired from flying also. He and Fran live in Perrysburg, OH and keep busy with volunteer work and traveling. He is past chairman of SCORE (Toledo Chapter).

LLOYD W. SHAMBUREK (ENS) was born in 1925 in Wisconsin. He attended the University of Wisconsin, 1942-1943, and entered the Navy V-12 Program on July 1, 1943, at the University of North Carolina. Transferred to Columbia University on July 1, 1944, and was commissioned an ensign in October 1944. Completed scouts and raider training at Ft. Pierce, FL in early 1945. His unit was assigned to Scout Intelligence School, Henry Hudson Hotel, NY. Upon completion, the unit was sent to Oahu, HI.

Entire unit received orders COMNAV Group, China, Washington, DC to be used as liaison couriers to Chiang, Peking, China. Before his turn came, V-J occurred and he was reassigned to Guadalcanal and back to Oahu in the spring of 1946. He returned to the U.S. on AOG and was discharged in June 1946.

Graduated from the University of Wisconsin with a BS degree in economics in 1948.

Shamburek has been married to Carol for 46 plus years and they have three sons. Was the owner of a school supply business for 43 1/2 years before retiring in 1992.

EDWARD S. SHERWOOD (LT) was born on May 29, 1925, in New York City, NY. While at Colby College, he enlisted in the USN and was sent to the V-12 unit at Bates College. After completing the pre-med programs, he served as corpsman at USN Hospital, Portsmouth, NH, be-

fore being sent to the University of Vermont, College of Medicine.

Received his MD in 1949 and was recalled as a medical officer during the Korean War, serving with the Marines at Quantico, VA. He was released from active duty in 1953, with the rank of lieutenant MC from the USNR.

After a pediatric residency, he practiced pediatrics in New Brunswick, NJ for over 30 years. He is now retired and lives with his wife, Lois, in East Corinth, VT.

DON HARVEY SHIRK (LT) was born on March 23, 1922, in Cleveland, OH. He enlisted in the USN on July 1, 1943, and entered the Navy V-12 Program at the Miami University, Oxford, OH. He transferred to Wellesley College, branch of Harvard, for Midshipman School. Served on aircraft carriers, one year in the Atlantic and one year in the Pacific. Participated in antisubmarine patrol in the North Atlantic and later in the Pacific Theater.

His memorable experience was pursuing and being pursued by German and Japanese submarines.

Discharged on Aug. 6, 1946, and served in the Reserves until 1957. He achieved the rank of lieutenant. His awards include the American Theater Medal, Asiatic Theater Medal, WWII Victory Medal and Atlantic Theater Medal.

Married Ann and they have a daughter Judy and two grandchildren. He is retired from Dun & Bradstreet Corp.

WALTER W. SHIRLEY entered the V-12 Program at Milligan College, Johnson City, TN on July 1, 1943. He finished Midshipman School on July 19, 1944, at Northwestern University and received amphibious training at Coronado and Morro Bay, CA. He served aboard the USS *Alamaack* (AKA-10) and the USS *Mattabesset* (AOG-52) in the Pacific until released on Sept. 10, 1946.

Received a BS degree from East Central State; MS degree from Oklahoma State and taught math at Oklahoma City University. Worked for Sohio in Oklahoma City and ARCO in Los Angeles as a petroleum engineer. He started a computer business in 1971 and retired in 1991.

He and his wife, Ruth, whom he met while in the USN, have three sons and nine grandchildren. They live in South Pasadena, CA.

DANIEL R. SIDOTI (LTJG) enlisted in the USN in November 1942 and entered the V-12 Program at Union College in July 1943. He at-

tended Midshipman School, Plattsburg, NY and was commissioned in June 1944. He trained at the USN Amphibious Base, Ft. Pierce, FL; assigned to the USS *Trousdale* (AKA-79), December 1944, and served 17 months as amphibious landing craft officer, watch and division officer, assistant communications officer and ship's secretary.

Transported troops and equipment to various areas of the Pacific Theater of Operations; participated in the Okinawa invasion. Post-war, he carried Chinese nationalist troops to northern China for combat with Communist troops.

Released to inactive duty in July 1946. His awards include the Asiatic-Pacific Campaign Medal with Bronze Star, Navy Occupation Service Medal with Asia Clasp, China Service Medal, American Campaign Medal and WWII Victory Medal.

Received his BA degree from Union College and MS degree from Stevens Institute of Technology. He worked and managed food product research and development at General Foods Corp., Monsanto and Anheuser-Busch Companies and was granted five patents. He is listed in *Who's Who in Technology* and active in professional societies.

Currently retired and working as consultant in Food Product R&D. He married Ginny and they have one daughter, Stephanie.

CHARLES E. SIGETY (ENS) entered the Cornell University V-12 Program on July 1, 1943. In 1944 he was sent to the Navy Supply Corp School at Harvard where he was commissioned an ensign and finished first in his class. He was sent to Newport, RI for the pre-commissioning of the heavy cruiser, USS *Fall River* as disbursing officer. When the atom bomb was dropped, Sigety left the ship and was decommissioned. He has degrees from Columbia University, Yale Law School and an MBA from Harvard Business School.

During his 50 year career in industry, government and education, he served as Rapporteur to the Committees on Federal Taxation for the U.S. Council of the International Chamber of Commerce and the American Institute of Accountants; as deputy federal housing commissioner of the U.S.; first assistant attorney general of New York State; director of the New York State Housing Finance Agency. He has practiced law in New York and Washington, DC and has taught at the Yale Law School, Yale College, Columbia University and Pratt Institute in New York City.

His primary career has been entrepreneurial. He founded the Florence Nightingale Health Center; a 968 bed nursing home in New York City; ski slopes in New York City on the Van Cortland Park Golf Course; Video Vittles, a food business for television, preparing foods that were seen on many nationally renowned television shows; in 1982 his family acquired the medical and surgical division of Parke Davis which became Professional Medical Products, Inc. a national disposable medical supply business, headquartered in Greenwood, SC. The businesses he has founded have grossed more than $2 billion in sales.

He has served on missions for the Overseas Private Investment Corp. in Hungary, Poland, Czechoslovakia, Russia and Germany; served on the Finance Committee of the World Games, organizing games in Santa Clara, London, Karlsruhe and the Hague.

Married to Katharine Kinne Snell, the nation's first network television cook on NBC; and has five children, four of them work in businesses he has founded. His activities are now in philanthropy and education. He has homes in Pennsylvania, South Carolina and Boca Raton, FL.

MITCHELL M. SIMMONS (CDR) was born on Dec. 18, 1925, in Mt. Airy, NC. He enlisted in the Aviation Cadet Program in March 1943 and entered the V-12A Program at Duke in July 1943. After two semesters, he was designated an Aviation Cadet, trained at various facilities and was commissioned a Naval Aviator in January 1946.

Flew SC-1 Seahawks float planes from USS *Portsmouth* (CL-102) during two cruises to the Mediterranean and became a flight instructor at Whiting Field, FL in July of 1948, joined VF-34 in F9F jets at Quonset Pt, in 1950 and made a cruise to Mediterranean aboard the USS *Essex*. When he carrier qualified in the F9F-2 aboard the *Tarawa* on April 12, 1951, he became the first Naval Aviator to make his first carrier landing in a jet aircraft, having initially qualified in seaplanes instead of carriers in operational flight training.

Following this tour, he served as assistant safety officer on ComAirLant staff, then went to Naval Air Special Weapons Facility in Albuquerque testing compatability of special weapons with the various naval aircraft. In 1956, he attended General Line School at Monterey and joined VA-113 at Miramar in December 1956. He had two tours to WestPac aboard the USS *Shangri-La* and attended the University of Colorado in 1960-1961 under the Hollaway Program, obtaining his BS degree.

Assigned as commanding officer of VA-192, he was present at the initial opening of NAS Lemoore, CA and had a tour to WestPac aboard the USS *Bonne Homme Richard*. After this successful tour, he retired in August 1963, and returned to Boulder, CO, where he went to work with the Flatiron Companies as an asphalt plant operator. These companies engaged in asphalt, concrete, sand and gravel, highway and bridge construction. In 1981, he was made CEO of the nine company organization and retired in 1986, returning to Coronado, CA.

His wife of 45 years died in 1991. He has two daughters, one in Jackson, WY and one in Spokane, WA with four grandchildren. He sails a Han Christan 41 foot cutter, plays tennis regularly and enjoys his computers.

MAXWELL OWENS SIMS (LTJG) was born on Dec. 19, 1924, in Delta, AL. He enlisted in the USN in December 1942, entered the V-12 Program at Howard College in Birmingham, AL and attended Midshipman School at Plattsburg, NY.

Stationed at Philippine Islands, Hong Kong, China and participated in action at Okinawa. H was discharged in September 1946.

Married Jo Anne and they have two daughters and seven grandchildren. He is a partner Dudley, Hopton-Jones, Sims and Freeman, CPA

WATSON S. SIMS was born on July 9, 192 in Pembroke, GA. He enlisted in the USN in 193 Served on the USS *Arkansas* before transferring Torpedo Boat Squadron Three, which was sent the Philippine Islands in September 1941. Aft his boat was disabled and sunk during evacuatic of Gen. Douglas MacArthur from Corregidor, h took part in four war patrols aboard submarine rising to radioman 1st class before being assigne to V-12 at Bates College in February 1944. Aft two terms at Bates, he was transferred to ROTC Tufts College and commissioned in June 1946.

Leaving the USN, he attended the graduat school of journalism at Columbia University an joined The Associated Press. He spent 24 years AP, with one year's time out for a Nieman Fellov ship at Harvard, serving as reporter, foreign corr spondent, bureau chief and world news editor. H served seven years as editor of the Battle Creel MI, *Enquirer* and eight years as editor of the Ne *Brunswick, NJ News* before retiring from journa ism.

He is now a resident of Rocky Hill, NJ and ha been general executive of The George H. Gallu International Institute since 1992. He and his wif the former Elisabeth Sturdivant, have a son who a medical doctor and a daughter who teaches New York State University.

FRANK S. SINISCALCHI (ENS) was bor on Feb. 14, 1923, in West Warwick, RI. He e listed in the USN on July 1, 1943, entered the V-1 Program at Brown University and attended Mi shipman School at Columbia University. He w stationed on the USS *Brookings* (APA-140), serve in the Asiatic-Pacific Campaign and Philippir Liberation.

His memorable experience was graduatic from Midshipman School.

Discharged on March 6, 1946. His awarc include the Philippine Liberation Ribbon, WW Victory Ribbon, American Area Service Ribbo and Asiatic-Pacific Area Service Ribbon.

Married Katherine and they have four chi dren, two step-children and four grandchildre He is a retired osteopathic physician.

MELVIN E. SINN (ENS) entered the Navy V 12 Program at Brown University on July 1, 194

Commissioned an ensign at Columbia University Midshipman School in 1944; served as instructor at Pre-Midshipman School, Asbury Park, NJ; and ASW training at Key West. He was assigned to the USS *Roper* (APD-20) which was hit by a suicide plane in Okinawa and decommissioned. Assigned to USS *Horace A. Bass*; served occupation duty in Japan and was released to inactive duty in March 1946.

Received his AB degree from Brown University; MA degree, Fletcher School of Law and Diplomacy; commissioned as a foreign service officer, U.S. State Dept. and member of the U.S. Delegations to conferences Geneva and Economic Commission for Latin America and Chile. He has had political, economic and consular assignments in Washington, Uruguay, Chile, Columbia, Spain and Guatemala; American Consul general, Barcelona; Charge D'Affaires, American Embassy Guatemala.

Retired in 1982 and resides in Alexandria, VA with wife, Lois K. Sinn. They have a son, Jeffrey, employed by Dunn and Bradstreet, Tucson, AZ; daughter, Dr. Leslie Sinn White, DVN, Associate Professor at Northern Virginia Community College, teaching veterinary technology.

CARL W. SLEMMER JR. was born on March 28, 1923, in Camden, NJ. He entered the V-12 Program at Muhlenberg College, went to Pre-Midshipman School at Asbury Park, NJ, then to Midshipman School at Northwestern University where he received his commission in January 1945. Following additional training at Advanced Line Officer's Training School at Hollywood Beach, FL and the Amphibious Warfare Base at Coronado, CA, he was sent to Okinawa for three months, then assigned to a mine sweeper until June 1946.

After being released from active duty, he received a BS degree from Muhlenberg in 1948. Subsequently, he studied law and received his law degree from Temple University. He then specialized in labor law and spent most of his career in that field. He retired in 1989. He and his wife, Renee, live in Moorestown, NJ. They have three married children.

CLYDE K. SMITH was born on Dec. 9, 1925. He joined the USN on June 30, 1943, and was assigned to active duty at NAS Glenview, IL on Nov. 24, 1943, with V-5 status. He was assigned to NTU V-12 Park College on March 1, 1944-June 18, 1945; to NROTC ND, Notre Dame, IN on July 1, 1945-July 1, 1946, and was discharged on July 1, 1946.

Received the following degrees from Michigan State University: BS degree in 1947; DVM in 1951; MS degree in 1953; and Ph.D. in microbiology from the University of Notre Dame in 1966. He was appointed to the faculty of the Dept. of Microbiology and Public Health, MSU on July 1, 1951, and taught veterinary microbiology, epidemiology and public health as well as conducting research in microbiology and gnotobiology (germ free life studies).

He joined the faculty of The Ohio State University in 1966 and remained there until he retired in 1986. He is a professor emeritus of The Ohio State University and currently lives at Sturgeon Bay, WI, where he is enjoying retirement with his lovely wife, Mary E. "Jinx," whom he met while attending MSU. They have four sons and 10 grandchildren.

DANA A. SMITH was born on Nov. 23, 1925, in Hollis, ME. He entered the Bates V-12 unit in July 1943 and transferred to Yale (M.E.) in March 1944. He was commissioned at Yale in February 1946 and with 2,000 other new ensigns, chipped paint on light cruisers (the *Cleveland*, in his case) until discharge in July 1946.

After his discharge, he moved back to Maine and completed his engineering studies at the University of Maine and immediately turned to teaching. He taught in various Maine high schools and was principal of a couple of small ones. He retired in 1986. His career was undistinguished, except for the day the state police arrived at the island high school of North Haven and arrested his fellow teacher, who turned out to be "The Great Impostor," Fred Demara.

Re-entered the Navy from 1952-1954, and served aboard the LST-1084, experiencing the offerings of San Diego, Pearl Harbor, Yokusuka, Sasebo, Inchon and Pusan. He participated in the POW exchange "Operation Big Switch" and again retreated to Maine.

Married to Flora, and has lived in Tenants Harbor since 1957. They have three daughters, one son and nine grandchildren. He enjoys boating, gardening, photography and doing volunteer work at the Marshall Point Lighthouse Museum in Port Clyde.

DANIEL B. SMITH (2/LT) was born on Nov. 13, 1922, in Texas. In 1943 he entered the V-12 Program at Louisiana Tech, was commissioned a 2nd lieutenant, in USMCR at Quantico, VA and served with the 4th Marine Div. in the South Pacific. He also served with a Marine MP unit on Guam and with the 2nd Pioneer Bn. of the 2nd Marine Div. out of Sasebo on Kyushu, Japan.

Placed on inactive duty in June 1946, he entered SMU (Dallas) and earned a BBA degree in August 1947. He was recalled to active duty in December 1950 for the Korean War and served as CO of a motor transport unit at Camp Pendleton, CA. He resigned from the USMCR in June 1952 and returned to Richardson, TX.

Married Gloria Mills in December 1950, and they have two sons and one daughter. Their older

son served in the USN in Japan and Vietnam. He is a self-employed commercial real estate appraiser.

DONALD H. SMITH (LT COL) enlisted in the Marine Corps in March 1943, and entered the V-12 Program on July 1. After WWII he returned to Yale for a BA degree and graduate studies in Chinese. He was recalled to active duty in 1950 for the Korean War, his assignments included serving as infantry company commander in combat, naval attaché to Iran, logistics officer in Okinawa, inspector-instructor with New Hampshire reservists, Marine aircraft wing plans officer, and Marine Corps project officer for automating the tactical intelligence functions of the armed forces. He retired in October 1968 with the rank of lieutenant colonel.

Started theological studies at Colgate Rochester/Bexley Hall/Crozer in Rochester, NY, graduating in 1971 with a master of divinity degree. Soon after ordination as a Episcopalian priest he began a circuit ministry in Abaco, Bahamas, serving six native churches. Retired from full time ministry in 1990, and he and his wife, Bernice, reside in Treasure Cay, Abaco, and have a Stateside home in north Ft. Myers, FL. They have four children and 10 grandchildren living in Green Turtle Cay, Bahamas; Sarasota, FL; and Rochester, NY.

GEORGE T. SMITH (CAPT) entered the V-12 Program on July 1, 1943, and was commissioned in 1945. Reported to USS *Shangri-La* (CV-38) and participated in the Atomic Bomb Tests at Bikini. Served on the USS *Bayfield* (APA-33), which evacuated Admiral Cook, his staff and military dependents from Tsingtao, China in 1948.

The majority of his career was in submarines. He was flag lieutenant COMSUBLANT and served on submarines *Argonaut*, *Flying Fish*, *Sablefish*, *Threadfin* and *Trutta*, which he commanded. His other commands were USS *Cadmus* (AR-14) and *SUBDEVGRU 2*. He served on the faculty, Submarine School and USNPGS, Monterey, CA and graduated from the Naval War College.

Retired from the USN in 1971, then began a second career in education from which he retired in 1986 and now devotes his time to volunteer work. He resides in Saunderstown, RI. He has four children and five grandchildren who live in New England.

JOHN L. SMITH, Milligan College, University of Wisconsin, Great Lakes NTC and USS *Cape Johnson* (AP-172). He graduated from the

University of Tennessee and was recalled to active duty in 1950 on the USS *James C. Owens* (DD-776). He received his doctor of jurisprudence degree from Nashville YMCA Night Law School. He retired in 1988 from TVA after 35 years in its electric power program.

Remembers Milligan College with the unique mixture of a small campus in the hills and the very active physical and mental training of an all Navy student body as his introduction to Navy life. The Navy V-12 experience greatly increased his aspiration for higher learning and served him well in his college career and family accomplishments.

Smith and his wife Jane live in Nashville, TN. These grandparents have a daughter Joni and three sons, Jeff, Jerry and Jay, with college degrees in electrical engineering, chemical engineering, doctor of medicine and business administration (marketing) respectively.

WAYLAND PATRICK SMITH was stationed in Oberlin from 1944-1945 and UW (Madison) from 1945-1946. He recieved his BS degree in 1947; MS degree in 1953; Ph.D. in 1960; and is a professional engineer.

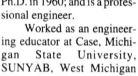

Worked as an engineering educator at Case, Michigan State University, SUNYAB, West Michigan University. He retired in 1989.

Enjoys full time motorhoming since 1991, scow skipper (MC898) and ultra light pilot (10WPS).

CORNELIUS E. SMYTH (NEIL) (CDR) was born on Aug. 20, 1926, in New York City, NY. He enlisted in December 1943 and entered the Navy V-12 Program at Yale University in March 1944; transferred to the University of Pennsylvania in July 1944; graduated ROTC in June 1946 and was commissioned an ensign with Supply Corps at the same time as receiving his BS degree. He served as assistant naval attaché, Cairo, Egypt, 1946-1949. He was the youngest assistant Naval Attaché in the history of the Navy; he was 20 years old.

Released to inactive duty in June 1949 and joined the Reserves. He served as commanding officer, Naval Intelligence Unit 10-1, 1969-1970; worked as an internal auditor, Pan American Airways, 1954-1959; as controller, El San Juan (Intercontinental) Hotel, 1959-1970; Caesar's World, Inc., 1970-1980 and 1983-1990 as vice president of finance, administrative vice president and ex-
138

ecutive vice president of Caesar's Palace, Las Vegas and executive vice president of international marketing. He was president of the Sands Hotel, Las Vegas, 1981-1982; and president of international association of hospitality accountants, 1969-1970. He retired in 1990.

He is an avid table tennis player and was U.S. National Doubles Champion (over 50) in 1985 and currently the U.S. Olympic representative for table tennis. He also enjoys body surfing. Married to Jeanne Dillingham and they have six children and 15 grandchildren.

FRANK SNYDER (CAPT) joined the V-12 unit at Cornell in July 1943. After two terms he transferred to the NROTC unit at Rennsalear for one term, then entered the Naval Academy and graduated in 1947. As a naval officer he served in nine ships and had four tours of duty in Washington, including one as the military assistant to the assistant secretary of defense, Eberhardt Rechtin (V-12. Cal Tech).

Retired in 1976 with the rank of captain and joined the staff of the committee at the National Academy of Sciences that studied telecommunications and computer issues.

In 1979, he and his wife, Margaret, moved to their present home near Newport, RI. As a professor at the Naval War College, he taught operations, command and control, and the military planning process.

GEORGE A. SOURIS (LCDR) entered Navy boot training at Faragut, ID in September 1943. He began V-12 training at the University of Oklahoma in July 1944, then transferred to OU NROTC in May 1946. He received a BA degree in journalism and was commissioned an ensign in June 1947.

He was the gunnery division and CIC watch officer on heavy cruiser *Columbus* (CA-74), until September 1949. During this time the ship transited the Panama Canal from Long Beach, CA to Norfolk, VA and made calls at ports from Oslo to Istanbul.

From 1952 until his retirement in 1976, he was Naval Intelligence Reservist in St. Louis, MO. He also worked at McDonnell Douglas from 1954 until retirement in 1987 as an engineering administrator on Gemini, Skylab and Cruise Missile programs.

Between trips, he and wife, Anne, take water aerobics and do volunteer work. Their daughter Renee was born in 1959 and son Paul in 1960.

RAYMOND CURTIS SOUTHWICK (LT) entered the Navy V-12 Program at Tufts College,

Medford, MA on July 1, 1943. He was commissioned an ensign from Columbia University Midshipman School in November 1945. He served overseas for six months at Radio Barrigada on Guam Island. He was in the Naval Reserve until 1952, with the rank of lieutenant.

Graduated from Tufts College with a BSCE degree in 1945. He later became a registered professional engineer and land surveyor in the New England states and spent 35 years in business as consultant in private civil and structural engineering practice. He is also licensed as a builder and real estate broker (realtor).

Currently resides in South Dartmouth, MA with wife, Patricia. They have two children: son and daughter, and six grandchildren.

NED A. SPENCER entered V-12 MIT, 1943-46; BSEE (also received MSCS in 1982 from George Washington University). V-12 training USS *Montpelier*. Joined Wheeler Laboratories a junior engineer, advancing to VP of engineering. Developed microwave devices and antennas, anti missile defense, air-to-air IFF, base-stations and mobile communication antennas.

In 1971 he joined the MITRE Corporation' Center for advanced aviation system develop ment, supporting the FAAs R&D into air traffic control. He led teams devoted to surveillance improvements; microwave landing system, ad vanced control concepts, and development of the Traffic Advisory and Collision Avoidance System (TCAS), which has now been implemented in al major commercial aircraft.

A private pilot and former radio amateur, he served on panels of the National Security Indus trial Association, on the Radio Technical Com mission for Aeronautics and on committees for the IEEE; he is a fellow of that organization and received citation for "Outstanding technical and managerial leadership in the development of an tennas for radar and communication." He retired in 1993, Potomac, MD.

He will never forget the time on the *Montpe lier* when they were doing gunnery practice in the Gulf Stream and inadvertently coupled a forward five-inch gun to an after gun-director" They blew out a porthole on the bridge, thus proving that V-12'ers were a force to be reckoned with.

DEAN ROY STAATS (LTJG) was born on Sept. 18, 1924, in Somerville, NJ. He entered the Drew University V-12 Program on July 1, 1943 transferred to Brown University ROTC Program on July 1, 1944, and was commissioned an ensign in February 1945 at Brown University.

Served in the Pacific Theater on the USS *Talita* (AKS), then as navigator and communications officer of LST 1043. They carried ammunition and later occupation troops to Sasebo and Wakayama, Japan in 1945. He was discharged to inactive Reserve in August 1946 with the rank of lieutenant junior grade.

Graduated from Brown University BSC in February 1945; received an MA degree in mathematics, education in June 1948; Fellow of the Society of Actuaries (FSA) in 1960; and a member of the American Academy of Actuaries (MAAA) in 1965. He worked for Metropolitan Life, then for North American Reassurance Company in New York City as a data processing officer, corporate actuary, and executive vice president; president of NARE Life Management Company and U.S. manager of Canadian Reassurance Company.

Retired in 1986 and resides in Newtown Square, PA with wife, Marilyn. He has a son, Barry and grandchild Patrick who live in Cresco, PA.

ERNEST N. STAMEY (LTJG) was born on Dec. 31, 1920, in Altamont, NC. Enlisted in January 1942 and entered the Navy V-12 Program at the University of North Carolina and transferred to North Western in Chicago for Midshipman School and was commissioned an ensign in 1945.

His memorable experience was his acceptance to the V-12 Program at University of North Carolina after two years as yeoman on the YMS-54 (yard mine sweeper) at York Town, VA. He was discharged in 1947 with the rank of lieutenant junior grade.

Practiced real estate law in Hialeah, FL from 1950-1983 while the city population increased from 4,500 to 150,000. He served on the Miami Springs, FL city council and always took an active part in politics.

His wife Patricia was his number one of six secretaries and office manager. She passed away in 1982. In 1983 he married Lula Belle, the great country singer and widow of Scotty; they were stars on The National Barn Dance in Chicago for 25 years. He and Lula Belle have had 11 wonderful years together. He is now a retired attorney on family farm in western North Carolina.

DONALD L. STARNER (MOMM2/c) was born on March 16, 1921, in Gardners, PA. He enlisted in the USNR on Oct. 14, 1942, and entered the V-12 Program at Yale University. He was stationed at the Naval Reserve Station, Los Angeles, CA; USNTS, Farragut, ID; Naval Training School (diesel), Cooper Bessemer Corp., Grove City, PA; RS, North Virginia; USS *Direct*; V-12 USN Training Station, Bainbridge, MD; Yale University, New Haven, CT; and PSC, Lido Beach, Long Island, NY.

His memorable experience was being selected from the Atlantic Fleet for the V-12 Program. He was discharged on Feb. 20, 1946. His awards include the Good Conduct Medal, WWII Victory Medal and American Theater Medal.

He married Mary M. Lake of Akron, OH on June 26, 1943, and they have one daughter. He is retired.

RICHARD P. STEIN (LT) was born on Sept. 2, 1925, in New Albany, IN. He enlisted in the USNR on May 18, 1943, and entered the V-12 Program. Attended Milligan College, TN; Duke University; and Notre Dame for Midshipman School. Stationed at Boston, Miami, Guam, Subic Bay, Kwajalein and Newport, RI.

Released to inactive duty on July 20, 1946; recalled for the Korean War from Aug. 1, 1950-May 20, 1952.

Floyd Co. prosecuting attorney, 1956-1961; U.S. Attorney S. District of Indiana, 1961-1967; chairman of Indiana Public Service Comm., 1967-1971; and senior vice president of the Public Service Co. of Indiana until 1989.

Married Charlotte Key and they have three children: Richard Jr., William and Patricia. He is semi-retired and practicing law.

JOHN M. STEINBRUGGE (LCDR) was born on Nov. 8, 1923, in Tucson, AZ. He entered the Navy V-12 Program at Colorado University, Boulder, CO, on July 1, 1943; volunteered for aerology studies at UCLA V-12 in November 1943 and graduated first in his class. Transferred to Midshipman School at Notre Dame in August 1944 and was commissioned an ensign USNR on Oct. 19, 1944. Served at NAS Opa Locka, Miami from October 1944-February 1945 and served as aerology officer NAS Bermuda from February 1945 to June 1946.

Received a Letter of Commendation at NAS Bermuda from ComAirLant for exceptional service. He retired from the Naval Reserve as LCDR in 1964.

Graduated from Oregon State University in 1948 in civil engineering with honors. Started his own engineering consulting firm in 1962 and is presently chairman of the board of Steinbrugge, Thomas and Bloom, Inc. (commonly known as STB Structural engineers) in Newport Beach, CA. Presently resides in Huntington Harbour, CA and has two daughters who live in Oregon.

JULES VERNE STEINHAUER (LCDR) was born on Aug. 25, 1925, in Brooklyn, NY. He enlisted in the USNR on May 5, 1943, and entered the Navy V-12 Program at Tufts College, Medford, MA; transferred to Midshipman School, Columbia University, New York (24th class).

(Limited) qualifications in submarines (diesel); stationed at Subic Bay and Luzon Philippines. He was discharged on July 1, 1970, with the rank of LCDR. Awarded six service medals.

Married, has no children. He works as volunteer in charge, Navy Retired Affairs Office, Naval Station, NY; recorder, New York Commandery, Naval Order of the U.S.

JAMES B. STEPHEN (LCDR) was born on May 17, 1925, in Spartanburg, SC. He entered the

University of South Carolina in 1942 and the University V-12 Program on July 1, 1943. Commissioned an ensign upon graduating from Carolina in February 1945; served in the Central Pacific in the USS *Melucta* (AK-129) during 1945, and thereafter on the USS *Pawkatuck* (AO) until separation as a Reservist in September 1946. He retired as LCDR in 1963.

Graduated from Duke University School of Law in 1949 and practiced law in Spartanburg for 30 years. He served in the South Carolina House of Representatives from 1961-1968; state senate, 1969-1979 and was elected to South Carolina Circuit Court bench in 1979. He retired in 1992, and still resides in Spartanburg with wife, Ginger, and holds court occasionally. They have one son, two daughters, one step-son, one step-daughter and five grandchildren.

HENRY LEONIDAS STEVENS III, (CAPT) was born on May 12, 1923, in Warsaw. NC. He graduated from the University of North Carolina with an AB degree and Wake Forest School of Law (JD). He joined the USMC in June 1943 and entered the V-12 Program at Chapel Hill; graduated from Quantico in 1945; and was commissioned as 2nd lieutenant. He served with the 3rd Mar. Div. (Guam) and with the 3rd Amphibious Corps (China) during WWII. He served with Base Legal Office, Camp Lejeune, as captain during the Korean Conflict.

Served as a North Carolina senior resident superior court judge; president, North Carolina, Conference of Superior Court Judges, 1993-1994; and was a member of the board of directors, North Carolina Judicial Conference. He practiced law in Kenansville, NC for 26 years, was a judge for 17 years and retired on January 1, 1995.

Judge Stevens and wife, Vernell Abernethy Stevens, reside in Warsaw. Their two children are Mildred Beasley (married to attorney, Allen C. Foster II) and Henry IV (served with 2nd Mar. Div., Amtracks, in Desert Storm, promoted to captain and attends law school); and one granddaughter, Christina Marie Foster.

ROY W. STEVENS remembers his V-12 experience as a special memory. The marvelous V-12 Program enabled him to escape from what, undoubtedly, would have been a very pedestrian life in Ottumwa, IA, to one of modest success in business including executive positions with major companies and world-wide responsibilities.

In the 50th anniversary V-12 report, April 1994, Meldrom F. Burrill, his CO at Illinois State was a special honoree. This recalled, what he believes to be the most memorable eight months of his life. From April 1944, as a V-12'er at ISU, he studied and was able to play baseball and football at Illinois State and was elected captain of the football team in the fall of 1944. It was a great two semester experience. Because he was busy playing football and traveling on many weekends, he did not complete a special course which was required of all V-12'ers to go on to Midshipman School. Lieutenant Burrill made a special exception on his part and kept him over for several days after all others had graduated to take a crash course. He passed the exam and was admitted to Midshipman School at Cornell University. This saved his life as far as his Navy career was concerned and he graduated with a commission in April 1945.

He shall always be indebted to Mr. Burrill for this special consideration and to the V-12 Program, which in retrospect had a tremendous impact on his life.

LAURENCE O. STINE (LTJG) was born in November 1924 in Williston, ND. He enlisted in June 1943 and entered the University of Minnesota V-12 Program, then transferred to Notre Dame for Midshipman School in 1945.

He served in the USS *Princeton* (CV-37) and was discharged in June 1946 with the rank of lieutenant junior grade.

Retired engineer; consulting; formerly senior vice president of UOP Inc. He has 50 patents and has written many papers.

RUSSELL A. STOKES (LTJG) was born on May 1, 1922, in Preston, MS. He entered the USNR on July 1, 1943, the V-12 Program at Mississippi College and Midshipman School at Columbia University, NY.

Served in the USS LST-566 from 1944-1946 and served in the Atlantic-Pacific and the Okinawa invasion.

Was discharged on June 30, 1947, with the rank of lieutenant junior grade.

Married Christine Boyd. Professor emeritus of mathematics at the University of Mississippi since 1988 and is now retired.

JAMES H. STONE (CAPT) was born on Dec. 20, 1925, in New York, NY. Enlisted in the USMC in June 1943; commissioned as 2nd lieutenant and assigned to the 2nd Bn., 1st Mar. Div. as platoon leader. Wounded on the Okinawa invasion and recieved the Navy Marine Corps Medal and Purple Heart. He went with the 1st Div. to North China and was put on inactive duty in 1946.

Graduated with a BA degree from Williams

College and studied geology at Texas A&M. He was recalled to active duty in 1950 for the Korean conflict and was company commander, 1st Mar. Div. Left the Marine Corps in 1952 to work for an independent oil operator and afterward started The Stone Oil Company, which is a predecessor company of Stone Energy Corp., listed on the New York Stock Exchange. He is currently chairman of the board and CEO.

Active in many activities in New Orleans and is a single father of six children.

THOMAS E. STOTT JR. (LTJG) was born on May 14, 1923, in Beverly, MA. He enlisted in September 1941 and entered the Navy V-12 Program at Tufts University and transferred to Tufts NROTC. He was stationed in the USS LST-669 during the Pacific Theater Operation and served in Leyte Gulf, Lingayen Gulf, Subic Bay and Okinawa.

His memorable experiences were being on a beach when MacArthur returned; and the kamikazes at Okinawa.

Discharged to inactive duty in February 1946. His awards include the Atlantic Theater, Pacific Theater with three stars, Philippine Medal with two stars and the WWII Victory Medal.

Married Mary Elizabeth Authelet on Feb. 26, 1944, and they have five children: Pamela, Randi, Wendy, Thomas E., and Diana. Received his BS degree from Tufts University in 1945. Worked as design engineer for Bethlehem Steel in Quincy, MA, 1956-1959; project engineer, 1959-1964; senior engineer, basic ship design, 1960-1963; project coordinator, 1963-1964. He then went to work for Stal-Laval, Inc, Elmsford, NY from 1964-1984; then Stott & Co., Cummaquid, MA, 1984-1988. Retired in 1988 and serves on the board of directors and as treasurer of Friends of Prisoners, Inc.; deacon of West Parish, Barnstable, MA; fellow ASME; and Soc. Naval Architects and Marine Engineers. He is retired.

JEHU JOSEPH STOUDENMIRE (JAKE) (2/LT) was born in December 1921 in Sumter, SC. He enlisted in the USMC on April 9, 1952, at Savannah, GA; entered the Navy V-12 Program at Duke University; and transferred to Quantico, VA for OCS. He was stationed at Parris Island and Camp Pendleton, CA and served at Kamuela, HI.

Discharged on Nov. 30, 1949, from the 6th Marine Corps Reserve.

Married Katie Shealy, Cola, SC in 1948. He worked as NCR branch manager in Hagerstown, MD and Baton Rouge, LA. Stoudenmire passed away on Jan. 25, 1992, his widow resides in Baton Rouge, LA and his daughter, Kate S. Allenbach, resides in West Chester, PA.

ALVIN L. STROH (CAPT) was born on Dec. 9, 1922, in Loveland, CO. He entered the Navy V-12 Program at Southwestern Louisiana Institute on July 1, 1943, and was commissioned a 2nd lieutenant, USMC, September 1944, at Special Officers Candidate School, Camp Lejeune, NC.

Joined the 2nd Mar. Div. on Saipan; served i 2nd Pioneer Bn; and went to Nagasaki and Saseb with the occupation forces. He was released fror active duty at Camp Pendleton, CA in May 194

Recalled to active duty for the Korean War i 1951; completed Transport QM School, Coronad CA and Army Engineers School, Ft. Belvoir, VA He was assigned to the 3rd Marine Div. when was reactivated at Camp Pendleton, CA. He wa released from active duty in 1953 at Cam Pendleton, CA with the rank of captain.

Currently resides in Downey, CA with hi wife Leonore. They have two children, Duane an Julie Ann. He is the owner of McMillin Wire & Plating, Inc. since 1968 and is still a happy work ing executive in September 1994.

EMIL GENE STUNZ (CAPT) entered the V 12 Program at the University of Washington i March 1944, followed by V-12 at Willamett University and NROTC at the University of Wash ington. Commissioned an ensign in June 194 active duty Korean War, Korean Theater, US *Logan* (APA-196) and navigator in the USS *Electr* (AKA-4) in the Alaskan area. Served in the USNI at Salem, Oregon and Boise, ID, including CO o Surface Div. and group commander. Retired fror the USNR in 1973 with the rank of captain.

Married Helen Sallee in 1946 and they resid in Nyssa, OR. They have four children and nin grandchildren. Received his law degree fror Willamette University in 1959 and practiced lav at Nyssa until retirement in 1973. Served 26 year on school boards, including tours as president o Oregon School Boards Assoc. and chairman of th State Board of Education. He is presently on th board of directors of several local corporations an does some motor home traveling.

JOHN F. SULLIVAN JR. entered the Nav V-12 Program on July 1, 1943, and was discharge on June 1, 1945. Returned to school in 1946 an received his BS and MS degrees in chemica engineering.

Joined General Electric Co. and spent 1 years in various positions; joined E.F. Houghto Co., a manufacturer of specialty chemicals, a president and CEO in 1968; joined the Hayes Albion Corp., an automotive equipment manufacturer, as executive vice-president of operations responsible for 13 operating divisions in 1972.

He joined the Bath Iron Works (BIW) in 1975 as president and CEO and chairman of the board and also elected vice-president of the parent company, Congoleum Corp. BIW is a builder of ships for the USN and has built a good portion of the guided missile frigates and cruisers presently in the fleet; BIW is the lead ship builder for the new Arlie Burke class of Aegis guided missile destroyers.

Retired in 1983 and resides in South Harpswell, ME and Naples, FL with his wife. He has two sons living in Maine and a daughter living in Boston.

RICHARD L. SULZER (LT) was born on March 21, 1925, in Englewood NJ. He enlisted in the USNR on Dec. 20, 1942; entered the Navy V-12 Program at Duke University and transferred to Cornell University for Midshipman School. He was stationed in the USS *Philadelphia* (CL-41) and USS *Helena* (CA-75).

His memorable experience was the long cruise on the *Helena* from Boston to China via Suez.

He was released to inactive duty on Aug. 15, 1946.

Married and has four children. He is retired from the Federal Aviation Administration.

JOHN H. SUTTON (CADET) was born on Feb. 17, 1926, in Royal Oak, MI. He enlisted in the USN in June 1944 and entered the Navy V-12 Program at Alma College and gradfuated from University of Michigan in 1948.

He was stationed at Brooklyn Navy Yard and Eagle Mountain, TX. Memorable experience was when the war was over before his training was completed.

Married Suzanne Davis in August 1947 and they have four daughters. He lived in Birmingham, MI from 1951-1987 and passed away in May 1992 in Palm Springs, CA.

HUGH MELVIN SWIFT JR. (LTJG) was born on Dec. 11, 1924, in Los Angeles, CA. He entered the Navy V-12 Program on July 1, 1943, at the University of Redlands, Redlands, CA. Was commissioned an ensign through the Northwestern Midshipman School (Tower Hall), Chicago, IL; then attended CIC School at Hollywood Beach, FL and Fighter Director's School at NAS St. Simon's Island, GA.

Served aboard the USS *Hornet* (CV-12) in the Pacific and the USS *Saidor* (CVE-117), the photographic ship at the atom bomb tests at Bikini.

He was discharged with the rank of lieutenant junior grade.

Graduated from Stanford University with an AB degree, received a JD degree from Stanford University School of Law and has practiced law in the greater Los Angeles area for 44 years. He is

now a senior partner in the law firm of Lagerlof, Senecal, Bradley & Swift, Pasadena, CA.

Married Jill Averill over 40 years ago, and they have three children: Julie Swift O'Connor, Diana and Molly. They reside in Tarzana, CA.

PAUL E. SWISHER was born on Aug. 21, 1922, in Alliance, OH. He attended Carnegie Institute of Technology, Pittsburgh, PA and Mt. Union College, Alliance, OH, before entering the Navy V-12 Program on July 1, 1943. He was commissioned on June 28, 1944, at Columbia University, NY, served as engineering officer in the USS *Papaya* and the USS *Corkwood* in the South Pacific until April 1946.

Graduated from Illinois Institute of Technology, Chicago, IL in January 1948 with a BS degree in electrical engineering. Spent 11 years at Bendix Corp., South Bend, IN in the Navy's Talos Missile Program, then moved to TRW and served in Los Angeles and San Bernardino, CA for 25 years on the Minuteman and Peacekeeper (MX) ICBM Programs. Received a master of engineering degree in 1970 from the University of Redlands, Redlands, CA. He retired in August 1984.

He and his wife Iona have done extensive traveling throughout the U.S. and Europe, as well as Canada and Hawaii, the Caribbean, Panama Canal, the Baltic and the Mediterranean. They are the parents of two daughters, Barbara and Carole, and two granddaughters, Jennifer and Brandy.

ALBERT EDWARD SYMONDS JR. (EM2/c) was born on Jan. 29, 1925, in Linwood, NJ. He enlisted in the USNR on Jan. 26, 1943, at Wilmington, DE,; attended boot camp at Newport, RI, then transferred to the V-12 prep unit at Newport in April 1943. Attended Dartmouth College from June 1943-March 1944 and Tufts College from March 1944 to November 1944 for V-12 service. Assigned to CUB-12 and UDT in November 1944 and served overseas in the Philippine Islands from January 1945 to March 1946 (NSD 3149).

Honorably discharged on April 25, 1946, at Bainbridge NTS, MD, with the rank of EM2/c. His awards include the Philippine Liberation Medal, USN Good Conduct Medal, American Theater of Operations Campaign, Asiatic Theater of Operations Campaign and WWII Victory Medal.

Recieved his BS and MS degrees in organic chemistry from the University of Delaware. Employed as research chemist at DuPont Experimental Station, Wilmington, DE in June 1951; transferred to the Savannah River Plant, Aiken, SC, in May 1955 where he worked as a radiation engineer and certified health physicist until retirement in 1986. He is presently a consultant and CEO of Radiological Consultants, Inc. He has one patent and numerous publications.

Married Lillian Helen on Sept. 9, 1950, and has two sons, one daughter and two grandsons. Currently resides in Aiken, SC.

FRANCIS C. TATEM was born on Nov. 9, 1924, in Westbury, NY. He joined the Navy V-12 unit at Swarthmore College from July 1943-1944. Served as secretary of the college student council and was a volunteer English tutor for members of the Chinese naval unit at Swarthmore. Attended USNR Midshipman School at Northwestern's Chicago Campus and served in the Pacific Theater aboard the USS LST-803 and a minesweeper in Japan 1946.

Graduated from Lehigh University, 1948; Seabury-Western Theological Seminary with M.Div. degree in 1951; was ordained deacon and priest in the Episcopal Church and served in parishes in New York, Wisconsin, Illinois, Kentucky and Virginia.

Married Ann Haldeman in July 1953 and they have five children and eight grandchildren. Currently retired and does volunteer work with hospice, the regional hospital and assists occasionally in nearby churches. He is a member of the county welfare board.

WILLIAM D. TAYLOR (LT, SC) was born on Sept. 12, 1925, in Nashville, NC. He entered the V-12 Program on May 15, 1943, and was released to inactive duty on Feb. 27, 1947. Attended the Navy Supply Corps Midshipman School at the Harvard Business School. Retired with the rank of lieutenant second class, USNR. After being commissioned, he served in the USS *Eldridge* (DE-173), USS *Atherton* (DE-169), USS *Carroll* (DE-170) and USS *Ft. Mandan* (LSD-21).

Returned to college in 1947 and received a BS degree in economics from the Wharton School, University of Pennsylvania and a MA degree in economics from the University of Virginia.

His business career included banking, 30 years in the textile industry and 13 years with the Charlotte, NC Board of Education.

Retired in 1994 and resides in Charlotte, NC with his wife. They have one daughter who lives in New York City.

THEO G. THEVAOS, MD (LT, MCUSNR) entered the V-12 Program from the V-1 Program. Stationed July 1, 1943, Emory University, Atlanta, GA; July 1944, Naval Hospital, Parris Island Marine Base, SC; hospital corpsman, September 1944, Medical College of Georgia, Augusta, GA; December 1945 unit deactivated to Reserve status; October 1950, Ft. Gordon, GA Army hospital. Suddenly reactivated as physician, when the Chinese entered the Korean War. Joined 1500 plus other former Navy V-12 physicians, mostly pulled from residency training, who were loaned to the Army in the emergency.

At the completion of WWII, the Army had

discharged its ASTP medical participants and had less than 24 physicians in Reserve Status. The medical draft was reinstated and Army doctors began replacing Navy personnel after about six months.

In May 1951 transferred to Breezy Point NAS, Norfolk, VA, and in July 1951 became the Destroyer Sqdn. 6 medical officer. Completed two tours of duty with the 6th Fleet and served as an official interpreter at welcoming ceremonies to Greece. In October 1952 deactivated to Reserves as LT MCUSNR.

Completed pediatric residency in pediatrics at MCG in July 1953. Has since served on medical staffs of Gracewood State School and Hospital (for the developmentally disabled) and the Pediatric and Pediatric Neurology Dept. at MCG. Now physician in chief and director medical education and research, GSSH, and professor of pediatric neurology, MCG.

In 1994 received the Faithful Service Appreciation Award for 45+ years of state service from Georgia Governor Zell Miller and the Plato Award from District I of American Hellenic Education Progressive Association for community service and professional excellence.

Married Artemisia Dennis Thevaos, associate professor, Fine Arts Department, Augusta College. A Magna Cum Laude graduate of Wesleyan Conservatory, Macon, GA and recipient of master's degree and performers certificate in piano from Indiana University. She performs professionally as a member of Thevaos-Porro duo-piano team. They have one son, George Philip, currently a senior at Medical College of Georgia and a recipient of an MCG Alumni Scholarship after graduating with highest honors at Georgia Tech. He has been a member of Atlanta Symphony Chorus under Robert Shaw for the past five years and is co-organist with his mother at the Greek Orthodox Church Holy Trinity, Augusta, GA.

ROBERT COLBY THOMPSON (2/LT) was
born on Oct. 10, 1922, in Woodhaven, NY. He enlisted in the USMCR on March 9, 1942, and entered the Navy V-12 Program at the University of Michigan. Stationed at Ann Arbor, MI; Parris Island and Camp Lejeune (SOCS).

Deactivated Jan. 27, 1945, and was medically discharged on May 24, 1945.

Thompson has been married for over 50 years and has three children. He is retired.

WILLIAM THOMPSON (RADM) attended
Wabash College, 1944-1945, and commissioned at Notre Dame on July 9, 1945. Served in the Korean War in the USS *Midway* and destroyer *John R. Craig* as operations officer. He became special duty only officer in public affairs and was special assistant to three secretaries of the Navy. He was the first public affairs officer selected for flag rank and became chief of Navy information.

Retired as rear admiral in 1975. He was awarded the Distinguished Service Medal for "...creation of the most effective and profes-

142

sional public affairs program in the history of the Navy."

In 1978, he became founding CEO and president of the USN Memorial Foundation and designed, built and funded the national Navy Memorial on Pennsylvania Avenue in Washington, DC. In the process, he received two Distinguished Public Service Awards, the highest citation the Navy can bestow on a civilian. He also received the foundation's Lone Sailor Award for "distinguishing himself in the private sector and accomplishing deeds of benefit to mankind." He retired in 1993.

He was a Wabash Sigma Chi and has been designated a Significant Sig by the fraternity.

Thompson and wife Dorothy have three children and three grandchildren. They reside in McLean, VA.

WILLIAM L. TIDWELL (CAPT) was born
on Jan. 14, 1926, in Greenville, SC. He entered the V-12 Program as NROTC on July 1, 1943, at the University of South Carolina. He graduated and was commissioned in February 1945. He served in the USS *Colorado* in gunfire control at Okinawa and later in the USS *Iowa* as administrative assistant to the gunnery officer.

Married Louise Van Hollebeke in 1946. Attended graduate schools in microbiology; MS degree from the University of Hawaii in 1948; Ph.D. from UCLA in 1951. He was a professor at Texas A&M from 1951-1955; and San Jose (CA) State University from 1955-1988.

Spent over 30 years in the USNR and retired as captain of the Medical Service Corps. Activist on campus and within church and the community before and after retirement.

Retired to Three Rivers, CA with wife Lou in 1990. Their son Paul, his wife Inka and a grandson Philip, live in Albuquerque, NM.

ALBERT W. TIEDEMANN JR. (CAPT)
was born on Nov. 7, 1924, in Baltimore, MD. He enlisted as a V-1 in December 1942 and was ordered to active duty with the V-12 Program at Mt. St. Mary's College in July 1943. Attended Midshipman School at Notre Dame in July 1944 and was commissioned an ensign on Oct. 26, 1944. Ordered to the 7th Amphibious Force and joined the USS LCI(L)-444 at Hollandia. Operated in New Guinea and the Philippines, including landing at Legaspi. The 444 rode out a typhoon returning to the U.S.

He was released to inactive duty as lieutenant (jg) on Aug. 25, 1946.

His Naval Reserve service included CO of a

research company, Surface Division, Mob Tea and System Analysis Div. (Sublant).

Awarded the Asiatic-Pacific, American, Phi ippine Liberation, WWII Victory, Naval Reserv and Armed Forces Reserve Medals.

As a Naval Reserve Assoc. founder, he he three national offices; was originator and chai man for five years of the Navy Sabbath Progra and transferred to the retired list in Novemb 1984.

Received his Ph.D. degree in chemistry an worked as chief chemist, Emerson Drug Co.; labo ratory director with Hercules Inc. at Allegan Ballistics Laboratory and Radford Army Ammu nition Plan; and VA Div. of Consolidated Labora tory Services. He currently resides in Richmon VA.

JAMES ROBERT TOMLINSON (LTJ(
was born on July 27, 1924, in Toledo, OH. Enliste in the USNR on June 5, 1943, and entered the V 12 Program at the University of Michigan in N vember 1943 and the University of Oklahom NROTC in March 1944. Commissioned in Ma 1945 and attended Destroyer School at Norfol Served aboard the USS *Dyess* (DD-880) in th Pacific, Japan and China.

His memorable experience was arriving i Tokyo Bay in December 1945 and the installatio of the British Governor General at Hong Kong i January 1946. The British were a grand host fo three days.

Transferred to inactive duty in August 194 and joined the USNR in March 1955. Award include the WWII Victory Ribbon, American The ater, Asiatic-Pacific, and Japan Occupation.

Retired as SW regional manager fro Ingersoll-Rand Co. after 40 years. He served in number of management positions and kept ir volved with the USN as a major supplier of pump and compressors for the fleet and new ship pr gram.

Married Carolyn Black and they have fou children and four grandchildren.

JAMES F. TOOLE entered the V-12 Progra
in 1943. Received his BA degree from Princeto University in 1947; MD degree from Cornell Un versity Medical College in 1949; and LLB fror LaSalle Extension University in 1963.

In October 1950, he was recalled to activ duty by the USN and assigned on TDY to th Army where he served as a battalion surgeon i Korea and Japan. In July 1951 he was reassigne to Mare Island Naval Hospital, then to fligh training in Pensacola, FL. He later became flight surgeon for the USN with assignments i North Africa, the Mediterranean, North Atlanti and the Gulf of Mexico, aboard the USS *Tarawe Mindoro*, *Saipan*, *Guadalcanal* and *Siboney*.

In 1951 he was awarded the Combat Medica Badge and in 1952 the Bronze Star with V. He wa discharged in 1945.

He is director of the Stroke Research Cente at Bowman Gray School of Medicine, where h also maintains his clinical practice. He resides in Winston-Salem, NC with his wife. He has three boys, two of whom live in North Carolina and the other in Baku Azerbaijan, and a daughter who also resides in North Carolina.

ROSCOE H. TURLINGTON (CADET) was born on Sept. 1, 1923, in Kenly, NC. He enlisted in the USN, V-5, V-6, V-1 on June 10, 1943, and entered the V-12 Program at the University of North Carolina.

Stationed at Chapel Hill, NC; Bainbridge; Norfolk; Quonset Point; Columbia, SC; Athens, GA; Jax, FL; Yellow Water, FL; Great Lakes, IL; CASU 21; CASU 22; and Cape Canaveral, FL. He served in the USS *Princeton*; USS *Charger* and USS *Wolverine*.

His memorable experience was V-J Day in downtown Chicago and his solo flight in Cadet School.

Discharged on April 18, 1946. He was awarded the American Area Campaign Medal, Good Conduct Medal, Point System, and WWII Victory Medal.

Married Louise Parker and they have twin sons, Roscoe and Lester, daughter, Wyn Page and three grandsons. He is the corporate director of East Coast Bank and a doctor of dental surgery in Clinton, NC.

HEYWOOD A. TURNER was born on March 6, 1926, in Richmond, VA and moved to Tampa, FL in 1936. Joined the USN on Dec. 17, 1943, reported to the V-12 unit at Duke University on March 1, 1944, transferred to the University of Louisville V-12 unit on Nov. 1, 1944, and attended Speed School. He pitched for the Cardinal baseball teams in 1945 and 1946. He was discharged from the USN on June 24, 1946.

Memorable experience: Graduated from Hillsborough HS, Tampa, Fl with Sam McMurray and Bill Patterson. The three of them were roommates at Duke University. After leaving Duke, it was 50 years before he next met Sam at their 50th HS reunion. He has no knowledge of Bill.

Received a bachelor's degree in electrical engineering from the University of Florida in 1949.

Employed by Tampa Electric Co. as an oiler in 1947; served in the U.S. Army from April 1953-April 1955; and retired from the Tampa Electric Co. as senior vice-president of production in May 1989.

Lives in Tampa, FL with wife Mary Louise. They have two daughters, one son and three grandsons.

COURTNEY R. TUSING entered the USMC in November 1943 at Parris Island; transferred to QM School, Camp Lejeune, NC and was scheduled to ship out with the replacement outfit in two weeks when his opportunity for OCS came.

Spent one term with the V-12 unit at Bucknell University in May 1944, then transferred to Dartmouth until the unit dissolved in July 1946. He lacked three credit hours for commission as a Reserve lieutenant, so took discharge instead of a four enlistment with commission.

Graduated from Bucknell University in June 1947 with an AB degree in economics and sold life insurance for one and a half years, taught school for one and a half years, then entered banking. Progressed through five banks during 38 years, finished Stonier Graduate School of Banking, retired as chairman of the board and president of First United, Oakland, MD.

Married Mattie Webster in 1949, and they have a daughter who teaches in Greenville, SC; a son who is a Navy commander in Norfolk (USNA 1975); and five grandchildren. Retired to a 200 acre farm in Baker, WV, and enjoys horses, golf, and travel. He has traveled all 50 states and most of Canada (62,000 miles in a motor home) and went to France in 1994. He is active in the Lutheran Church.

ELMO F. VESTAL (AVN CADET) was born on Dec. 28, 1926, in Whitewright, TX. He enlisted in the USN in April 1944 and entered the V-12 Program at Southwestern Louisiana Institute. Other schools and stations included Tulane University; Bunker Hill NAS; Glenview NAS; and Ottumwa PFS.

The leadership training he received has been invaluable to him. He was discharged in August 1946.

Married and has eight children and 11 grandchildren. He works in sales for office systems and real estate investments; served as governor from 1992-1993; and is a member of Optimist International, Pennsylvania and Upper Delaware District.

WILLIAM V. VICTOR (BILL) (CDR) was born and raised a Nebraska cornhusker. After high school the Navy sent him to Minnesota University College of St. Thomas (St. Paul) and Northwestern University under the V-12 Program where he earned his BS degree. He completed his education at the Navy Supply Corps School, Harvard Business School (MBA degree) and Navy Postgraduate School.

Victor was a purchasing specialist in the USN. His most memorable and satisfying duties

were at the Navy Purchasing Office, London and the USS *Lexington* (CVA-16).

Retired from the USN in 1965 and began employment with Teledyne Systems Co. (Aerospace) in the Los Angeles area, retiring for good 20 years later.

Special Memories: The friendships made which have lasted a lifetime. The dazzling attractions of the "Big Shoulders" City, a short "EL" ride away were an education in themselves. Best of all, he met a beautiful, blond co-ed (Rosalie) on his first day at Northwestern. This romance blossomed into a 46 year marriage which only death could diminish. Rosalie passed away in August 1991. He has four children and five grandchildren and resides in Northridge, CA..

JAMES N. WAGGONER (LT) was born on Nov. 10, 1925, in Elgin, IL. He enlisted in the USNR in July 1943 and entered the V-12 Program. Attended the University of North Carolina and Indiana University. Stationed at the USN Ship Yards, Norfolk, VA; as a surgeon in the USS *Princeton* from 1952-1953; and served at Inchon, Korea.

Earned his BS degree in 1946; MD degree in 1949 from the Indiana University School of Medicine.

He was discharged on Nov. 10, 1952, with the rank of lieutenant (MC).

Married Patricia Ann Dougherty on Feb. 1, 1947, and they have two children, Steven and Kimberly. He retired from his last position as medical directory of The Garrett Corp. in Los Angeles, CA.

JUNIOR S. WAGNER (CAPT) was born on Oct. 25, 1922, in Smith Center, KS. He entered Bowling Green State University's V-12 Program on July 1, 1943; was commissioned a 2nd lieutenant USMCR in March 1945 at Marine Corps School, Quantico, VA. Served in 12th Marines in Guam and 1st MP Bn., 3rd Amphibious Corps in Tientsin, China, 1945-1946.

Released to inactive duty and recalled in July 1951. Served in the 11th Marines as CO HQ Btry. in Korea, 1952. He was discharged as captain, USMCR in 1953.

Received his BS degree in education in 1948 at Bowling Green, OH; MED and Ed.S. from the University of Colorado, Boulder in 1955 and 1960.

Served as high school instructor at Lebanon, KS, 1948-1954; superintendent of schools at Kensington, KS, 1955-1987. Retired in 1987 and was president of the Kansas Retired Teachers Association, 1994-1995. He resides in Kensington, KS with wife Betty. They have three children: James (employed with Frito-Lay of Topeka, KS); Joel (vocational director of Three Lakes Special Education Co-op of Lyndon, KS and resides in Silver Lakes, KS); and daughter Nancy Weber (living in Oakley, KS and employed by Weber Battery of Oakley); and two grandsons, Gregory Wagner and Joseph Weber.

EDWARD S. WAJDA was born on Oct. 31, 1924, in Schenectady, NY. Entered the Navy V-12 Program at Union College in July 1943; commissioned ensign in November 1945 from Columbia Midshipman School. He joined the Pacific Fleet at Pearl Harbor and was assigned as communications officer aboard USS PCE-902 in December 1945. Served in the USS LST-461 (April 1946) and worked with the Seabees in decommissioning naval bases in the South Pacific Theater.

He was honorably discharged in June 1946 and awarded the Asiatic-Pacific Campaign Medal, WWII Victory Medal and American Theater Ribbons.

Received BS in physics from Union College in 1945; MS from Cornell University, 1948; Ph.D in physics from Rensselaer Polytechnic Institute, 1953. He taught physics at Union College, 1953-1955, and was senior research physicist with IBM, 1955-1986.

Retired in 1986 and is noted travelogue lecturer on world-wide travel experiences and the author of several books.

He and wife Sophie live in Poughkeepsie, NY and have two daughters, Mary-Jane Hahn and Susan-Ann Siebl.

FRANCIS J. WALKER entered the V-12 Program on July 1, 1943, at Gonzaga University, Spokane, WA; graduated Columbia University Midshipman School, 1944, as ensign USNR and assigned to the USS *Clay* (APA-39). Participated in the invasions of Leyte Gulf and Lingayen Gulf, Philippines and Okinawa.

Received his BA degree from St. Martin's College, Lacey, WA, 1947; JD degree from the University of Washington School of Law, 1950; and admitted to the Washington State Bar in 1950.

Served as assistant attorney general for the state of Washington, 1950-1951; started his private practice in 1951; general counsel Washington Catholic conference, 1967-1976; listed in *Who's Who in West* and *Who's Who in American Law*.

Married Julia O'Brien (awarded B.Sc (M.T.) Seattle University 1950) in 1951 and they have six children: Vincent, Monica Hylton, Jill Nudell, John, Michael and Thomas; and 12 grandchildren. His wife was awarded a BS (M.T.) degree from the Seattle University in 1950.

GRANT J. WALKER (CAPT) was born in Ramey, PA. He entered the V-12 Program on July 1, 1943, and attended Midshipman School at Columbia. He remained in the USN until 1976 and attained the rank of captain.

Participated in the Korean and Vietnam con-

flicts and was awarded three Bronze Stars and two Meritorious Service Medals.

Immediately after WWII he was in Bikini for the first A-Bomb tests and was in initial group of ensigns to attend GLS (Newport). Subsequent tours were mostly in cruisers and destroyers and sea going staffs, having several DD commands. His last sea command was an LPD (*Coronado*) which he was commissioning CO.

Shore tours included staffs of DESLANT, PHIBLANT and strike warfare Ops. in the Pentagon. He was a graduate of the Senior Crs., Naval War College. He currently resides in Virginia Beach, VA, with his wife Dorothy.

KENDALL L. WALKER (LTJG) was born on July 22, 1924, in Martin, PA. He enlisted in the USN in November 1943 and entered the V-12 Program. Attended Case School, Yale University, where he earned his BSME degree. He had no active service after graduation.

Discharged in June 1946, with the rank of lieutenant junior grade in the USNR

Married and has three children. He is retired.

ROBERT KIRK WALKER was born on May 22, 1925, in Jasper, TN. He entered the University of South (Sewanee) V-12 Program on July 1, 1943. Commissioned ensign on May 24, 1945, at Northwestern University Midshipman School, Chicago, IL. Served in the Atlantic Fleet as CIC/Intercept Officer aboard USS *Fechteler* (DD-870). Released to inactive duty May 8, 1946.

Served Surface Div. 6-75, Chattanooga on Jan. 25, 1949; recalled during the Korean Conflict with duty in the Atlantic Fleet in the USS *Botetourt* (APA-136) from January 1951-May 1951 and USS *Pocono* (AGC-16) from June 1951-August 20, 1952.

Attended the University of Virginia Law School and received his JD degree in 1948. Engaged in practice of law with Strang, Fletcher, Carriger, Walker, Hodge and Smith of Chattanooga, TN, currently managing partner. Served as president of Chattanooga Bar Association, 1962-1963; and of Tennessee Bar Association, 1965-1966; and as mayor of Chattanooga, 1971-1975.

Married Joy Holt in August 1945 and they have three children: Kirk, head master of Ensworth School in Nashville; Marilyn, teacher at Girl's Preparatory School in Chattanooga; and James Holt, attorney in Nashville; and four grandchildren.

PAUL R. WALLER was born in Hinckley, MN. He enlisted in the USN on May 26, 1943, and

entered the V-12 Program. Stationed at the USNA Wold-Chamberlain; St. Mary's College, Winon, MN; St. Mary's College, Orenda, CA and USNA Glenview, IL.

Discharged on June 6, 1946, as a license airport control tower operator.

Married Eva and they have four childre Cynthia, Alan, Richard and Martha. He is a minin and processing consultant in his own compan Paul R. Waller, P.E.

BILLY WALLIS (LT COL) was born in Feb ruary 1923, in Navasota, TX. He joined the USMC in November 1942 and was assigned to the V-1 Program at Southwestern Louisiana Institute July 1943. Commissioned at Quantico, VA May 1945; assigned to 1st MP BN and went to th Pacific, then to China when the war ended.

Re-entered Southwestern Louisiana Inst tute and graduated in 1949. Taught in the Iberi Parish School System and was recalled to activ duty and served in Korea with the 7th Mot Trans. Bn. in 1951. Returned to New Iberia, L and taught. He received his master's degree fro Louisiana State University in 1954 and later be came principal then supervisor in the centr office.

Retired a lieutenant colonel from th USMCR and then retired from the school sys tem in 1979.

Resides with his wife Josephine Ware Wall in New Iberia, LA. They have two childre William T. Wallis (employed by Gas Company New Mexico) and Caroline Wallis Goddar (Methodist minister in Tennessee). They hav two grandchildren, Ben and Catherine Goddar

GEORGE WILLIAM WATSON (LTJG was born on March 1, 1926, in Eaton Rapids, M He entered the V-12a Program at Milligan Co lege, TN on March 1, 1944; commissioned a ensign in June 1946 from the University of Okl homa NROTC and served in LST-1138 in th Pacific Theater. Discharged in September 194 with the rank of lieutenant junior grade.

Graduated from the University of Michiga with an AB degree in 1947 and a JD degree 1950. Practiced as an attorney in Charlotte an Kalamazoo, MI, 1950-1962; served as assista prosecuting attorney; friend of the court Kalamazoo County; served as attorney, associat general counsel and general counsel with Feder Emergency Management Agency (FEMA) an predecessor agencies. Retired from the feder service in 1991 and has been a lawyer and leg consultant on administrative law and governmer tal affairs since 1991.

Active in community service and sailbo racing and cruising; president of Mt. Vernon L Enterprises, 1989-1994; past commodore, Na tional Yacht Club; and past president of Stratfor Recreation Association.

Resides in Alexandria, VA with his wi Ruth Murphy Watson. They have five childre and four grandchildren.

PHILIP B. WATSON JR. was born on June 8, 1926, in Louisville, KY. He attended Speed Scientific School and graduated with bachelor's degree in chemical engineering in 1947; received his master's degree in chemical engineering in 1973; attended the University of Louisville Law School graduating with a bachelor of law degree in 1950 and juris doctor degree in 1969. He was admitted to the Kentucky Bar Association in 1950 and to practice before Kentucky and the U.S. Supreme Courts in 1950 and 1960.

Joined the Navy V-12 unit at the University of Louisville, Louisville, KY in 1944 and served aboard the USS *Belet* and USS *Hopping*. He was discharged in June 1946.

Worked as legal counsel, 1952-1954, to E.I. DuPont de Nemours, Wilmington, DE; 1954-1972, house counsel for Girdler Corp., Louisville, KY, counsel and corporate secretary to C/I Girdler Co., Cincinnati, OH and manager of international operations, Tube Turns, Inc., Louisville, KY; 1972-1980, vice president, Industrial Clean Air, Inc., Berkeley, CA; and 1980-1987, western legal counsel, Joy Mfg. Co., Los Angeles, CA. He retired to Whispering Pines, NC in 1987.

STEN-ERIC E. WEIDLER (LTJG) was born on Feb. 26, 1925, in Milwaukee, WI. He enlisted in the USNR on Dec. 7, 1942, in the V-1 Program. Entered the V-12 Program at the University of Wisconsin from July 1, 1943-July 1945 and was commissioned from the University of Notre Dame Midshipman School in November 1945.

Married Audrey and they have three married children: Patricia, Peter and Faith; and an assortment of grandchildren. Retired in 1990 and enjoys oils, P.C. Church, friends, etc.

L. BOURKE WELCH served on destroyer duty in China and Indo-China from 1945-1946. He was discharged in Charleston, SC. Forthwith, he and George Carr, Dartmouth '46, bought a 1930 Pierce Arrow and drove it from New York to California—fabulous trip. Has lived in California since 1946.

Entered the University of California (Berkeley) in September 1947 and graduated in 1949. Recalled to Korea in 1951 and served on more destroyers until 1953.

Married and has two children and one stepchild. He was active in big volume printing for 33 years, during which time he lost a son, a marriage, and gained a new superlative life partner, Ruth. Following cardiac setbacks they retired to California's garden spot, Santa Maria.

Serves part time as a marketing manager for an agricultural laboratory and his wife enters her final year as a transcriber for a neurologist. They delight in breeding and showing top flight English setters.

ROBERT D. WELCH (LTJG) enlisted in the USNR in 1942 to enter the V-12 Program at Dartmouth College. Commissioned an ensign from Notre Dame Midshipman School, October 1944. Served with AdComPhibsPac as officer-in-charge aboard the USS LCT(6)-837 and XO LCT(6) Group 69. Later became a deck officer in the USS *Blackhawk*, destroyer tender, in Tsingtao, China. After WWII he participated in Reserve assign-

ments, two at the Little Creek Virginia Amphibious Intelligence School and one aboard the destroyer USS *Coney* on antisubmarine exercises in the Caribbean.

Graduated from Dartmouth, Phi Beta Kappa, Summa Cum Laude in 1948 and received an MA degree in English language and literature from the University of Chicago in 1952. A 43 year career as an educator included teaching at St. Paul's School in Garden City, NY, teaching and work as director of secondary curriculum for the Grosse Pointe, MI Public Schools, nine summers as a master teacher of English in the Harvard-Newton graduate MAT Program, and teaching English at Wayne County Community College. He retired in 1991.

Currently travels extensively with his wife of 38 years, the former Susanne Dengler of Saginaw, MI, often accompanied with son Robert's wife, Terry, and granddaughters, Lauren and Dana. Son Robert Dean is an emergency physician and assistant professor in the Wayne State University School of Medicine and son Richard Allen is a supervisor in the Wayne State University Computer Department.

BIDDLE A. WHIGHAM (LCDR) was born on Nov. 9, 1925, in Pittsburgh, PA. Entered the V-12 Program in July 1943 at Bloomsburg STC. In March 1944 he transferred to RPI NOTC Unit in Troy, NY; commissioned there in October 1945 and served in the USS *Vincennes* (CL-64) both in the South Pacific and at Mare Island, CA. Separated from active duty in June 1946.

Graduated from Lehigh University with a BSME in 1948 and joined Armstrong World Industries as a project engineer. He was recalled to active duty during the Korean War and served two years in the USS *Salem* (CA-139) flagship of the 6th Fleet. He was able to complete his 20 years of service via the organized Naval Reserve.

Entire working career was spent with Armstrong and split between engineering and manufacturing. Noteworthy were 11 years as Pittsburgh plant manager and eight years in international operations as director of manufacturing. He retired in 1988.

He and his wife Pat live in Lancaster, PA but spend winters in Florida. They both do volunteer work, enjoy travel, antiquing, and "elderhosteling."

EDGAR W. WHITE was born on Sept. 7, 1923, in Pretty Prairie, KS. Graduated from Elkhart HS, Elkhart, KS in 1941 and attended Sterling College, Sterling, KS for two years where he enlisted in the Navy V-12 Program in December 1942. Went to Southeast Missouri State Teachers

College, Cape Girardeau, MO, July 1, 1943, to begin his V-12 training. Completed training there in February 1944 and went to pre-Midshipman School at Asbury Park, NJ, then entered Midshipman School at Columbia University in NYC, graduating in October 1944 with commission of ensign.

Married Doris Caywood of Sterling, KS at the Riverside Church in NYC on Oct. 26, 1944. They later had two children, Lana White Willimon and Kevin White, both born at Boulder, CO, while he was in Law School.

Sent to Navy Communication School at Harvard University, which he completed in February 1945; spent some time at the 8th Naval District HQ in New Orleans, LA; went to San Francisco where he shipped out to Ford Island, on Oahu, and worked in the communications issuing office. Codes were received in wholesale lots and distributed out to individual planes and ships. In June 1945 he went to Eniwetok Island and boarded the USS *Denebola*, where he and Carl Lindstrom set up a mobile issuing office.

In October he was sent to Okinawa and boarded the USS *Inca* and helped close out that mobile issuing office as the war had ended. When this office was closed out, he had a short trip to Shanghai, China. Then back to the Navy base on Guam, where he boarded a troop ship for the States. Mustered out of service in San Francisco, CA in March 1945. Doris met him in San Francisco where they purchased a car and drove home to Elkhart, KS.

Attended Law School at the University of Colorado at Boulder and received a JD degree. He practiced law in Elkhart and later took in his nephew, Darrell Johnson, and now practices under the name White and Johnson.

ROBERT S. WHITE (LCDR) was born on Oct. 29, 1924, in Wakefield, MA. Enlisted in the USN on Dec. 7, 1942, and entered the V-12 Program at Tufts University. Called to active duty on July 1, 1943, transferred to Columbia University in October 1945 and commissioned ensign Nov. 2, 1945, Midshipman School, Columbia University.

Served during WWII from November 1945-June 1946: USS *Briscoe* (APA-65), LCI(L)-1064 and LST-748. Released to inactive duty and recalled from February 1951-August 1953 for the Korean War. Served in the USS LST-603, February-September 1951, and Boat Unit 2 until discharge in August 1953. Transferred to Retired Reserve in July 1971 and retired from USN in October 1984 as LCDR. He was awarded the Naval Reserve Medal, WWII Victory Medal, Asiatic-Pacific Campaign Medal, National Defense Service Medal and Armed Forces Reserve Medal.

Graduated from Tufts University in June 1945 with a BSME. Taught mechanical engineering at Northeastern University, September 1946-February 1951, and September 1953-June 1955. Received his MME degree in June 1955 and transferred to BF Goodrich Co. in July 1955 as a design engineer. Retired on Jan. 1, 1987, as supervising engineer.

Married Beatrice, who passed away in August 1993. They have two daughters, Helen Stacy and Janet Smith, and two sons, Robert White and Donald White.

ROBERT P. WICKINS (1/LT) was born on Jan. 8, 1925, in Rochester, NY. He enlisted in the USMC on Jan. 1, 1943; entered the Navy V-12 Program, University of Rochester (New York) Colgate and was stationed at Parris Island, Quantico and New River.

His memorable experience was seeing Nagasaki so soon after the H Bomb hit it.

Married Elsie Keirsbilek on March 31, 1951, and they have four children and seven grandchildren. He is retired.

EDWARD A. WILBANKS (LTJG) enlisted in May 1943 and entered the Navy V-12 Program at Newberry College, SC. Graduated from Midshipman School, Columbia University in October 1944 and attended Communications School at Harvard University. Assigned in February 1945 to staff COMFAIRWING 18 (RADM M.R. Greer) Tinian, Guam and Iwo Jima.

After WWII he was recalled to active duty in communications aboard the USS *F.D. Roosevelt* (CVB-42) and honorably discharged in 1949 with the rank of lieutenant junior grade. Received BS from Furman University, Greenville, SC and MS from Troy State University, Dothan, AL. Postgraduate studies in math/physics at University of North Carolina, Chapel Hill and NYU School of Engineering (Bronx).

Employed and later retired, as a physicist at the Navy Mine Countermeasures Station (later Navy Coastal Systems Center) Panama City, FL. Did research in marine physics related to naval operations, mine countermeasures and tactical instruction. Organized and headed the first Tactics and Doctrine Division and received two patent awards for naval systems. He is the author of 25 scientific papers, member of Sigma Xi, the Scientific Research Society and the American Physical Society (SE Sec).

Retired from Federal service, where he was adjunct Gulf Coast Community College, self employed consultant, community and state advocate in the field of human potential and mental health.

Widower, he has one daughter, three sons and four grandchildren.

ARTHUR CLYDE WILLHELM JR. (LCDR) was born on Oct. 10, 1921, in Galveston, TX. Enlisted in the USNR on June 26, 1942; entered the Navy V-12 Program at Case Western Reserve and graduated from Midshipman School, USNA, Annapolis, IN. Served aboard the USS *Wolverine* (1X-64), May 1944-November 1945 and the USS *Randolph* (CV-15) in January 1946 and participated in the Battle of
146

Randolph Street (Chicago main drag).

His most memorable experience was the celebration ceremony for the one millionth student landing. Discharged on Feb. 21, 1946, with the rank of LCDR. He was awarded the American Theater Medal and WWII Victory Medal.

Married Mary Louise Coburn in Cleveland, OH on April 29, 1944 (three days after receiving his commission). They have three children, eight grandchildren and four great-grandchildren, all of whom live within a 25 mile radius of the Crestline community in Birmingham, AL. Retired from the U.S. Pipe and Foundry Company on May 1, 1989, and his wife, Mary Lou, retired from the Children's Hospital in Birmingham on the same date.

BERNIE ARTHUR WINKLER was born on Dec. 5, 1925, at The Grove, TX. Entered the service on Feb. 7, 1944, and the V-12 Program at Millsaps College. Assigned to the V-5 Program at North Carolina Preflight, Chapel Hill as aviation cadet. Released to inactive duty at Camp Shelton, Norfolk, VA on Sept. 18, 1945.

Returned to Texas Tech University, Lubbock, TX, graduating with a BS degree in dairy science and all border conference honors as a tackle on the 1947 football team which met Miami of Ohio in the Sun Bowl. He played two years of professional football for the Los Angeles Dons of the All-American Football Conference. After retiring from football he began a 40 year career in the dairy industry in various management and plant engineering positions with major producers in Pasadena, CA, Dallas, San Antonio and Austin, TX.

Following his retirement he began a small agricultural services company and actively pursued his interests in cattle ranching and raising quarter horses up until his death on June 30, 1990. He is survived by his wife Carolyn Winkler of New Braunfels, TX, five children and five grandchildren.

PAUL H. WINTER (LTJG) was born on July 28, 1922, in Los Angeles, CA. Enlisted in the V-12 Program on June 5, 1942 and entered the Program on active service on July 1, 1943. Graduated from Cal Tech in February 1944 and attended RMS, Camp Perry, VA. Commissioned an ensign, CEC, USNR on May 6, 1944, and sent to O.T., Camp Endicott, RI.

Shipped out with the 111th Amphibious Construction Bn. in October 1944 to the Philippines

and assigned to an LST in charge of a pontoo causeway platoon. Participated in the landing Balikpapen, Borneo and later served with th 143rd Construction Bn. as XO in the Philippine

Left active service on Oct. 18, 1946, an resigned from the Navy as a lieutenant junior grad CEC on Dec. 10, 1953. He was awarded th Asiatic-Pacific Medal, American Theater Meda Philippine Liberation Medal, Expert Rifle an Expert Pistol Ribbons.

Married 46+ years ago and has two sons an three grandchildren. Worked with internation engineering firms and as a private consultant Kabul, Afghanistan; Tokyo; Tehran; Aucklan NZ; as well as the U.S. He and his wife sailed wit another couple in their own boat to New Zealan and back. He is still active as an engineerin consultant.

WILLIAM K. WITHERSPOON (LTJC was born on Aug. 28, 1924, Hot Springs, AF Attended Phillips University, Enid, OK, and vol unteered for the Navy V-1 Program; called t active duty on July 1, 1943, and assigned to Milliga College V-12 unit, in Tennessee. Commissione an ensign in October 1944, Columbia Universit Midshipman School, New York, NY. Attende USN Minesweeping School, Yorktown, VA, as signed to the USS *Hogan* (DMS-6) operating the Pacific and served as assistant communica tions officer. When the ship was decommissione he was assigned to Port Director's Office at Guam He was promoted to lieutenant junior grade.

Attended East Tennessee State University received a BS degree in chemistry in 1947 and M degree in chemistry at Tulane University in 194 Spent two years at the University of Tennesse working in the Chemistry Department. For 4 years he worked with major chemical companie Retired in 1984 and became a consultant, teachin statistical process control.

Lives with his wife Christine in Seabrook TX. They have two sons and a grandson in Seabroo and a daughter with two grandchildren in Austi TX.

GEORGE C. WITTERIED (AVN CADE1 V-5) was born in Chicago, IL. Enlisted in the USI in May 1943 and entered the Navy V-12 Progra at the University of Notre Dame. He was statione at Hunter's Point, CA. Discharged in May 194 with the rank of aviation cadet V-5.

Married Joann Nockels in August 1950 an they have six children (three served in the mili tary). All six children are college graduates and a in various professional positions (four are eng neers). Retired as associate professor emeritu Management and Indo Religion, University o Missouri, St. Louis, MO.

GLENN M. WOOD (LT) entered the Navy V 12 Program at George Williams College on July 1943, transferred to the Illinois Institute of Tech nology and graduated with a BS degree in me chanical engineering. Commissioned an ensign i

February 1946. After completing Steam Engineering School at Newport, RI he reported aboard the USS *Gillette* (DE-681) in San Diego in June 1946. Sea duty ended with decommissioning of the ship.

Released to inactive duty in June 1947 and was recalled in June 1951 to recommission the USS *H.D. Crow* (DE-252). Attended Damage Control School at Philadelphia in the fall of 1951. Served as engineering officer until his discharge in May 1953. He is currently a lieutenant in inactive Reserves.

Graduated with a master's degree in mechanical engineering from Massachusetts Institute of Technology in 1954. Joined Pratt & Whitney the same year for an engineering career in development of jet engines for aircraft, including work on high temperature liquid metal components for nuclear powered engine, compressor development, and engineering support of various engine models. Earned his master's degree in engineering science from Rensselaer Polytechnic Institute in June 1960. Elected Fellow American Society of Mechanical Engineers in April 1979.

Retired in 1987 and resides in East Hampton, CT with wife, Eleanor. They have three sons: Daniel, Thomas and Roger.

HAROLD W. WOODS JR. (LTJG) enlisted Dec. 12, 1942, as an apprentice seaman in Class V-1, USNR. Attended Tulane University, New Orleans V-12 Program, A/S USNR (July 1943-February 1944); Midshipman Indoctrination (March 1944) at Camp McDonough, Plattsburg, NY and graduated (June 1944) to ensign, Co. L; (August 1944) appointed to Crew 5685; (July-August 1944) amphibious warfare, Solomon, MD; (September 1944) USS LCT-6 Group 72, San Francisco, CA; (November 1944) assumed command of LCT-1126.

Served (1945-1946) in Manila, Philippines, New Guinea, Shikokee, Mitsuhama; flew over Hiroshima in a Piper Cub with the 218th FA, 41st Div. (November 1945). Japanese Naval Academy, Koyo (November 1945-February 1946); graduated from Loyola University, New Orleans (June 1947); promoted to lieutenant (jg) in 1948. Organized Intelligence Reserve and attended refresher course in Monterey, CA (1949); served inactive Reserve status (April 1951).

Honorably discharged from the Reserves on Oct. 1, 1956. Awarded the American Theater of War, Asiatic-Pacific Area, WWII Victory Medal, Philippine Liberation Medal and Naval Service Occupational Medal.

Married Elaine Grundmann Woods on June 25, 1949, and they have four children: Stephen, Ronald, Charlene Landry and Dianne Caro; and seven grandsons: Scott, Philip, Eric and Jared Landry; Patrick and Ryan Woods; and Nicholas Caro. Retired in 1990 from various sales work. He passed away on June 12, 1993.

C. BRUCE WRIGHT was born on May 11, 1924, in Bath, ME. He worked as a shipfitter and tinsmith helper at the Bath Iron Works, ME. Entered military service on Jan. 25, 1943, joined the Navy V-12 Program on July 1, 1944, and served overseas as a seabee mail censor and in the States with other Navy units.

After service he returned to college at the University of Maine and Boston University, receiving a BS degree in journalism with graduate study in public relations.

Began 30 years in public relations as PR director for Maine Civil Defense. Later he wrote speeches in the executive office of the president, office of civil and defense mobilization, and for Air Force generals. He also headed PR and/or development offices at colleges, hospitals and the federal government.

Listed in several biographical references and is a 21 year member of the National Defense Executive Reserve and is active in church and community. In retirement he is a full time special assignment high school teacher. He and wife Phyllis live in Rowley, MA. Their daughter Lynne was killed in 1980 at the age of 18 in a car accident.

RICHARD O. WRIGHT (LCDR) was born on July 6, 1925, in Webster, IA. Enlisted in the USN in May 1943 and started active duty July 1, 1943 as apprentice seaman V-12, Midshipman School at the University of Notre Dame. Commissioned an ensign in USNR on Nov. 2, 1945. Served as a training instructor at Puget Sound Naval Shipyard and communications watch officer, CINCPAC Staff, Pearl Harbor. He was released to inactive duty in August 1946. Received his BSEE degree from Iowa State University in June 1947. Joined Republic Steel Corp., Chicago District, as mgmt. trainee in July 1947.

Recalled to active duty as a gunnery officer aboard the USS *Charles H. Roan* (DD-853) in June 1951 and served with the 6th Fleet in the Mediterranean. Released to inactive duty in June 1953 and transferred to Retired Reserve on July 1, 1966, then to Naval Reserve Retired List on July 6, 1985.

Married Lauretta Rose La Ponte on April 19, 1952. They have six children: Michael, Kathleen, Richard Jr., James, Nancy and Rae Ann. Retired from Republic Steel on Dec. 31, 1984. Since August 1985 has worked for Amoco Oil as an electrical construction advisor. Member of the National Society Sons of American Revolution.

JOHN H. YOCUM (LTJG) was born on June 14, 1925, in El Dorado, AR. Enlisted in the USNR

on July 1, 1943; entered the Navy V-12 Program at the University of North Carolina; transferred to the University of Notre Dame Midshipman School where he was commissioned.

Stationed at NTS, Miami, FL; Officers Gunnery School, Washington, DC; and served aboard the USS *Oregon City* (CA-122). Discharged on July 22, 1946, with the rank of lieutenant junior grade.

Married and has two children. He works as a controller for an oil company in El Dorado, AR.

BRADLEY W. YOUNG JR. (BRAD) (LCDR) was born on March 28, 1926, in Seattle, WA. He was selected for the Navy V-12 Program while still in high school and sent to Willamette University, Salem, OR in June 1944; transferred to the University of Washington for NROTC in Seattle, WA, July 1945. Released from active duty in June 1946 and immediately signed back into the Reserves to complete his training at the University of Washington.

While in V-12, NROTC, he fought in Navy intramural boxing matches and was the light-heavyweight boxing champion at the University of Washington in 1947. He also played piano with the Navy Stage Band.

Commissioned ensign in June 1947 and received a BA in journalism in 1950. Rejoined Active Reserves in 1952. Naval Reserve career included harbor defense, MSTS, Military Sealift Cmd. and recruiting.

Promoted to LCDR in September 1963. He was awarded the WWII Victory Medal, American Theater of Action, and Naval Reserve Medal.

Married Lucille and they have two sons, two daughters and 14 grandchildren. Civilian career in sales and management for 34 years with Sears. Retired, but still plays piano part-time, works in dance bands and jazz combos.

ROBERT E. YUNG (LTJG) was born in 1924 in Toronto, Ontario, Canada. He enlisted in the USN in May 1943, entered the Navy V-12 Program at Union College and Dartmouth College, then transferred to Ft. Schuyler, Columbia University for Midshipman School. Served in the Pacific in the USS *Roi* (CVE-103) and participated in "Magic Carpet", returning men from the Pacific.

His memorable experience was visiting China to return the Marines.

Separated from active service in April 1946 and graduated from Union College in 1948. He achieved the rank of lieutenant junior grade.

Married Betty Roth in 1950 and they have a son, daughter and three grandchildren. He has been associated with the Mutual of New York for over 42 years and is now semi-retired.

JOHN F. ZAMPARELLI (LTJG) was born Dec. 13, 1922, Boston, MA and grew up in the city of Medford, Middlesex County, MA. Enlisted in the USN on Dec. 2, 1942, and entered active duty in the Navy V-12 Program on July 1, 1943, at Tufts University where he won the Welterweight Boxing (Gold Glove) championship and the middleweight (Silver Glove) boxing championship. Ordered to Notre Dame University Midshipman School, South Bend, IN, where he was commissioned an ensign and won the middleweight boxing championship.

Served in the Amphibious Corps of the Navy in the Philippine Liberation stationed in Leyte Gulf from which the Flotilla supplied the battleships, aircraft carriers and cruisers with ammunition for combat.

Received the Unit Commendation with star in the name of the President of the U.S. for taking the first LCT across the Pacific on its own power. In August 1946 he separated as a lieutenant junior grade. Other awards include the American Theater Ribbon, Asiatic-Pacific Theater Ribbon, Philippine Liberation, WWII Victory Ribbon and Unit Commendation w/star.

Returned to Tufts University for his AB degree in history government in 1947. Graduated from Boston College Law School, LLB degree in 1950 and practiced law with litigation his primary source. Served 30 years in elective office on the Medford School Committee, Massachusetts House of Representatives and Middlesex County Registry of Deeds (was never defeated in any of the offices after being elected).

Recipient of an Honorary Degree of Doctor of Humane Letters and has been honored by the highest awards of the undergraduate school, law school, fraternity, Bar Associations, Chamber of Commerce, charitable associations, Veteran's Organization, social and political associations and religious affiliations.

Of lasting memory to those present in Cousens Gymnasium of Tufts College was an incident after the Welterweight Boxing bout between Tim Canty and John Zamparelli which Zamp won but which neither V-12er would give up with blood from each on the ring floor. Captain Preston B. Haines (USN), the commanding officer of the Naval Unit, went into the ring and spontaneously addressed the assemblage saying, "There's blood on this deck, blood that has been shed by both fighting to win." He stated the courage displayed was typical of our Navy and predicted victory in WWII.

He is still practicing law. Married the former Betty M. Doane, whom he met at Tufts, and is still living in Medford. They have three children: John Jr., Lisa and Laurie.

HOWARD DAVIS ZELLER (WHITEY) was born on Dec. 24, 1922, at Cleveland, OH. He enlisted in the USN on Dec. 11, 1942, and reported to Baldwin Wallace College on July 1, 1943, for the first V-12 class. In February 1944 he reported to NTS Camp McDonough, Plattsburg, NY and was commissioned an ensign June 27, 1944.

First duty was in Charleston, SC, he became plankowner of LSM-164 commissioned on Aug. 26, 1944. Transferred to LSM-126 at Little Creek, VA, Sept. 13, 1944, and left for the Pacific to see action at Iwo Jima as one of the first LSM's to beach in assault landing. He was then involved in Philippine liberation until V-J Day, then made two trips to Japan, taking units of the 1st Cav. to Yokohama and Air Corps personnel to Hokkaido.

Assigned to LSM-144 as CO at Guam in November 1945 and decommissioned it at Norfolk, VA, in March 1946. He was then sent to Portland, OR and LSM-96 for "mothballing" and released to inactive duty in July 1946 at Great Lakes.

Employed as Geolgist (Mineral Fuels) by the U.S. Geological Survey 1950-1981. Married Betty in 1958 in Boulder, CO and they have two children, Howard and Betsy. Now retired and enjoys 4-wheeling and photography in the Rocky Mountains.

FREDERICK ZIEGLER, MD (LT) was born on May 12, 1924, in Henry, IL. He enlisted in the USNR on July 1, 1943, and entered the Navy V-12 Program at DePauw University at Greencastle, IN and Johns Hopkins University School of Medicine. Served at the USN Gun Factory, Washington, DC, 1949-1950 and USN Hospital, Camp Pendleton, CA, 1953-1954.

Memorable experience was being on the full time faculty at Johns Hopkins University (in USNR "Ready Reserve"); V-12 pre-med and medical school made his professional career possible.

Discharged in 1956 with the rank of lieutenant (MC).

Married Julie Richards in 1953 and they have three children and three grandchildren. He is a practicing physician in Monterey, CA.

JOHN W. ZIEGLER (LT COL) was born in Denver, CO. Graduated from East HS in 1942; participated in baseball and excelled in track and football and attained all-city and all-state status in both track and football. He still holds the Colorado State record for the 220 yard hurdles set in 1942. Received a four year scholarship to the University of Colorado and at the end of his first year, he enlisted in the USMC V-12 Program and was transferred to Colorado College, excelling in football and winning honorable mention All American Recognition; he played with the Philadelphia Eagles. Was president of the Beta Theta Pi fraternity at Colorado College and received the coveted

VanDiest Award as the school's outstanding scholar and athlete.

Received a BA degree from Colorado College; MA degree from Denver University; and graduated from the U.S. Army Command and General Staff College. Served four years in the USMC in WWII and the U. S. Army during the Korean Conflict as a lieutenant colonel.

His biographical record is included in the Centennial Edition of *Who's Who in Colorado.* During his career in Denver, he was affiliated with CF&I Steel Corp., was vice-president of Thompson Pipe and Steel and retired in 1987 as vice president of marketing for Rocky Mountain Orthodontics.

His wife Suzanne is a retired DPS Librarian and his two daughters, Jennifer and Suzette, live in California with their families. He passed away in May 1993. His family and friends have established the East High School Memorial Scholarship Fund and the John W. Ziegler Memorial Fund at the University of Colorado.

JOHN L. ZORACK (LT COL) was born Feb 23, 1925, in Colorado Springs, CO. Enlisted in the USMC and was assigned to the Colorado College V-12 unit on July 1, 1943. Appointed 2nd lieutenant in June 1945, served in Japan and was discharged in March 1946.

Graduated from Colorado College (BA), American University (MA) and Denver University (LLD). He was recalled for the Korean Conflict and retired after 20 years as a lieutenant colonel with tours in Vietnam (Bronze Star), Senior Staff and Command College, and HQMC.

Represented associations and corporations as a lawyer/lobbyist for over 25 years and was the subject of articles in *News Week, Inc.* magazine, *The Washington Post*, and the *Wall Street Journal* and called "the lobbyist for Mom and Pop" by *Business Week.* CBS' *60 Minutes* reporter, Morley Safer, extolled Zorack's lobbying credentials. He has been interviewed by international print and television media and lectures at colleges and universities, including Harvard University's Kennedy School of Government. Author of the highly acclaimed *Lobbying Handbook*, Zorack chairs a course on Lobbying and Legislative Practice for the Georgetown Law Center, and he is President and Founder of the National Association of Registered Lobbyists.

Members of the University of Pennsylvania Navy V-12 program march at Franklin Field.

A lmost everything was scarce on the home front in World War II: sugar, coffee, gasoline, red meat, shoes — you name it. For colleges, the war brought another shortage: young men.

Many colleges were almost forced to close because so many students had been drafted or had enlisted in the armed services. Fifty years ago today, the U.S. Navy came to the rescue.

On this date, 70,000 young men reported to 131 colleges selected to participate in the Navy College Training Program better known as V-12. The Army had a similar program. The idea was to use college campuses and college professors to help train military officers.

The program seemed a good idea from all angles. The young men gained several semesters of education on their way to the ranks. The Navy gained educated officers. And the colleges gained the students they needed to stay in business.

USS Cleveland. (Courtesy of LB. Mann, Jr.)

Gunnery Sgt. John E. Lilley (right) and Sgt. Onuska (left) outstanding V-12 Marine Corps instructors, Colorado College. (Courtesy of John L. Zorack)

Outside Stratford Arms Hotel at Marquette University in Milwaukee, WI, July 1943. Back row, L to R: Al Rigling, Leon Schweda, Bob Thoresen. Front row, L to R: Bill Reed, John Rivera, Bin Randolph. (Courtesy of Leon J. Schweda)

2 Platoon 557, CC and Bucknell University V-12 students, Paris Island, S.C., Dec. 1944. (Courtesy of J. L. Zorack)

Printed in the USA
CPSIA information can be obtained
at www.ICGtesting.com
JSHW051825011124
72840JS00004B/30